Plant Breeding and Genetics

NIPA® GENX ELECTRONIC RESOURCES & SOLUTIONS P. LTD.
New Delhi-110 034

About the Author

Dr. Hari Har Ram with an excellent academic career (I position in UP in Intermediate Exam, I position in Gorakhpur Univ. in B. Sc. Ag. Exam, I position in Masters and Ph. D. programmes in Pantnagar Univ. and I position in ICAR Senior Fellowship awardees on all India basis), completed his Masters and Ph. D. (Plant Breeding) from Pantnagar Univ. and Pot-Doc from Germany. He served for 33 years at Pantnagar University in various positions including Prof. and Head Vegetable Science, Joint Director Vegetable Research Centre, Coordinator Seed Science and Technology, and Vegetable Research, Nodal Officer Plant Genetic Resources Management, and General Manager Seed Production. After leaving Pantnagar, he served as Vice-President and Chief R&D for eight years in a vegetable seed company in Pune. Later on, he worked for two years as Consultant to UP Council of Agricultural Research, Lucknow. He has served several times FAO as International Consultant in Myanmar, Libya and DPR Korea. There are 9 released varieties of soybean, 26 varieties/hybrids of vegetable crops and two NBPGR registered novel vegetable germplasm lines to his credit during the university service. He has commercialized 39 cultivars/hybrids of vegetable crops in seed industry. He has authored 18 books, 260 research papers/conference proceeding papers/ abstracts, 16 chapters in the books, and guided 25 Masters and 26 Ph. D. students. He has served as member in RACs and QRTs of several ICAR Institutes. His two books, Vegetable Breeding, and Crop Breeding and Biotechnology are immensely popular among students in State Agricultural Universities. He has visited Nepal, Germany, UK, Netherlands, France, Belgium, Luxembourg, Denmark, Mongolia, Myanmar, Libya, North Korea, Egypt, Colombia, Taiwan, Thailand, Hong Kong, Malaysia and China.

Plant Breeding and Genetics

Hari Har Ram
Former Professor
Department of Plant Breeding and Genetics
Former Head
Department of Vegetable Science
G.B.Pant University of Agriculture and Technology
Pantnagar, Udham Singh Nagar
Uttarakhand, India
Former Vice President
Krishidhan Vegetable Seeds, Pune
Former FAO Consultant (Myanmar, Libya, DPR Korea)

NIPA® GENX ELECTRONIC RESOURCES & SOLUTIONS P. LTD.
New Delhi-110 034

NIPA® GENX ELECTRONIC RESOURCES & SOLUTIONS P. LTD.

101,103, Vikas Surya Plaza, CU Block
L.S.C. Market, Pitam Pura, New Delhi-110 034
Ph : +91 11 27341616, 27341717, 27341718
E-mail: newindiapublishingagency@gmail.com
web: www.nipabooks.com

For customer assistance, please contact
Phone: + 91-11-27 34 17 17
Fax: + 91-11- 27 34 16 16
E-Mail: feedbacks@nipabooks.com

ISBN: 978-81-19254-18-7

This is the Paperback edition of Harback edition

Composed and Designed by NIPA®.

Dedicated to

Late Smt. Chandra Yadava

10 February 1950 - 31 July 2019

My wife Chandra Yadava, an epitome of honesty, self-discipline and self-sacrifice with commitment to ethical values of highest standards who left us suddenly forever on 31/07/2019 while being out cheerfully on morning walk and coincidentally at that time I was correcting proofs of this book. She had taken complete responsibility of entire household activities making me absolutely free to fully devote myself to the teaching and research work which I cherish the most. I remain indebted to the noble soul forever and will endeavor to follow her footsteps.

Hari Har Ram
31/07/2019

Preface

Presently, all under-graduate students pursuing degree programme for B. Sc. Agriculture., B. Sc. Horticulture, B. Sc. Forestry, B. Sc. Fisheries, B. V. Sc. and A. H., B. Sc. Biotechnology, and B. Sc. Biology have to go through at least one basic course on genetics and few of them have to take additionally one basic course on plant breeding as well. Several chapters especially those related to statistical and biotech applications overlap also between these two courses. Therefore, it was intended to prepare an introductory book covering fundamentals of plant breeding and genetics as one book to serve as a text book for a wide section of students as listed above. While doing so, efforts have been made to start a particular chapter with definitions of the most relevant terms, moving towards basic principles with reasonable details and illustrating the difficult propositions with simple solved numerical problems so that the students are fully equipped to understand the underlying concepts. Emerging areas such as molecular markers, genomics and GM crops have been adequately dealt with. I am sure the intended students will feel comfortable while going through the chapters and will comprehend the basic principles of plant breeding and genetics easily.

Prof. P. K. Gupta, Professor Emeritus, CCS University, Meerut helped quite liberally making available soft copies of his outstanding reviews on molecular markers. I sincerely thank him. Dr. Dinesh Yadav, Professor and Head, Biotechnology, DDU, Gorakhpur and Dr. Gautam Anand, Hebrew University, Jerusalem, Israel made substantial contributions in the chapter on Plant Genomics for which they deserve appreciation.

Some family members, particularly, my wife Chandra Yadava, for providing hassle-free office like environment at home, Dr. Rakesh Yadava (elder son, agricultural professional), Ranjan Yadava (younger son, computer specialist), Vaibhav Yadava (nephew, civil engineer) for offering some professional help including removal of few technical and computer related glitches while preparing the manuscript, deserve appreciation. I am thankful to them. I have pleasure to mention names of few kids, namely, Aakash, Dhara, Gauri, Vidhi and Krishnav some of whom might be induced to make a professional career in the field of genetics and plant breeding which is going to be far more exciting and thrilling than ever before.

My sincere appreciation is to Sri Sumit Jain and Ms. Rita Verma of New India Publishing Agency for their cooperation.

Hari Har Ram
Lucknow
31 July 2019

Contents

Glossary

Abscissa. The horizontal scale on a graph.

Acentric chromosome. Chromosome fragment lacking a centromere.

Accesion. A distinct, uniquely identified sample of seeds, plants or other germplasm materials that is maintained as an integral part of a germplasm collection.

Acridine dyes. A class of positively charged polycyclic molecules that intercalate into DNA and induce frameshift mutations.

Acrocentric. A modifying term for a chromosome or chromatid that has its centromere near the end.

Adaptation. Adjustment of an organism or a population to an environment.

Adaptedness. The degree or capacity of an individual to survive in a local environment and to transmit its genotype to the next generation.

Adaptor. A synthetic oligonucleotide that contains two or more restriction enzyme cleavage sites.

Additive allelic effects. Genetic factors that raise or lower the value of a phenotype on a linear scale of measurement.

Adenine. A purine base found in DNA and RNA.

A-DNA. A right handed DNA double helix that has 11 base-pairs per turn. DNA exists in this form when partially dehydrated.

Agrobacterium. A type of soil inhabiting bacteria that is capable of introducing DNA from plasmids in the bacteria into the genome of plant cells. Often used in genetic transformation of plants.

***Agrobacterium tumefaciens*-mediated transformation.** A naturally occurring process of DNA transfer from the bacterium *A. tumefaciens* to plants.

AIDS (acquired immunodeficiency syndrome). The usually fatal human disease in which the immunc system is destroyed by the human immunodeficiency virus (HIV).

Albinism. Absence of pigment in skin, hair and eyes of an animal. Absence of chlorophyll in plants.

Aleurone. The outermost layer of the endosperm in a seed.

Alkaptonuria. An inherited metabolic disorder. Alkaptonurics excrete excessive amounts of homogentisic acid (alkapton) in the urine.

Alkylating agents. Chemicals that transfer alkyl (methyl, ethyl and so on) groups to the bases in DNA.

Allelic exclusion. A phenomenon whereby only one functional allele of an antibody gene can be assembled in a given B lymphocyte. The "allele" on the other homologous chromosome in a diploid mammalian cell cannot undergo a functional rearrangement which would result in the production of two different antibodies by a single plasma cell.

Allele (allelomorph; adj. allelic, allelomorphic). One of a pair or series of alternative forms of a gene that occur at a given locus in a chromosome. Alleles are symbolized with the same basic symbol (for example, *D* for tall peas and *d* for dwarf).

Allele frequency. The proportion of one allele relative to all alleles at a locus in a population.

Allogamy. Alternative term for cross-pollination.

Allopatric speciation. Speciation occurring at least in part because of geographic isolation.

Allopolyploid. A polyploidy having chromosome sets from different species, a polyploidy containing genetically different chromosome sets derived from two or more species.

Allosteric transition. A reversible interaction of a small molecule with a protein molecule that causes a change in the shape of the protein and a consequent alteration of the interaction of that protein with a third molecule.

Allotetraploid. An organism with four genomes derived from hybridization of different species. Usually in forms that become established, two of the four genomes are from one species and two are from another species.

Allozygote. A diploid individual in which the two genes of a particular locus are not identical by descent from a common ancestor.

Amino acid. Any one of a class of organic compounds containing an amino (NH_2) group and a carboxyl (COOH) group. Amino acids are the building blocks of proteins. Alanine, proline, threonine, histidine, lysine, glutamine, phenlalanine, tryptophan, valine, arginine, tyrosine and leucine are among the common amino acids.

Aminoacyl (A) site. The ribosome binding site that contains the incoming aminoacyl-tRNA.

Aminoacyl-tRNA synthetases. Enzymes that catalyze the formation of high energy bonds between amino acids and tRNA molecules.

Amniocentesis. A procedure for obtaining amniotic fluid from a pregnant woman. Chemical contents of the fluid are studied directly for the diagnosis of some diseases. Cells are cultured and metaphase chromosomes are examined for irregularities (for example Down syndrome).

Amorph. A mutation that obliterates gene function, a null mutation.

Amphidiploid. A species or type of plant derived from doubling the chromosomes in the F_1 hybrid of the two species, an allopolyploid. In an amphidiploid, the two species are known whereas in other allopolyploids they may not be known.

Anabolic pathway. A pathway by which a metabolite is synthesized, a biosynthetic pathway.

Anaphase. The stage of mitosis or meiosis during which the daughter chromosomes pass from the equatorial plate to opposite poles of the cell (toward the ends of the spindle). Anaphase follows metaphase and precedes telophase.

Anchor gene. A gene that has been positioned on both the physical map and the genetic map of a chromosome.

Aneuploid. An organism or cell having a chromosome number that is not an exact multiple of the monoploid (n) with one genome, that is hyperploids, higher (for example, 2n +1) or hypoploid, lower (for example, 2n – 1). Also applied to cases where part of a chromosome is duplicated or deficient.

Antibody. Substance is a tissue or fluid of the body that acts in antagonism to a foreign substance (antigen).

Antocodon. Three bases in transfer RNA molecule that are complementary to the three bases of a specific codon in messenger RNA.

Antigen. A substance, usually a protein, that is bound by an antibody or a T cell receptor when introduced into a vertebrate organism.

Antisense gene. A gene that produces a transcript that is complementary to pre-mRNA or mRNA produced from a gene.

Apomixis. Asexual reproduction in plants through the formation of seeds without fertilization (agamospermy).

Artificial selection. The practice of choosing individuals from a population for reproduction, usually because these individuals possess one or more desirable traits.

Ascospore. One of the spores contained in the ascus of certain fungi such as *Neurospora*.

Ascus (pl. asci). Reproductive sac in the sexual stage of a type of fungi (Ascomycetes) in which ascospores are produced.

Asexual reproduction. Any process of reproduction that does not involve formation and union of gametes from the different sexes or mating types.

Assortative mating. Mating in which the partners are chosen because they are phenotypically similar.

Asynapsis. The failure or partial failure in the pairing of homologous chromosomes during the meiotic prophase.

ATP. Adenosine triphosphate: an energy rich compound that promotes certain activities in the cell.

Attenuation. A mechanism for controlling gene expression in prokaryotes that involves premature termination of transcription.

Attenuator. A nucleotide sequence in the 5' region of a prokaryotic gene (or in its RNA) that causes premature termination of transcription, possibly by forming a secondary structure.

Autocatalytic reaction. A reaction catalyzed by a substrate without the involvement of any other catalytic agent.

Autogamy. Self-fertilization.

Autopolyploid. A polyploidy that has multiple and identical or nearly identical sets of chromosomes (genomes). A polyploidy with species with genomes derived from the same original species.

Autoradiograph. A record or photograph prepared by labeling a substance such as DNA with a radioactive material such as tritiated thymidine and allowing the image produced by decay radiations to develop on a film over a period of time.

Autosome. Any chromosome that is not a sex chromosome.

Auxotroph. A mutant microorganism (for example bacteria or yeast) that will not grow on a minimal medium but that requires the addition of some compound such as an amino acid or a vitamin.

Backcross. A cross of an F_1 to one of the parental types. The offspring of such a cross are referred to as the backcross generation or backcross progeny.

Back mutation. A second mutation at the same site in a gene as the original mutation which restores the wild type nucleotide sequence.

BACs (bacterial artificial chromosome). Cloning vectors constructed from bacterial fertility (F) factors, like YAC vectors. They accept large inserts of size 200 to 500 kb.

Bacteriophage. A virus that attacks bacteria. Such viruses are called bacteriophages because they destroy their bacterial hosts.

Balanced lethal. Lethal mutations in different genes on the same pair of chromosomes that remain in repulsion because of close linkage or crossover suppression. In a close population, only the trans-heterozygotes (l_1 +/ + l_2) for the lethal mutations survive.

Base analogs. Unnatural purine or pyrimidine bases that differ slightly from the normal bases and can be incorporated into nucleic acids. They are often mutagenic.

Base collection. A comprehensive collection of germplasm accessions held for the purpose of long term conservation.

Base substitution. A single base change in a DNA molecule.

Basic number. The number of chromosomes in ancestral diploid species of polyploids, represented by x.

Binomial expansion. Exponential multiplication of an expression consisting of two terms connected by a plus (+) or minus (–) sign such as $(a + b)^n$.

Bioinformatics. A broad term to describe application of computer technology and information science to organize, interpret, and predict biological structure and function. Bioinformatics is usually applied in the context of analyzing DNA sequence data.

Biometry. Application of statistical methods to the study of biological problems.

Biopharming. The use of genetically transformed crop plants and livestock animals to produce valuable compounds, especially pharmaceuticals. Also called pharming.

Bioremediation. The use of biological organism to render hazardous wastes non-hazardous or less hazardous.

Biotechnology. A set of biological techniques developed through basic research and now applied to research and product development.

Biotype. A group of individuals with the same genotype. Biotypes may be homozygous or heterozygous.

Bivalent. A pair of synapsed or associated homologous chromosomes that have undergone the duplication process to form a group of four chromatids.

Breeding. The science and art of manipulating the heredity of an organism for a specific purpose.

Breeding line. A genetic group that has been selected and bred for its special combinations of traits.

Breeding value. For quantitative genetics, the part of the deviation of an individual phenotype from the population mean that is due to additive effects of alleles.

Broad sense heritability. In quantitative genetics, the proportion of the total phenotypic variance that is the genotypic variance.

Bt (*Bacillus thuringiensis*). A naturally occurring bacterium that produces a protein toxic to certain lepidopteran insects.

Bulk breeding. The growing of genetically diverse populations of self-pollinated crops in a bulk plot with or without mass selection followed by single-plant selection.

CAAT box. A conserved nucleotide sequence in eukaryotic promoters involved in the initiation of transcription.

Callus. A cluster of undifferentiated plant cells that have the capacity to regenerate a whole plant in some species.

Carcinogen. A substance capable of inducing cancer in an organism.

Carrier. An individual who carries a recessive allele that is not expressed (that is, is obscured by a dominant allele).

Catabolic pathway. A pathway by which an organic molecule is degraded in order to obtain energy for growth and other cellular processes; degradative pathway.

Catabolite repression. Glucose mediated reduction in the rates of transcription of operons that specify enzymes involved in catabolic pathways (such as *lac* operon).

cDNA (complementary DNA). A DNA molecule synthesized *in vitro* from an RNA template.

cDNA library. A collection of cDNA clones containing copies of the RNAs isolated from an organism or a specific tissue or cell type of an organism.

Cell. The fundamental level of structural organization in complex organisms. Cells contain a nucleus (with chromosomes), and cytoplasm with the protein synthesis machinery, bounded by a membrane.

Cell culture. A technique for growing cells under laboratory conditions.

Cell cycle. The cyclical events that occur during the divisions of mitotic cells. The cell cycle oscillates between mitosis and the interphase which is divided into G_1, S and G_2.

Cell fusion. The formation of hybrid cell produced by fusing two different cells.

Centimorgan (cM). A unit of measure of recombination frequency. One centimorgan is equal to a 1% chance that a marker at one genetic locus will be separated from a marker at a second locus due to crossing over in a single generation.

Central dogma. The underlying model for describing gene structure and function. It states that genes are transcribed in the nucleus into messenger RNA molecules, which are then translated into proteins on ribosomes.

Centriole. An organelle in many animal cells that appears to be involved in the formation of spindle during mitosis.

Centromere. Spindle-fiber attachment region of a chromosome.

Character (contraction of the word **characteristic**). One of the many details of structure, form, substance or function that make up an individual organism.

Chiasma (pl. **Chiasmata**). A visible change of partners or a crossover in two of a group of four chromatids during the first meiotic prophase. In the diplotene stage of meiosis, the four chromatids of a bivalent are associated in pairs but in such a way that one part of two chromatids is exchanged. This point of "change of partner" is the chiasma.

Chimera (plant). Part of a plant with genetically different constitution as compared with other parts of the same plant. It may result from different zygotes that grow together or from artificial fusion (grafting); it may either be pernical , with parallel layers of genetically different tissues or sectorial.

Chimeric selectable marker gene. A gene that is constructed from parts of two or more different genes and allows the host cell to survive under conditions where it would otherwise die.

Chi-square. A statistic used to test the goodness of fit of data to the predictions of any hypothesis.

Chloroplastids. A green organelle in the cytoplasm of plants that contains chlorophyll and in which starch is synthesized. A mode of cytoplasmic inheritance, independent of nuclear genes, has associated with these cytoplasmic organelles.

Chromatid. In mitosis or meiosis, one of the two identical strands resulting from self-duplication of a chromosome.

Chromatin. The deoxyribonucleohistone in the chromosomes; originally named because of the readiness with which it stains with certain dyes (chromaticity).

Chromatin fibers. A basic organizational unit of eukaryotic chromosomes that consists of DNA and associated proteins assembled into strands of average diameter 30 nm.

Chromatography. A method for separating and identifying the components from mixtures of molecules having similar chemical and physical properties.

Chromocentre. Body produced by fusion of the heterochromatic regions of the chromosomes in the polytene tissues (for example, the salivary glands) of certain *Diptera.*

Chromomeres. Small bodies, described by J. Belling, that are identified by their characteristic size and linear arrangement along a chromosome.

Chromonema (pl. **chromonemata**). An optically single thread forming an axial structure within each chromosome.

Chromosome. A condensed structure found in the cell nucleus that contains the genes of that cell.

Chromosome aberration. Abnormal structure or number of chromosomes; includes deficiency, duplication, inversion, translocation, aneuploidy, polyploidy or any othe changes from the normal pattern.

Chromosome banding. Staining of chromosomes in such a way that light and dark areas occur along the length of the chromosomes. Lateral comparisons identify pairs. Each human chromosome can be identified by its banding pattern.

Chromosome jumping. A procedure that uses large DNA fragments to "jump" along a chromosome from one site to another site.

Chromosomes. Darkly staining nucleoprotein bodies that are observed in cells during division. Each chromosome carries a linear array of genes.

Chromosome theory of inheritance. The theory that chromosomes carry the genetic information and that their behaviour during meiosis provides the physical basis for segregation and independent assortment.

Chromosome walking. A procedure that uses overlapping clones to "walk" down a chromosome from one site to another site.

Cis heterozygote. A heterozygote that contains two mutants arranged in the *cis configuration*, for example, $a^+ b^+/a\ b$.

Clonal selection. The production of a population of plasma cells all producing the same antibody in response to the interaction between a B lymphocyte producing that antibody and the antigen bound by that antibody.

Clonal propagation. The reproduction of plans through asexual means, such as cuttings, grafts, or tissue culture.

Clone. All the individuals derived by vegetative propagation from a single original individual. In molecular biology, a population of identical DNA molecules all carrying a particular DNA sequence from another organism.

Cloning vector. A small, self-replicating DNA molecule, usually a plasmid or viral chromosome, into which foreign DNAs are inserted in the process of cloning genes or other DNA sequences of interest.

Codominant alleles. Alleles that produce independent effects when heterozygous.

Codon. A set of three adjacent nucleotides in an mRNA molecule that specifies the incorporation of an amino acid into a polypeptide chain or that signals the end of polypeptide synthesis. Codons with the latter function are called termination codons.

Coefficient. A number expressing the amount of some change or effect under certain conditions (for example coefficient of inbreeding).

Coenzyme. A substance necessary for the activity of an enzyme.

Coincidence. The ratio of the observed frequency of double crossovers to the expected frequency where expected frequency is calculated by assuming that the two crossover events occur independently of each other.

Cointegrate. A DNA molecule formed by the fusion of two different DNA molecules usually mediated by a transposable element.

Colchicine. An alkaloid derived from the autumn crocus that is used as an agent to arrest spindle formation and interrupt mitosis.

Colinearity. A relationship in which the units in one molecule occur in the same sequence as the units in another molecule which they specify, for example, the nucleotides in a gene are collinear with the amino acids in the polypeptide encoded by that gene.

Combining ability. The performance of a line with others in a cross.

Competence. Ability of a bacterial cell to incorporate DNA and become genetically transformed.

Complementarity. The relationship between the two strands of a double helix of DNA. Thymine in one strand pairs with adenine in the other strand, and cytosine in one strand pairs with guanine in the other strand.

Complementation test (trans **test**). Introduction of two recessive mutants into the same cell to determine they are alleles of the same gene, that is whether they affect the same genetic function. If the mutants are allelic, the genotype m_1 +/+ m_2 will exhibit a mutant phenotype, whereas if they are nonallelic, it will exhibit the wild phenotype.

Compound chromosome. A chromosome formed by the union of two separate chromosomes, as in attached-X-chromosome or attached X-Y chromosomes.

Conditional lethal mutation. A mutation that is lethal under one set of environmental conditions, the restrictive conditions, but is viable under another set of conditions, the permissive conditions.

Conjugation. Union of sex cells (gametes) or unicellular organisms during fertilization, in *E. coli*, a one-way transfer of genetic material from a donor ("male" cell) to a recipient ("female" cell).

Consanguinity. Related by descent from a common ancestor.

Consensus sequence. The nucleotide sequence that is present in the majority of genetic signals or elements that perform a specific function.

Conserved sequence. A base sequence in a DNA molecule (or an amino acid sequence in a protein) that has remained essentially unchanged throughout evolution.

Constitutive enzyme. An enzyme that is synthesized continually regardless of growth conditions.

Constitutive gene. A gene that is continually expressed in all cells of an organism.

Contig. A set of overlapping clones that provide a physical map of a portion of a chromosome.

Continuous variation. Variation not represented by distinct classes. Individuals grade into each other and measurement data are required for analysis. Multiple genes are usually responsible for this type of variation.

Controlling elements. In maize, a transposable element such as Ac or Ds that is capable of influencing the expression of a nearby gene.

Coordinate repression. Correlated regulation of the structural genes in an operon by a molecule that interacts with the operator sequence.

Copolymers. Mixtures consisting of more than one monomer; for example, polymers of two kinds of organic bases such as uracil and cytosine (poly UC), have been combined for studies of genetic code.

Co-repressor. An effector molecule that forms a complex with a repressor and turns off the expression of a gene or set of genes.

Correlation. A statistical association between variables.

Cosmids. Cloning vectors that are hybrids between phage λ chromosomes; they contain λ *cos* sites and plasmid origins of replication.

Coupling (*cis* configuration). The condition in which a double heterozygote has received two linked mutations from one parent and their wild type alleles from the other parent (for example, a b/a b x + +/+ + produces a b/+ +).

Covalent bond. A bond in which an electron pair is equally shared by protons in two adjacent atoms.

Covariance. A measure of the statistical association between variables. The mean of the product of deviation of two variates from their individual means.

cpDNA. The DNA of plant plastids, including chloroplasts.

Crossbreeding. Mating between members of different races or species.

Crossing over. A process in which chromosomes exchange material through the breakage and reunion of their DNA molecules.

Crossover unit. A measure of distance on genetic maps that is based on average number of crossing-over events that take place during meiosis. A map interval that is one crossover unit in length (sometimes called a centimorgan) implies that only one in every hundred chromatids recovered from meiosis will have undergone a cross-over event in this interval.

Cultivar. A product of plant breeding that is released for access to producers.

Cytogenetics. Area of biology concerned with chromosomes and their implications in genetics.

Cytokinesis. Cytoplasmic division and other changes exclusive of nuclear division that are a part of mitosis or meiosis.

Cytology. The study of structure and function of cells.

Cytoplasm. The protoplasm of a cell outside the nucleus in which cell organelles (mitochondria, plastids and alike) reside; all living parts of the cell except the nucleus.

Cytoplasmic inheritance. Hereditary transmission dependent on the cytoplasm or structures in the cytoplasm rather than the nuclear genes; extrachromosomal inheritance. For example, plastid characteristics in plants may be inherited by a mechanism independent of nuclear genes.

Cytosine. A pyrimidine base found in RNA and DNA.

Cytotype. Maternally inherited cellular condition in *Drosophila* that regulates the activity of transposable P elements.

Dalton. Mass of a hydrogen atom.

Deficiency (deletion). Absence of a segment of a chromosome, reducing the number of loci.

Degeneracy (of the genetic code). The specification of an amino acid by more than one codon.

Degrees of freedom. The number of independent comparisons that can be made in a set of data.

Deme. A local population of organisms.

Denaturation. Loss of native configuration of a macromolecule, usually accompanied by loss of biological activity. Denatured proteins often unfold their polypeptide chains and express changed properties of solubility.

de novo. Arising anew, afresh, once more.

Deoxyribonuclease (DNase). An enzyme that hydrolyzes DNA.

Derepression. The process of turning on the expression of a gene or set of genes whose expression has been repressed (turned off).

Deviation. As used in statistics, a departure from an expected value.

Diakinesis. A stage of meiosis just before metaphase I in which the bivalents are shortened and thickened.

Diallel cross (complete). The crossing in all possible combinations of a series of genotypes.

Dicentric chromosome. One chromosome having two centromeres.

Dicot. A plant with two cotyledons or seed leaves.

Differentiation. A process in which unspecialized cells develop characteristic structures and functions.

Dihybrid. An individual that is heterozygous for two pairs of alleles; the progeny of a cross between homozygous parents differing in two respects.

Dimer. A compound having the same percentage composition as another but twice the molecular weight; one formed by polymerization.

Dimorphism. Two different forms in a group as determined by such characteristics as sex, size or colouration.

Dioecious. Plants in which staminate and pistillate flowers occur on different individuals.

Diploid. An organism or cell with two sets of chromosomes (2n) or two genomes. Somatic tissues of higher plants or animals are ordinarily diploid in chromosome constitution in contrast with haploid (monoploid) gametes.

Diplonema (adj. **diplotene**). The stage in prophase of meiosis I following the pachytene stage, but preceding diakinesis, in which the chromosomes of bivalents separate from each other at and around their centromeres.

Discontinuous variation. Phenotypic variability involving distinct classes such as red versus white, tall versus dwarf.

Disjunction. Separation of homologous chromosomes during anaphase of mitotic or meiotic divisions.

Dizygotic twins. Two-egg or fraternal twins.

DNA. Deoxyribonucleic acid; the information-carrying genetic material that comprises the genes. DNA is a macromolecule composed of long chain of deoxyribonuclotides joined by phosphodiester linkages. Each deoxyribonucleotide contains a phosphate group, the five-carbon sugar 2-deoxyribose and a nitrogen-containing base.

DNA fingerprint. The banding pattern on a Southern blot of an individual's genomic DNA that has been cleaved with a restriction enzyme(s) and hybridized to an appropriate nucleic acid probe(s).

DNA gyrase. An enzyme in bacteria that catalyzes the formation of negative supercoils in DNA.

DNA helicase. An enzyme that catalyzes the unwinding of the complementary strands of a DNA double helix.

DNA ligase. An enzyme that catalyzes covalent closure of nicks in DNA double helices.

DNA polymerase. An enzyme that catalyzes the synthesis of DNA.

DNA primase. An enzyme that catalyzes the synthesis of short strands of RNA that initiate the synthesis of DNA strands.

DNA profile. A distinctive pattern of DNA restriction fragments or PCR products that can be used to identify, with great certainty, any person, biological sample from a person, or organism from the environment.

DNA repair enzymes. Enzymes that catalyze the repair of damaged DNA.

DNA topoisomerase. An enzyme that catalyzes the introduction or removal of supercoils from DNA.

Domestication. The process of bringing wild plants under cultivation to produce crops under the supervision of humans.

Dominance. A condition in which one member of an allele pair is manifested to the exclusion of the other.

Dominant selectable marker gene. A gene that allows the host cell to survive under conditions where it would otherwise die.

Donor parent. The parent from which one or a few genes are transferred to the recurrent parent in backcross breeding.

Dosage compensation. A phenomenon in which the activity of a gene is increased or decreased according to the number of copies of that gene in the cell.

Double cross. A cross between two F_1 hybrids.

Double haploid (DH). An individual that is produced by doubling its gametic (n) chromosome numbers into 2n.

Duplication. The occurrence of a segment more than once in the same chromosome or genome, also, the multiplication of cells.

Ecotype. A population or strain of organism that is adapted to a particular habitat.

Effector molecule. A molecule that influences the behaviour of a regulatory molecule, such as a repressor protein, thereby influencing gene expression.

Egg (ovum). A germ cell produced by a female organism.

Electrophoresis. The migration of suspended particles in an electric field.

Electroporation. A process whereby cell membranes are made permeable to DNA by applying an intense electric current.

Elongation factors. Soluble proteins that are required for polypeptide chain elongation.

Embryo. An organism in the early stages of development.

Embryo rescue. The removal and culture of an immature embryo to produce a plant, often following a wide cross.

Embryo sac. A large thin-walled space within the ovule of the seed plant in which the egg and after fertilization, the embryo develop; the mature female gametophyte in higher plants.

Endomitosis. Duplication of chromosomes without division of the nucleus, resulting in increased chromosome number within a cell. Chromosomes strands separate but the cell does not divide.

Endonuclease. An enzyme that breaks strands of DNA at internal positions; some are involved in recombination of DNA.

Endoplasmic reticulum. Network of membranes in the cytoplasm to which ribosomes adhere.

Endopolyploidy. A state in which the cells of a diploid organism contain multiples of the diploid chromosome number (that is 4n, 8n and so on).

Endosperm. The nutritive tissue that develops in the embryo sac of most angiosperms. It usually forms after the fertilization of two fused primary endosperm nuclei of the embryo sac with one of the two male gametes. In most diploid plants, the endosperm is triploid (3n), but in some (for example, the lily) it is 5n.

Enhancement. The process of improving a germplasm accession by breeding while retaining the important genetic contributions of the accession.

Enhancer. A substance or object that increases a chemical activity or a physiological process; a major or modifier gene that increases a physiological process; a DNA sequence that influences transcription of a nearby gene.

Environment. The aggregate of all external conditions and influences affecting the life and development of an organism.

Enzyme. A protein that accelerates a specific chemical reaction in a living system.

Epigenetic. A term referring to the non-genetic causes of a phenotype.

Episome. A genetic element that may be present or absent in different cells and that may be inserted in a chromosome or independent in the cytoplasm (for example, the fertility factor (F) in *Escherichia coli*).

Epistasis. Interactions between products of non-allelic genes. Genes suppressed are said to be hypostatic. Dominance is associated with members of allelic pairs, whereas epistasis is interaction among products of non-alleles.

Epitope. A specific structural feature of an antigen that stimulates the production of antibody.

Equational division. Mitotic-type division that is usually the second division in the meiotic sequence; somatic mitosis and nonreductional division of meiosis.

Equatorial plate. The figure formed by the chromosomes in the centre (equatorial plane) of the spindle in mitosis.

Equilibrium density gradient centrifugation. A procedure used to separate macromolecules based on their density (mass per unit volume).

Error variance. Variance arising from unrecognized or uncontrolled factors in an experiment with which the variance of recognized factors is compared in test of significance.

ESTs (expressed sequence tags). Short cDNA sequences that are used to link physical maps and genetic (RFLP) maps.

Euchromatin. Genetic material that is not stained so intensely by certain dyes during interphase and that comprises many different kinds of genes.

Eugenics. The application of the principles of genetics to the improvement of humankind.

Eukaryote. A member of the large group of organisms that have nuclei enclosed by a membrane within their cells.

Euploid. An organism or cell having a chromosome number that is an exact multiple of the monoploid (n) or haploid number. Terms used to identify different levels in an euploid series are diploid, triploid, tetraploid and so on.

Excinuclease. The endonuclease-containing protein complex that excises a segment of damaged DNA during excision repair.

Excision repair. DNA repair process that involves the removal of the damaged segment of DNA and its replacement by the synthesis of a new strand using the complementary strand of DNA as template.

Exit (E) site. The ribosome binding site that contains the free tRNA prior to its release.

Exogenote. Chromosomal fragment homologous to an endogenote and donated to a merozygote.

Exon amplification. A procedure that is used to identify coding regions (exons) that are flanked by 5' and 3' intron splice sites.

Exons. The segments of a eukaryotic gene that correspond to the sequences in the final processed RNA transcript of that gene.

Exonuclease. An enzyme that digests DNA or RNA, beginning at the ends of strands .

Expressivity. Degree of expression of a trait controlled by a gene. A particular gene may produce different degrees of expression in different individuals.

Extrachromosomal. Structures that are not part of the chromosomes; DNA units in the cytoplasm that control cytoplasmic inheritance.

F1. The first filial generation; the first generation of descent from a given mating.

F2. The second filial generation produced by crossing *inter se* or by self-pollinating the F_1. The inbred "grand children" of a given mating, but in controlled genetic experimentation, self-fertilization of the F_1 (or equivalent) is implied.

Factor. Same as gene.

Family. A group of individuals directly related by descent from a common ancestor.

F factor. A bacterial episome that confers the ability to function as a genetic donor ("male") in conjugation; the fertility factor in bacteria.

Feedback inhibition. The accumulated end product of a biochemical pathway stops synthesis of that product. A late metabolite of a synthetic pathway regulates synthesis at an earlier step of the pathway.

Fertilization. The fusion of a male gamete (sperm) with a female gamete (egg) to form a zygote.

Fitness. The number of offspring left by an individual, often compared with the average of the population or with some other standard, such as the number left by a particular genotype.

Fixation. An event that occurs when all the alleles at a locus except one are eliminated from a population. The remaining allele, with frequency 100 percent, is said to have been fixed.

Folded genome. The condensed intracellular state of the DNA in the nucleoid of a bacterium. The DNA is segregated into domains, and each domain is independently negatively supercoiled.

Forced cloning. The insertion of a foreign DNA into a cloning vector in a predetermined orientation.

Founder principle. The possibility that a new, small, isolated population may diverge genetically because the founding individuals are a random sample from a large, main population.

Frameshift mutation. A mutation that changes the reading frame of an mRNA, either by inserting or deleting nucleotides.

Frequency distribution. A graph showing either the relative or absolute incidence of classes in a population. The classes may be defined by either a discrete or a continuous variable; in the latter case, each class represents a different interval on the scale of measurement.

Functional genomics. The field of study that attempts to determine the function of all genes (and gene products) largely based on knowing the entire DNA sequences of an organism.

Fusion protein. A polypeptide made from a recombinant gene that contains portions of two or more different genes. The different genes are joined so that their coding sequences are in the same reading frame.

Gall. A tumorous growth in plants.

Gamete. A mature male or female reproductive cell (sperm or egg).

Gametogenesis. The formation of gametes.

Gametophyte. That phase of the plant life cycle that bears the gametes, the cells have n chromosomes.

Gametophytic incompatibility. A botanical phenomenon controlled by the complex *S* locus in which a pollen grain cannot fertilize an ovule produced by a plant that carries the same *S* allele as the pollen grain. For example, S_1 pollen cannot fertilize an ovule made by an S_1/S_2 plant.

Gap gene. A gene that controls the formation of adjacent segments in the body of *Drosophila*.

Gene. A hereditary determinant of a specific biological function; a unit of inheritance (DNA) located in a fixed position on a chromosome; a segment of DNA encoding one polypeptide and defined operationally by *cis-trans* or complementation test.

Gene addition. The addition of a functional copy of a gene to the genome of an organism.

Gene cloning. The incorporation of a gene of interest into a self-replicating DNA molecule and the amplification of the resulting recombinant DNA molecule into an appropriate host cell.

Gene conversion. A process, often associated with recombination, during which one allele is replicated at the expense of another, leading to non-

Mendelian segregation ratios. In whole tetrads, for example, the ratio may be 6: 2 or 5: 3 instead of the expected 4: 4.

Gene expression. The process by which genes produce RNAs and proteins and exert their effect on the phenotype of an organism.

Gene flow. The spread of genes from one breeding population to another by migration, possibly leading to allele frequency changes.

Gene library. A collection of DNA fragments (carried on vector molecules) that, taken together, represents the total DNA of a certain cell type or organism.

Gene pool. The sum total of all different alleles in the breeding members of a population at a given time.

Gene regulation. The process of controlling the synthesis or suppression of gene products in specific cells or tissues.

Gene replacement. The incorporation of a transgene into a chromosome at its normal location by homologous recombination, thus replacing the copy of the gene originally present at the locus.

Gene splicing. Joining pieces of DNA from different sources using gene recombinant technology.

Gene therapy. The treatment of inherited diseases by introducing wild-type copies of the defective gene causing the disorder into the cells of affected individuals. If reproductive cells are modified, the procedure is called *germ-line* or *heritable gene therapy.* If cells other than reproductive cells are modified, the procedure is called *somatic cell* or *noninheritable gene therapy.*

Genetic code. The set of 64 nucleotide triplets that specify the 20 amino acids and polypeptide chain initiation and termination.

Genetic engineering. The manipulation of genes, composed of DNA, to create heritable changes in biological organisms and products that are useful to people, living things, or the environment.

Genetic equilibrium. Condition in a group of interbreeding organisms in which the allele frequencies remain constant over time.

Genetic erosion. The loss of genetic diversity caused by either natural or manmade processes.

Genetic marker. A genetic factor that can be identified and thus acts to determine the presence of genes or traits linked with it but not easily identified.

Genetic stocks. Accessions that typically possess one or more special genetic traits that make them of interest for research.

Genetic vulnerability. The condition that results when a crop or plant species is genetically and uniformly susceptible to a pest, pathogen, or environmental hazard.

Genetically modified (GM) organism. An organism whose genetic makeup has been changed by genetic engineering.

Genetics. The science of heredity and variation.

Genome. A complete set (n) of chromosomes (hence, of genes) inherited as a unit from one parent.

Genomic DNA library. A collection of clones containing the genomic DNA sequences of an organism.

Genomics. The field of study that seeks to understand the structure and function of all genes in an organism based on knowing the organism's entire DNA sequence, with an extensive reliance on powerful computer technologies.

Genotype. The genetic constitution (gene makeup) of an organism.

Germ cell. A reproductive cell capable when mature of being fertilized and reproducing an entire organism.

Germ plasm. The hereditary material transmitted to the offspring through the germ cells. The sum total of all hereditary material in a single (interbreeding) species.

Green revolution. A quantum yield jump obtained in wheat and rice in mid-sixties to mid-seventies specially in south and south east Asia due to deployment of dwarfing genes through the aggressive efforts of CIMMYT-Mexico (under the leadership of Borlaug) and IRRI-Manila, Philippines under the leadership of Khush. In Indian context, Swaminathan is credited as the "Father of Green Revolution".

Guanine. A purine base found in DNA and RNA.

Guide RNAs. RNA molecules that contain sequences that function as templates during RNA editing.

Gynandromorph. An individual in which one part of the body is female and another part is male; a sex mosaic.

Haploid (monoploid). An organism or cell having only one complete set (n) of chromosomes or one genome.

Hardy-Weinberg equilibrium. Mathematical relationship that allows the frequencies of genotypes in a population to be predicted from their constituent allele frequencies; a consequence of random mating.

Helix. Any structure with a spiral shape. The Watson and Crick model of DNA is in the form of double helix.

Hemizygous. The condition in which only one allele of a pair is present, as in sex linkage or as a result of deletion.

Hemoglobin. Conjugated protein compound containing iron, located in erythrocytes of vertebrates; important in the transportation of oxygen to the cells of the body.

Hemophilia. A bleeder's disease; tendency to bleed freely from even a slight wound; hereditary condition dependent on a sex-linked recessive gene.

Heredity. Resemblance among individuals related by descent; transmission of trait from parent to offspring.

Heritability. Degree to which a given trait is controlled by inheritance.

Hermaphrodite. An individual with both male and female reproductive organs.

Heteroalleles. Mutations that are functionally allelic but structurally non-allelic; mutations at different site in a gene.

Heterochromatin. Chromatin staining darkly even during interphase, often containing repetitive DNA with few genes.

Heteroduplex. A double-stranded nucleic acid containing one or more mismatched (non-complementary) base pairs.

Heterogametic sex. Producing unlike gametes with regard to the sex chromosomes. In humans, the XY male is heterogametic, and XX female is homogametic.

Heterogeneous nuclear RNA (hnRNA). The population of primary transcripts in the nucleus of a eukaryotic cell.

Heterokaryon. A cell containing two or more different nuclei.

Heteroplasmy. A cellular condition in which two genetically different types of an organelle are present.

Heteropyknosis (adj. **heteropyknotic**). Property of certain chromosomes , or of their parts, to remain more dense and to stain more intensely than other chromosomes or parts during the cell cycle.

Heterosis. Superiority of heterozygous genotypes in respect to one or more traits in comparison with corresponding homozygotes.

Heterozygote (adj. **heterozygous**). An organism with unlike members of any given pair or series of alleles that consequently produce unlike gametes.

Hfr. High frequency recombination strain of *Escherichia coli*; in such strains, the F episome is integrated into the bacterial chromosome.

Histones. Group of proteins rich in basic amino acids. They function in the coiling of DNA in chromosomes and in regulation of gene activity.

HIV (human immunodeficiency virus). The retrovirus that causes AIDS in humans.

Holoenzyme. The form of multimeric enzyme in which all of the component polypeptides are present.

Homeobox. A DNA sequence found in several genes that are involved in the specification of organs in different body parts in animals; characteristics of genes that influence segmentation in animals. The homeobox corresponds to an amino acid sequence in the polypeptide encoded by these genes; this sequence is called the homeodomain.

Homeotic mutation. A mutation that causes a body part to develop in an inappropriate position in an organism; for example, a mutation in *Drosophila* that causes legs to develop on the head in place of antennae.

Homoalleles. Mutations that are both functionally and structurally allelic; mutations at the same site in the same gene.

Homogametic sex. Producing like gametes with regard to the sex chromosomes.

Homologous chromosomes. Chromosomes that occur in pairs and are generally similar in size and shape, one having come from the male parent and the other from the female parent. Such chromosomes carry the same array of genes.

Homoplasmy. A cellular condition in which all copies of an organelle are genetically identical.

Hormone. An organic product of cells of one part of the body that is transported to the body fluids to another part where it influences activity or serves as a coordinating agent.

Human genome project. A huge international effort to map and sequence the entire human genome.

Hybrid. An offspring of homozygous parents differing in one or more genes; more generally, an offspring of a cross between unrelated strains.

Hybridization. Interbreeding of species, races, varieties, and so on, among plants or animals, a process of forming a hybrid by cross pollination of plants or by mating animals of different types.

Hybrid vigour (heterosis). Unusual growth, strength, and health of heterozygous hybrids from two less vigorous homozygous parents.

Hydrogen bonds. Weak interactions between electronegative atoms and hydrogen atoms (electropositive) that are linked to other electronegative atoms.

Hydrophobic interactions. Association of nonpolar groups with each other when present in aqueous solutions because of their insolubility in water.

Hyperploid. A genetic condition in which a chromosome or a segment of a chromosome is overrepresented in the genotype.

Hypersensitive sites. Regions in the DNA that are highly susceptible to digestion with endonucleases.

Hypomorph. A mutation that reduces but does not completely abolish gene expression.

Hypoploid. A genetic condition in which a chromosome or segment of a chromosome is underrepresented in the genotype.

I1, I2, I3,.... Symbols used to designate first, second, etc. inbred generations.

Inbreeding. Mating between related individuals.

Incomplete dominance. Expression of two alleles in a heterozygote that allows the heterozygote to be distinguishable from either of its homozygous parents.

Independent assortment. The random distribution of alleles to the gametes that occurs when genes are located in different chromosomes. The distribution of one pair of alleles is independent of the other genes located in nonhomologous chromosomes.

Inducer. A substance of low molecular weight that is bound by a repressor to produce a complex that can no longer bind to the operator; thus, the presence of the inducer turns on the expression of the gene(s) controlled by the operator.

Inducible enzyme. An enzyme that is synthesized only in the presence of the substrate that acts as an inducer.

Inducible gene. A gene that is expressed only in the presence of a specific metabolite, the inducer.

Induction. The process of turning on the expression of a gene or set of genes by an inducer.

Inhibitor. Any substance or object that retards a chemical reaction; a major modifier gene that interferes with a reaction.

Initiation factors. Soluble proteins required for the initiation of translation.

In situ. From the Latin, meaning in the natural place; refers to the experimental treatments performed on cells or tissues rather than on extracts from them.

***In situ* colony or plaque hybridization.** A procedure for screening colonies or plaques growing on plates or membranes for the presence of specific DNA sequences by the hybridization of nucleic acid probes to the DNA molecules present in these colonies or plaques.

Interaction. In statistics, an effect that cannot be explained by the additive action of contributing factors; a departure from strict additivity.

Intercalating agent. A chemical capable of inserting between adjacent base pairs in a DNA molecule.

Interference. Crossing over at one point that reduces the chance of another crossover nearby; detected by studying the pattern of crossing over with three or more linked genes.

Interphase. The stage in the cell cycle when the cell is not dividing; the metabolic stage during which DNA replication occurs; the stage following telophase of one division and extending to the beginning of prophase in the next division.

Intersex. An organism displaying secondary sexual characters intermediate between male and female; a type that shows some phenotypic characteristics of both males and females.

Intragenic complementation. Complementation that occurs between two mutant alleles of a gene; common only when product of the gene functions as a homomultimer.

Introns. Intervening sequences of DNA bases within eukaryotic genes that are not represented in the mature RNA transcript because they are spliced out of the primary RNA transcript.

Inversion. A rearrangement that reverses the order of a linear array of genes in a chromosome.

Inverted repeat. A sequence present twice in a DNA molecule but in reverse orientation.

In vitro. From the Latin meaning "within glass"; biological processes made to occur experimentally outside the organism in a test tube or other container.

In vivo. From the Latin meaning "within the living organism".

Ionic bonds. Attraction between oppositely charged chemical groups.

Ionizing radiation. The portion of the electromagnetic spectrum that results in the production of positive and negative charges (ion pairs) in molecule. X rays and gamma rays are examples of ionizing radiation.

IS element (insertion sequence). A short (800-1400 nucleotide pairs) DNA sequence found in bacteria that is capable of transposing to a new genomic location; other DNA sequences that are bounded by IS elements may also be transposed.

Isoalleles. Different forms of a gene that produce the same phenotype or very similar phenotypes.

Isochromosomes. A chromosome with two identical arms and identical genes. The arms are mirror images of each other.

Isoform. A member of a family of closely related proteins-proteins that have some amino acid sequences in common and some different.

Isogenic stocks. Strains of organisms that are genetically uniform except one gene and are homozygous.

Isoenzyme (isozyme). Different chemical forms of the same enzyme that can be distinguished from one another by electrophoresis because of difference in molecular weight.

Karyotype. The chromosome constitution of a cell or an individual; chromosomes arranged in order of length and according to position of centromere; also, the abbreviated formula for the chromosome constitution, such as 47, XX + 21 for human trisomy 21.

Lagging strand. The strand of DNA that is synthesized discontinuously during replication.

Lampbrush chromosome. Large diplotene chromosomes present in oocyte nuclei, particularly conspicuous in amphibians. These chromosomes have extended regions called loops, which are active sites of transcription.

Landrace. A population of plants, typically genetically heterogeneous, commonly developed in traditional agriculture from many years of farmer-directed section and which is specifically adapted to local conditions.

Leader sequence. The segment of an mRNA molecule from the 5' terminus to the translation initiation codon.

Leading strand. The strand of DNA that is synthesized continuously during replication.

Leptonema (adj. **leptotene).** Stage in meiosis immediately preceding synapsis in which the chromosomes appear as single, fine, threadlike structures (but they are really double because DNA replication has already taken place).

Lethal mutation. A mutation that renders an organism or cell possessing it inviable.

Ligase. An enzyme that joins the ends of two strands of nucleic acid.

Ligation. The joining of two or more DNA molecules by covalent bonds.

Line breeding. A system of breeding in which a number of genotypes which have been progeny tested in respect to some character or group of characters are composited to form a variety.

LINEs (long interspersed nuclear elements). Families of long (average length = 6500 bp) moderately repetitive transposable elements in eukaryotes.

Linkage. A relationship among genes in the same chromosome. Such genes tend to be inherited together.

Linkage equilibrium. A state in which the alleles of linked loci are randomized with respect to each other on the chromosomes of a population.

Linkage map. A linear or circular diagramme that shows the relative positions of genes on a chromosome as determined by genetic analysis.

Linker. A short, double-stranded DNA that contains one or more restriction enzyme cleavage sites.

Locus (pl. **loci**). A fixed position on a chromosome that is occupied by a given gene or one of its alleles.

Lod score. The logarithm of the ratio of odds for pedigree data, where the odds are calculated under the assumption of linkage with a specified frequency of recombination and under assumption of no linkage, that is 50 percent recombination.

Lysis. Bursting of a cell by the destruction of the cell membrane following infection by a virus.

Lysogenic bacteria. Those harboring temperate bacteriophages.

Macromolecule. A large molecule; term used to identify molecules of proteins and nucleic acids.

Mass pedigree method. A system of breeding in which a population is propagated in mass until conditions favourable for selection occur, after which pedigree selection is practiced.

Mass selection. As practiced in plant and animal breeding, the choosing of individuals for reproduction from the entire population based on the basis of the individual's phenotype rather than on the phenotypes of their relatives.

Maternal effect. Trait controlled by a gene of the mother but expressed in the progeny.

Maternal inheritance. Inheritance controlled by extrachromosomal (that is, cytoplasmic) factors that are transmitted through the egg.

Maturation. Formation of gametes or spores.

Mean. The arithmetic average; the sum of all measurements or values in a sample divided by the sample size.

Median. In a set of measurements, the central value above and below which there are equal number of measurements.

Meiosis. The process by which the chromosome number of a reproductive cell becomes reduced to half the diploid (2n) or somatic number; results in formation of gametes in animals or of spores in plants; important source of variability through recombination.

Meiotic drive. Any mechanism that causes alleles to be recovered unequally in the gametes of a heterozygote.

Melanin. Brown or black pigment.

Mendelian population. A natural interbreeding unit of sexually reproducing plants or animals sharing a common gene pool.

Merozygote. Partial zygote produced by a process of partial genetic exchange; such as transformation in bacteria. An exogenote may be introduced into a bacterial cell in the formation of a merozygote.

Messenger RNA (mRNA). RNA that carries information necessary for protein synthesis from the DNA to the ribosomes.

Metabolic cell. A cell that is not dividing.

Metabolism. Sum total of all chemical processes in living cells by which energy is produced and used.

Metacentric chromosome. A chromosome with the centromere near the middle and two arms of about equal length.

Metafemale (superfemale). In *Drosophila*, abnormal female, usually sterile, with an excess of X chromosomes compared with sets of autosomes (for example, XXX; AA).

Metaphase. The stage of cell division in which the chromosomes are most discrete and arranged in an equatorial plate; stage following prophase and preceding anaphase.

Metaxenia. Influence of pollen on maternal tissues of the fruit.

Microarray. A large set of cloned DNA molecules spotted onto a solid matrix (such as microscope slide for use in probing a biological sample to determine the gene expression, marker pattern, or nucleotide sequence of DNA/RNA.

Microprojectile bombardment. A procedure for transforming plant cells by shooting DNA-coated tungsten or gold particles into the cell.

Microsatellite. A repeated motif of nucleotides, usually only two or three bases in length, where the number of repeats frequently differ from different members of a species.

Midparent value. In quantitative genetics, average of the phenotypes of two mates.

Mismatch repair. DNA repair processes that correct base pairs that are not properly hydrogen-bonded.

Missense mutation. A mutation that changes a codon specifying an amino acid to a codon specifying a different amino acid.

Mitochondria. Organelles in the cytoplasm of plant and animal cells where oxidative phosphorylation takes place to produce ATP.

Mitosis. Disjunction of duplicated chromosomes and division of the cytoplasm to produce two genetically identical daughter cells.

Modal class. In a frequency distribution, the class having the greatest frequency.

Model. A mathematical description of a biological phenomenon.

Modifier (modifying gene). A gene that effects the expression of some other gene.

Molecular marker. An identifiable physical location on a chromosome (e. g. restriction enzyme cutting site, gene) whose inheritance can be monitored.

Monohybrid. An offspring of two homozygous parents differing in only one trait or in which only one trait is being considered.

Monomer. A single molecular entity that may combine with others to form more complex structures.

Monoploid (haploid). Organism or cell having a single set of chromosomes or one genome (chromosome number n).

Monosomic. A diploid organism lacking one chromosome of its proper complement (2n-1), an aneuploid. Monosome refers to the single chromosome, disome to two chromosomes of a kind, and trisome to three chromosomes of a kind.

Monozygotic twins. One-egg or identical twins.

Morphogen. A substance that stimulates the development of form or structure in an organism.

Morphology. Study of the form of an organism; developmental history of viable structures and comparative relation of similar structures in different organisms.

Mosaic. An organism or part of an organism that is composed of cells of different genotypes.

mtDNA. The DNA of mitochondria.

Multigene family. A group of genes that are similar in nucleotide sequence or that produce polypeptides with amino acid sequence.

Multiline. A mixture of isolines, each of which is different for a single gene conditioning different forms of the same gene.

Multiple alleles. A condition in which a particular gene occurs in three or more allelic forms in a population of organisms.

Mutable genes. Genes with an unusually high mutation rate.

Mutagen. An environmental agent, either physical or chemical, that is capable of inducing mutations.

Mutant. A cell of individual organism that shows a change brought about by mutation; a changed gene.

Mutation. A change in the DNA at a particular locus in an organism. The term is used loosely to include point mutations involving a single gene change as well as a chromosomal change.

Mutation pressure. A constant mutation rate that adds mutant genes to a population; repeated occurrences of mutations in a population.

Narrow sense heritability. In quantitative genetics, the proportion of the phenotypic variance that is due to the additive effects of alleles.

Natural selection. Differential survival and reproduction in nature that favours individuals that are better adapted to their environment; elimination of less fit organisms.

Negative autogenous regulation (negative self-regulation). Inhibition of the expression of a gene or set of coordinately regulated genes by the product of the gene or the product of one of the genes.

Negative control system. A mechanism in which the regulatory protein(s) is required to turn off gene expression.

Neutral mutation. A mutation changes the nucleotide sequence of a gene but has no effect on the fitness of the organism.

Neutral theory. The theory that the evolution of trait with little or no effect on fitness is a random process involving mutation and genetic drift.

Nick-translation. A procedure for labeling DNA by nicking it with an endonuclease and "translating" the nick along the DNA molecule in the presence of labeled deoxyribonucleoside triphosphates by the concerted action of 5'—3' exonuclease and 5'—3'polymerase activities of *E. coli* DNA polymerase I.

Nondisjunction. Failure of disjunction or separation of homologous chromosomes in mitosis or meiosis, resulting in too many chromosomes in some daughter cells and too few in others. Examples: In meiosis, both members of a pair of chromosome go to one pole so that the other pole does not receive either of them; in mitosis, both sister chromatids go to the same pole.

Nonsense mutation. A mutation that changes a codon specifying an amino acid to a termination codon.

Nontemplate strand. In transcription, the nontranscribed strand of DNA. It will have the same sequence as the RNA transcript, except that T is present at positions where U is present in the RNA transcript.

Northern blot. The transfer of RNA molecules from an electrophoretic gel to a cellulose or nylon membrane by capillary action.

Nuclease. An enzyme that catalyzes the degradation of nucleic acids.

Nucleic acid. A macromolecule composed of phosphoric acid, pentose sugar, and organic bases; DNA and RNA.

Nucleolar organizer (NO). A chromosomal segment containing genes that control synthesis of ribosomal RNA, located at the secondary constriction of some chromosome.

Nucleolus. An RNA-rich, spherical sac in the nucleus of metabolic cells; associated with the nucleolar organizer; storage place for ribosomes and ribosome precursors.

Nucleoprotein. Conjugated protein composed of nucleic acid and protein; the material of which the chromosomes are made.

Nucleoside. An organic compound consisting of a base covalently linked to ribose or deoxyribose.

Nucleosome. Spherical subunit of eukaryotic chromatin that are composed of a core particle consisting of an octamer of histones and 146 nucleotide pairs.

Nucleotide. A subunit of DNA and RNA molecule containing a phosphate group, a sugar, and a nitrogen-containing organic base.

Nucleus. The part of a eukaryotic cell that contains the chromosomes; separated from the cytoplasn by a memebrane.

Null hypothesis. Hypothesis that there is no discrepancy between observation and expectation based on some set of postulates.

Nulliplex. The condition in which a polyploid is recessive in all chromosomes in respect to a particular gene. Simplex denotes recessive at all loci except one, duplex two, triplex three, quadruplex four, etc.

Nullisomic. An otherwise diploid cell or organism lacking both members of a chromosome pair (chromosome formula 2n – 2).

Null mutation. A mutation that abolishes the expression of a gene.

Octoploid. Cell or organism with eight genomes or sets of chromosomes (chromosome number 8n).

Oligonucleotide-directed site-specific mutagenesis. A procedure by which specific nucleotide sequence can be changed to another predetermined sequence.

Oncogene. A gene that can cause neoplastic transformation in animal cells growing in culture and tumor formation in animals themselves.

Oocyte. The egg mother cell; the cell that undergoes two meiotic divisions (oogenesis) to form the egg cell. Primary oocyte—before completion of the first meiotic division; secondary oocyte—after completion of the first meiotic division.

Oogenesis. The formation of the egg or ovum in the animals.

Open reading frame. A sequence of nucleotide triplets that lacks a termination codon.

Operational definition. An operation or procedure that can be carried out to define or delimit something.

Operator. A part of an operon that controls the expression of one or more structural genes by serving as the binding site for one or more regulatory proteins.

Operon. A group of genes making up a regulatory or control unit. The unit includes an operator, a promoter, and structural genes.

Ordinate. The vertical axis in a graph.

Organelle. Specialized part of a cell with a particular function or functions (for example, the cilium of a protozoa).

Outbreeding. Mating of unrelated individuals.

Ovary. The swollen part of the pistil of a plat flower that contains the ovules; the female reproductive organ or gonad in animals.

Overdominance. A condition in which heterozygotes are superior (on some scale of measurement) to either of the associated homozygotes.

Ovule. The macrosporangium of a flowering plant that becomes the seed. It includes the nucellus and the integuments.

P. Symbol for the parental generation or parents of a given individual.

Pachynema (adj. **pachytene**). A mid-prophase stage in the meiosis immediately following zygonema and preceding diplonema. In favourable microscopic preparations, the chromosomes are visible as long, paired threads. Rarely, four chromatids are detectable.

Palindrome. A segment of DNA in which the base-pair sequence reads the same in both directions from a central point of symmetry.

Panmictic population. A population in which mating occurs at random.

Panmixis. Random mating in a population.

Paracentric inversion. An inversion that is entirely within one arm of a chromosome and does not include centromere.

Parameter. A value or constant based on an entire population.

Parthenogenesis. The development of a new individual from an egg without fertilization.

Paternal. Pertaining to the father.

Pathogen. An organism that causes a disease.

Pedigree. A table, chart, or diagramme representing the ancestry of an individual.

Pedigree breeding. A system of breeding in which individual plants are selected in the segregating generations from a cross on the basis of their desirability judged individually and on the basis of a pedigree record.

Penetrance. The percentage of individuals that show a particular phenotype among those capable of showing it.

Peptide. A compound containing amino acids; a breakdown or buildup unit in protein metabolism.

Peptide bond. A chemical bond holding amino acid subunits together in proteins.

Peptidyl (P) site. The ribosome binding site that contains the tRNA to which the growing polypeptide chain is attached.

Peptidyl transferase. An enzyme activity built into the large subunit of the ribosome—that catalyzes the formation of peptide bonds between amino acids during translation.

Pericentric inversion. An inversion including the centromere, hence involving both arms of a chromosome.

Petite mutant. A respiration-deficient yeast mutant that produces small colonies when grown on glucose-containing medium.

Phasmids. Cloning vectors that contain components derived from both phage chromosomes and plasmids.

Phenocopy. An organism whose phenotype (but not genotype) has been changed by the environment to resemble the phenotype of a different (mutant) organism.

Phenotype. The observable characteristics of an organism.

Phenylketonuria. Metabolic disorder resulting in mental retardation; transmitted as a Mendelian recessive and treated in early childhood by special diet.

Photoreactivation. A DNA repair process that is light-dependent.

Phylogeny. Evolutionary history of populations of related organisms .

Physical map. A map of location of identifiable landmarks on DNA (e. g. restriction enzyme cutting sites, genes) regardless of inheritance. Distance is measured in base pairs.

Physiological races. Pathogens of the same species with similar or identical morphology but differing pathogenic capabilities.

Plaque. Clear area on otherwise opaque culture plate of bacteria where the bacteria have been killed by a virus.

Plasmid. An extrachromosomal hereditary determinant that exists in an autonomous state and is transferred independently of chromosomes.

Plastid. A cytoplasmic body found in the cells of plants and some protozoa. Chloroplasts, for example, produce chlorophyll that is involved in photosynthesis.

Pleotropy (adj. **pleiotropic**). Condition in which a single gene influences more than one trait.

Point mutations. Changes that occur at specific sites in genes. They include nucleotide-pair substitutions and insertion or deletion of one or few nucleotide pairs.

Polar mutation. A mutation that influences the functioning of genes that are downstream in the same transcription unit.

Polyadenylation. The addition of poly(A) tails to the eukaryotic gene transcripts (RNAs).

Poly(A) polymerase. An enzyme that adds the poly(A) tails to the 3' termini of eukaryotic gene transcripts (RNAs).

Polycross. Open pollination of a group of genotypes (generally selected) in isolation from other compatible genotypes in such a way as to promote random pollination *inter se.*

Polygene (adj. **polygenic).** One of many genes involved in quantitative inheritance.

Polymer. A compound composed of many smaller subunits; results from the process of polymerization.

Polymerase. An enzyme that catalyzes the formation of DNA or RNA.

Polymerase chain reaction (PCR). A procedure involving multiple cycles of denaturation, hybridization to oligonucleotide primers, and polynucleotide synthesis that amplifies a particular DNA sequence.

Polymerization. Chemical union of two or more molecules of the same kind to form a new compound having the same elements in the same proportions but a higher molecular weight and different physical properties.

Polymorphism. Two or more kinds of individuals maintained in a breeding population.

Polynucleotide. A linear sequence of joined nucleotides in DNA or RNA.

Polypeptide. A linear molecule with two or more amino acids and one or more peptide groups. They are called dipeptides, tripeptides, and so on according to the number of amino acids present.

Polyploid. An organism with more than two sets of chromosomes (2n diploid) or genomes—for example, triploid (3n), tetraploid (4n), pentaploid (5n), hexaploid (6n), heptaploid (7n), octoploid (8n).

Polysaccharide capsules. Carbohydrate coverings with antigenic specificity that are present on same types of bacteria.

Polytene chromosome. Giant chromosomes produced by interphase replication without division and consisting of many identical chromatids arranged side by side in a cablelike pattern.

Population. Entire group of organisms of one kind; an interbreeding group of plants or animals; the extensive group from which a sample might be drawn.

Population (effective). Breeding members of the population.

Population genetics. The branch of genetics that deals with frequencies of alleles and genotypes in breeding populations.

Positional cloning. The isolation of a clone of a gene or other DNA sequence based on its map position in the genome.

Position effect. A difference in phenotype that is dependent on the position of a gene or group of genes, often caused by heterochromatin that is nearby.

Positive control system. A mechanism in which the regulatory protein(s) is required to turn on gene expression.

Postreplication repair. A recombination-dependent mechanism for repairing damaged DNA.

Prepotency. The capacity of a parent to impress characteristics on its offspring so they are more alike than usual.

Primary transcript. RNA molecule produced by transcription prior to any post-transcriptional modifications; also called a pre-mRNA in eukaryotes.

Primer. A short nucleotide sequence with a reactive 3' OH that can initiate DNA synthesis along a template.

Promosome. A protein replication complex that catalyzes the initiation of Okazaki fragments during discontinuous synthesis. It contains DNA primase and DNA helicase activities.

Probability. The frequency of occurrence of an event.

Proband. The individual in a family in whom an inherited trait is first identified.

Probe. Single-stranded DNA or RNA molecules of a specific base sequence, labeled either radioactively or immunologically, that are used to detect the complementary base sequence by hybridization.

Progeny testing. The practice of ascertaining the genotype of an individual by mating it to an individual of known genotype and examining the progeny.

Prokaryote. A member of a large group of organisms (including bacteria and blue green algae) that lack true nuclei in their cells and that do not undergo meiosis.

Promoter. A nucleotide sequence to which RNA polymerase binds and initiates transcription; also, a chemical substance that enhances the transformation of benign cells into cancerous cells.

Proofreading. The enzymatic scanning of DNA for structural defects such as mismatched base pairs.

Prophage (provirus). The genome of a temperate bacteriophage integrated into the chromosome of a lysogenic bacterium and replicated along with the host chromosome.

Prophase. The stage of mitosis between interphase and metaphase. During this phase, the centriole divides and the two daughter centrioles move apart. Each sister DNA strand from interphase replication becomes coiled, and the chromosome is longitudinally double except in the region of the centromere. Each partially separated chromosome is called a chromatid. The two chromatids of a chromosome are sister chromatids.

Protandry. Maturation of anthers before pistils.

Protease. An enzyme that hydrolyzes proteins.

Protein. A macromolecule composed of one to several polypeptides. Each polypeptide consists of a chain of amino acids linked together by peptide bonds.

Protogyny. Maturation of pistils before anthers (reverse of protandry).

Proto-oncogene. A normal cellular gene that can be changed to an oncogene by mutation.

Protoplast. A plant or bacterial cell from which the wall has been removed.

Prototroph. An organism such as bacterium that will grow on a minimal medium.

Provirus. A viral chromosome that has integrated into a host—either prokaryotic or eukaryotic genome.

Pseudoautosomal gene. A gene located on both the X and Y chromosomes.

Pseudogene. An inactive but stable component of a genome resembling a gene; apparently derived from active genes by mutation.

Pulsed-field gel electrophoresis. A procedure used to separate very large DNA molecules by alternating the direction of electric currents across a semisolid gel in a pulsed manner.

Punctuated equilibrium. The occurrence of speciation events in bursts separated by long intervals of species stability.

Purine. A double-ring nitrogen-containing base present in nucleic acids; adenine and guanine are the two purines present in most DNA and RNA molecules.

Pure-line. The progeny of a single homozygous individual by repeated selfing.

Pyrimidine. A single-ring nitrogen-containing base present in nucleic acids; cystine and thymine are commonly present in DNA and uracil usually replaces thymine in RNA.

Quantitative inheritance. Inheritance of measurable traits (height, weight, colour intensity) that depend on cumulative action of many genes, each producing a small effect on the phenotype.

Quantitative trait loci (QTL). Two or more genes that affect a single quantitative trait.

Quantum speciation. The formation of a new species in one or a few generations by selection and genetic drift.

Race. A distinguishable group of organisms of a particular species.

Radioactive isotope. An unstable isotope (form of an atom) that emits ionizing radiation.

Random genetic drift. Changes in allele frequency in small breeding populations due to chance fluctuations.

Reading frame. The series of nucleotide triplets that are sequentially positioned in the *A* site of ribosome during translation of an mRNA; also, the sequence of nucleotide-pair triplets in DNA that correspond to these codons in mRNA.

Receptor. A molecule that can accept the binding of a ligand.

Recessive. A term applied to one member of an allele pair lacking the ability to manifest itself when the other or dominant member is present.

Reciprocal crosses. Crosses between different strains with the sexes reversed; for example, female A x male B and male A x female B are reciprocal crosses.

Recombinant DNA molecule. A DNA molecule constructed *in vitro* by joining all or parts of two different DNA molecules.

Recombination. The production of gene combinations not found in the parents by the assortment of nonhomologous chromosomes and crossing over between homologous chromosomes during meiosis. For linked genes, the frequency of recombination can be used to estimate the genetic map distance; however, high frequencies (approaching 50%) do not yield accurate estimates.

Recurrent selection. A breeding method whereby plants are repeatedly selected and intercrossed to increase the frequency of favourable alleles.

Reduction division. Phase of meiosis in which the maternal and paternal chromosomes of the bivalent separate.

Regeneration. The process of growing an entire plant from a single cell or group of cells.

Regression coefficient. A numerical measure of the rate of change of the dependent on the independent variable.

Regulator gene. A gene that controls the rate of expression of another gene or genes. Example: The *lacI* gene produces a protein that controls the expression of the structural genes of the *lac* operon in *Escherichia coli.*

Release factors. Soluble proteins that recognize termination codons in mRNAs and terminate translation in response to these codons.

Renaturation. The restoration of molecule to its native form. In nucleic acid biochemistry, the term usually refers to the formation of double-stranded helix from complementary single-stranded molecules.

Repetitive DNA. DNA sequences that are present in a genome in multiple copies—sometimes a million times or more.

Replication. A duplication process that is accomplished by copying from a template (for example, reproduction at the level of DNA).

Replicon. A unit of replication. In bacteria, replicons are associated with segments of the cell membrane that control replication and coordinate it with cell division.

Replisome. The complete replication apparatus—present at a replication fork—that carries out the semiconservative replication of DNA.

Repressible enzyme. An enzyme whose synthesis is diminished by a regulatory molecule.

Repression. The process of turning off the expression of a gene or set of genes in response to an environmental signal.

Repressor. A protein that binds to DNA and turns off gene expression.

Repressor gene. A gene that encodes a repressor.

Repulsion (*trans* configuration). The condition in which a double heterozygote has received a mutant and a wild-type allele from each parent; for example, $a^{+}/a + x + b/+ b$ produces a +/+ b.

Restitution nucleus. A nucleus with unreduced or doubled chromosome number that results from the failure of a meiotic or mitotic division.

Restriction enzyme. An endonuclease that recognizes a specific short sequence in DNA and cleaves the DNA molecule at or near that site.

Restriction fragment. A fragment of DNA produced by cleaving a DNA molecule with one or more restriction endonucleases.

Restriction map. A liner or circular physical map of a DNA molecule showing the sites that are cleaved by different restriction enzymes.

Restriction site. A DNA sequence that is cleaved by a restriction enzyme.

Retroposon. A transposable element that moves via reverse transcription of RNA into DNA but lacks the long terminal repeat sequences.

Retrovirus. A virus that stores its genetic information in RNA and replicates by using reverse transcriptase to synthesize a DNA copy of its RNA genome.

Reverse transcriptase. An enzyme that catalyzes the synthesis of DNA using an RNA template.

Reversion (reverse mutation). Restitution of a mutant gene to the wild-type condition, or at least to a form that gives the wild phenotype; more generally, the appearance of a trait expressed by a remote ancestor.

RFLP (Restriction Fragment Length Polymorphism). A genetic difference among individuals that is detected by comparing DNA fragments released by digestion with one or more restriction enzymes.

Rh factor. Antigen in the red blood corpuscles of certain people. A pregnant Rh negative woman carrying an Rh positive child may develop antibodies, causing the child to develop a hemolytic disease.

Ribonuclease. An enzyme that hydrolyses RNA.

Ribosomal RNAs (rRNAs). The RNA molecules that are structural components of ribosomes.

R-loops. Single-stranded DNA regions in RNA-DNA hybrids formed *in vitro* under conditions where RNA-DNA duplexes are more stable than DNA-DNA duplexes.

RNA. Ribonucleic acid, the information carrying material in some viruses; more generally, a molecule derived from DNA by transcription that may carry information (messenger or mRNA), provide sub-cellular structure (ribosomal or rRNA), transport amino acids (transfer or tRNA) or facilitate the biochemical modification of itself or other RNA molecules.

RNA editing. Post-transcriptional processes that alter the information encoded in gene transcripts (RNAs).

RNA polymerase. An enzyme that catalyzes the synthesis of RNA.

Roentgen (r). Unit of ionizing radiation.

Sample. A finite series of observations taken from a population.

Sampling error. Deviation of a sample value from the true value owing to the limited size of sample.

Satellite DNA. A component of the genome that can be isolated from the rest of the DNA by density gradient centrifugation. Usually, it consists of short, highly repetitious sequences.

Scaffold. The central core structure of condensed chromosomes. The scaffold is composed of nonhistone chromosomal proteins.

Segregation. The separation of paternal and maternal chromosomes from each other at meiosis; the separation of alleles from each other in heterozygotes; the occurrence of different phenotypes among progeny, resulting from chromosome or allele separation in their heterozygous parents; Mendel's first principle of inheritance.

Selection. Differential survival and reproduction among genotypes; the most important of the factors that change allele frequencies in large populations.

Selection coefficient. A number that measures the fitness of genotype relative to a standard.

Selection differential. In plant and animal breeding, the difference between the means of the individuals selected to be parents and the mean of the overall population.

Selection pressure. Effectiveness of differential survival and reproduction in changing the frequency of alleles in a population.

Selection response. In plant and animal breeding, the difference between the mean of individuals selected to be parents and mean of their offspring.

Selector gene. A gene that influences the development of specific body segments in *Drosophila*; a homeotic gene.

Self-fertilization. The process by which pollen of a given plant fertilizes the ovules of the same plant. Plants fertilized in this way are said to have been selfed. An analogous process occurs in some animals, such as nematodes and molluscs.

Semisterility. A condition of only partial fertility in plant zygotes (for example, maize); usually associated with translocations.

Sense RNA. A primary transcript or mRNA that contains a coding region (contiguous sequence of codons) that is translated to produce a polypeptide.

Serology (Adj. **serological).** The study of interactions between antigens and antibodies.

Sex chromosomes. Chromosomes that are connected with the determination of sex.

Sexduction. The incorporation of bacterial genes into F factors and their subsequent transfer by conjugation to a recipient cell.

Sex factor. A bacterial episome (for example, the F plasmid in *E. coli*) that enables the cell to be a donor of genetic material. The sex factor may be propagated in the cytoplasm, or it may be integrated into the bacterial chromosome.

Sex-influenced dominance. A dominant expression that depends on the sex of the individual. For example, horns in some breeds of sheep are dominant in males and recessive in females.

Sex-limited. Expression of a trait in only one sex. Examples: milk production in mammals and egg production in chickens.

Sex linkage. Association or linkage of a hereditary trait with sex; the gene is in a sex chromosome, usually the X; often used synonymously with X-linkage.

Sexual reproduction. Reproduction involving the formation of mature sperm cells (that is, eggs and sperm).

Shine-Dalgarno sequence. A conserved sequence in prokaryotic mRNA that is complementary to a sequence near the 5' terminus of the 16S ribosomal RNA and is involved in the initiation of translation.

Shuttle vector. A plasmid capable of replicating in two different organisms, such as yeast and *E. coli.*

Sib-mating (crossing of siblings). Matings involving two individuals of the same parentage; brother-sister matings.

Sibs. Progeny of the same parents derived from different gametes. Half-sibs, progeny with one parent in common.

Sigma factor. The subunit of prokaryotic RNA polymerases that is responsible for the initiation of transcription at specific initiation sequences.

Silencer. A DNA sequence that helps to reduce or shut off the expression of a nearby gene.

Single cross. A cross between two genotypes, usually two inbred lines, in plant breeding.

Single-stranded DNA-binding protein. A protein that coats DNA single strands, keeping them in an extended state.

Small nuclear ribonucleoproteins (snRNPs). RNA-protein complexes that are components of spliceosomes.

Somatic cell. A cell that is a component of the body; in contrast with a germ cell that is capable, when fertilized, of reproducing the organism.

Southern blot. The transfer of DNA fragments from an electrophoretic gel to a cellulose or nylon membrane by capillary action.

Species. Interbreeding natural populations that are reproductively isolated from such groups.

Sperm (abbreviation of spermatozoon, pl. **spermatozoa).** A mature germ cell.

Spermatids. The four cells formed by the meiotic divisions in spermatogenesis. Spermatids become mature spermatozoa or sperm.

Splicing. The process that covalently joins exon sequences of RNA and eliminates the intervening intron sequences.

Sporophyte. The diploid generation in the life cycle of a plant that produces haploid spores by meiosis.

Standard deviation. A measure of variability in a set of data; the square root of variance.

Standard error. A measure of variation among a population of means. A statistic which is the estimated value of the standard deviation (a parameter).

Statistic. A value based on a sample or samples of a population from which estimates of a population value or parameter may be obtained.

Stem cell. A mitotically active somatic cell from which other cells arise by differentiation.

Sterility. Inability to produce offspring.

Structural gene. A gene that specifies the synthesis of a polypeptide.

STSs (sequence-tagged sites). Short, unique DNA sequences (usually 200-500 bp) that are amplified by PCR and used to link physical maps and genetic maps.

Subspecies. One of two or more morphologically or geographically distinct but interbreeding populations of a species.

Supercoil. A DNA molecule that contains extra twists as a result of overwinding (positive supercoils) or underwinding (negative supercoils).

Suppressor mutation. A mutation that partially or completely cancels the phenotypic effect of another mutation.

Sympatric speciation. The formation of new species by populations that inhabit the same or overlapping geographic regions.

Synapsis. The pairing of homologous chromosomes in the meiotic prophase.

Syndrome. A group of symptoms that occur together and represent a particular disease.

Synkaryon. A nucleus formed by the fusion of nuclei from two different somatic cells during somatic cell hybridization.

Synteny. The occurrence of two loci on the same chromosome, without regard to the distance between them.

Synthetic variety. A variety produced by crossing *inter se* a number of genotypes selected for good combining ability in all possible hybrid combinations, with subsequent maintenance of the variety by open pollination.

Taq polymerase. A heat-stable DNA polymerase isolated from the thermophylic bacterium *Thermus aquaticus.*

TATA box. A conserved promoter sequence that determines the transcription start site.

Tautomeric shift. The transfer of a hydrogen atom from one position in an organic molecule to another position.

T-DNA. The segment of DNA in the Ti plasmid of *Agrobacterium tumefaciens* that is transferred to plant cells and inserted into the chromosome of the plant.

Telomerase. An enzyme that adds telomere sequences to the ends of eukaryotic chromosomes.

Telomere. The unique structure found at the end of eukaryotic chromosomes.

Telophase. The last stage in each mitotic or meiotic division in which the chromosomes are assembled at the poles of the division spindle.

Temperate phage. A phage (virus)that invades but may not destroy (lyse) the host (bacterial cell). However, it may subsequently enter the lytic cycle.

Template. A pattern or mold. DNA stores coded information and acts as a model or template from which information is copied into complementary strand of DNA or transcribed into messenger RNA.

Template strand. In transcription, the DNA strand that is copied to produce a complementary strand of RNA.

Terminalization. Repelling movement of the centromeres of bivalents in the diplotene stages of meiotic prophase that tends to move the visible chiasmata toward the ends of the bivalents.

Terminal transferase. An enzyme that adds nucleotides to the 3' termini of DNA molecules.

Terminal signal. In transcription, a nucleotide sequence that specifies RNA chain termination.

Testcross. Backcross to the recessive parental type, or a cross between genetically unknown individuals with a fully recessive tester to determine whether an individual in question is heterozygous or homozygous for a certain allele. It is also used as a test for linkage.

Tetrad. The four cells arising from the second meiotic division in plants (pollen tetrads) or fungi (ascospores). The term also used to identify the quadruple group of chromatids that is formed by the association of duplicated homologous chromosomes during meiosis.

Tetraploid. An organism whose cells contain four haploid (4n) sets of chromosomes or genomes.

Tetrasomic (noun, **tetrasome**). Pertaining to a nucleus or an organism with four members of one of its chromosomes whereas the remainder of its chromosome complement is diploid. (Chromosome formula: $2n + 2$).

Tetratype. In fungi, a tetrad of spores that contains four different types, for example, AB, Ab, aB and ab.

Thymine. A pyrimidine base found in DNA. The other three organic bases –adenine, cytosine and guanine—are found in both RNA and DNA, but in RNA, thymine is replaced by uracil.

Ti plasmid. The large plasmid in *Agrobacterium tumefaciens.* It is responsible for the induction of tumors in plants with crown gall disease and is an important vector for transferring genes into plants, specially dicots.

Tissue culture. Growing cells, tissues, or tissue fragments (from complex, multi-cellular organism) on a nutrient medium in a dish, test tube, or flask.

Top cross. A cross between a selection, line, clone, etc., and a common pollen parent which may be a variety, inbred line, single cross, etc. The common pollen parent is called the top cross or tester parent. In corn, a top cross is commonly an inbred-variety cross.

Totipotent cell (or **nucleus**). An undifferentiated cell (or nucleus) such as blastomere that when isolated or suitably transplanted can develop into a complete embryo.

Transcription. Process through which RNA is formed along a DNA template. The enzyme RNA polymerase catalyzes the formation of RNA from ribonucleoside triphosphates.

Transcriptional antiterminator. A protein that prevents RNA polymerase from terminating transcription at specific transcription-termination sequences.

Transcription factor. A protein that regulates the transcription of genes.

Transcription unit. A segment of DNA that contains transcription initiation and termination signals and is transcribed into one RNA molecule.

Transcription vector. A cloning vector that allows the foreign gene or DNA sequence to be transcribed *in vitro.*

Transduction (t). Genetic recombination in bacteria mediated by bacteriophage. Abortive t: Bacterial DNA is injected by a phage into a bacterium but it does not replicate. Generalized t: Any bacterial gene may be transferred by a phage to a recipient bacterium. Restricted t: Transfer of bacterial DNA by a temperate phage is restricted to only one site on the bacterial chromosome.

Transfection. The uptake of DNA by a eukaryotic cell, followed by incorporation of genetic markers present in the DNA into the cell's genome.

Transfer RNAs (tRNAs). RNAs that transport amino acids to the ribosomes, where the amino acids are assembled into proteins.

Transformation (bacterial). Genetic alteration of bacteria brought about by the incorporation of foreign DNA in the bacterial cells.

Transformation (eukaryotic cells). The conversion of eukaryotic cells growing in culture to a state of uncontrolled cell growth (similar to tumor cell growth).

Transgene. A foreign or modified gene that has been introduced into an organism.

Transgenic. A term applied to organisms that have been altered by introducing DNA molecules into them.

Transgressive variation. The appearance in the F_2 (or later) generation of individuals showing more extreme development of a trait than either of the original parent.

Trans heterozygote. A heterozygote that contains two mutations arranged in the *trans* configuration—for example, a $b^+/a^+ b$.

Transition. A mutation caused by the substitution of one purine by another purine or one pyrimidine by another pyrimidine in DNA or RNA.

Translation. Protein (polypeptide) synthesis directed by a specific messenger RNA; occurs on ribosomes.

Translocation. Change in position of a segment of a chromosome to another part of the same chromosome or to a different chromosome.

Transposable genetic element. A DNA element that can move from one location to another in the genome.

Transposons. DNA elements that can move ("transpose themselves") from one position in a DNA molecule to another.

Transversion. A mutation caused by the substitution of a purine by a pyrimidine or a pyrimidine by a purine in DNA or RNA.

Trihybrid. The offspring from homozygous parents differing in three pairs of genes.

Trinucleotide repeats. Tandem repeats of three nucleotides that are present in many human genes. In several cases, these trinucleotide repeats have undergone expansions in copy number that have resulted in inherited diseases.

Trisomic. An otherwise diploid cell or organism that has an extra chromosome of one pair (chromosome formula 2n + 1).

Ultraviolet (UV) radiation. The portion of electromagnetic spectrum –wavelengths from about 1 to 350 nm—between ionizing radiation and visible light. UV is absorbed by DNA and is highly mutagenic to unicellular organisms and to the epidermal cells of multicellular organisms.

Univalent. An unpaired chromosome at meiosis.

Universality (of the genetic code). The codes have the same meaning with minor exceptions, in virtually all species.

Uracil. A pyrimidine base found in RNA but not in DNA. In DNA, uracil is replaced by thymine.

Variance. A measure of variation in a population; the square of the standard deviation.

Variation. In biology, the occurrence of differences among individuals.

Vector. A plasmid or viral chromosome that may be used to construct recombinant DNA molecules for introduction into living cells.

Viability. The capability to live and develop normally.

Virulence. Capacity of a pathogen to incite a disease.

Virulent phage. A phage (virus) that destroys the host (bacterial) cell.

VNTR (variable number tandem repeat). A short DNA sequence that is present as tandem repeats and in highly variable copy number.

Western blot. The transfer of proteins from an electrophoretic gel to a cellulose or nylon membrane by means of an electric force.

Wild type. The customary phenotype or standard for comparison.

Wobble hypothesis. Hypothesis to explain how one tRNA may recognize two codons. The first two bases of mRNA codon and anticodon pair properly, but the third base in the anticodon has some play (or wobble) that permits it to pair with more than one base.

X chromosome. A chromosome associated with sex determination. In most animals, the female has two, and the male has one x chromosome.

Xenia. Effect of pollen on the embryo and endosperm.

YACs (yeast artificial chromosomes). Linear cloning vectors constructed from essential elements of yeast chromosomes. They can accommodate foreign DNA inserts of 200 to 500 kb in size.

Y chromosome. The partner of x chromosome in the male of many animal species.

Z-DNA. The left-handed double helix that forms in GC-rich DNA molecules. The Z refers to the zig-zagged path of the sugar-phosphate backbones in this form of DNA.

Zygonema (adj. **zygotene**). Stage in meiosis during which synapsis occurs; after the leptotene stage and before the pachytene stage in the meiotic prophase.

Zygote. The cell produced by the union of two mature sex cells (gametes) in reproduction; also used in genetics to designate the individuals developing from such a cell.

Part-I

PLANT BREEDING

1

Plant Breeding and Breeding Systems

Definition

Based on various standard sources, plant breeding has been defined in various ways. A few important and standard definitions are:

1. The art and science of changing plants genetically.
2. Plant breeding is the genetic adjustment of plants to the service of man.
3. Improvement of crop plants as the result of process of evolution directed by man for his own ends. The basic principles of plant breeding are thus the principles of evolution. Directed evolution as imposed in plant breeding differs from natural evolution chiefly in a greatly reduced time-scale, and in a more precise control over the factors which govern hybridization and selection.
4. The science and art of manipulating the heredity of plants for a specific purpose.
5. Plant breeding is a deliberate effort by human to nudge nature, with respect to the heredity of plants, to an advantage. The changes made in plants are permanent and heritable and the professionals who carry out this task are called plant breeders. This effort at adjusting the status quo is instigated by a desire of humans to improve certain aspects of plants to perform new roles or enhance existing ones. As a result, the term "plant breeding" is often used synonymously with "plant improvement" in recent years. Normally the term plant breeding connotes the involvement of sexual process in effecting a desired change; however, it is also true that modern plant breeding also includes the manipulation of asexually reproducing plants (plants that do not reproduce through the sexual process). Breeding is therefore about manipulating plant attributes, structure, and composition, to make them more useful to humans.
6. Gepts and Hanock (2006) have defined plant breeding as an applied, multidisciplinary science. It is the application of genetic principles and practices associated with the development of cultivars more suited to the

needs of human than the ability to survive in the wild; it uses knowledge from agronomy, botany, genetics, physiology, pathology, entomology, biochemistry and statistics. Of particular importance is the ability to transfer, in addition to major genes, large suites of genes conditioning quantitative traits such as productivity and other traits of interest to humans. The ultimate outcome of plant breeding is mainly improved cultivars. Therefore, plant breeding is primarily an organismal science even though it is eminently suited to translate information at molecular level (DNA sequences, protein products) into economically important phenotypes. The traditional definition of plant breeding includes only those scientists who develop new cultivars and improved germplasm, however, many feel this definition should be expanded to include scientists who contribute to crop improvement through breeding research.

As a science, plant breeding started soon after the rediscovery of Mendel's laws at the beginning of the 20th century. Since then, plant breeding has evolved by absorbing approaches from different areas of science, allowing breeders to increase their efficiency and exploit genetic resources more thoroughly. Over the years, it has put to productive use progress in crop evolution, population and quantitative genetics, statistical genetics and biometry, molecular biology, and genomics. Thus, plant breeding has remained a vibrant science, with continued success in developing and deploying new cultivars on a worldwide basis. On average, around 50% of productivity increase can be attributed to genetic improvement.

Plant Breeding Goals

High yield, appropriate maturity, modification in plant architecture to suit mechanical harvesting and resist lodging, resistance to diseases and insect-pests, resistance/tolerance to abiotic stresses, namely, drought, water-logging, saline-alkaline soils and change in quality traits to meet the requirement of farmers and consumers, improved product appeal and adaptation to production environment, improvement in plant traits to resist climate change and also to engineer plants genetically to enable plants to be used as bioreactors to produce certain pharmaceuticals (called as biopharming or simply pharming).

Selected Milestones in Plant Breeding

Year	Milestone
9000 BC	First evidence of plant domestication in the hills above the Tigris river
3000 BC	Domestication of all important food crops in the Old World completed
1000 BC	Domestication of all important food crops in the New World completed
700 BC	Assyrians and Babylonians hand pollinate date palms

Year	Milestone
1694	Camerarius of Germany first to demonstrate sex in plants and suggested crossing as a method to obtain new plant types
1716	Mather of USA observed natural crossing in maize
1719	Fairchild created first artificial hybrid (carnation x sweet William)
1727	Vilmorin Company of France introduced the pedigree method of breeding
1753	Linnaeus published Species Plantarium. Binomial nomenclature born
1761-1766	Koelreuter of Germany demonstrated that hybrid offspring received traits from both parents and were intermediate in most traits, produced first scientific hybrid using tobacco
1847	Reid's Yellow Dent maize developed
1866	Mendel published his discoveries in *Experiments in Plant Hybridization* culminating in the formulation of laws of inheritance and discovery of unit factors (genes)
1876	The first inbreeding experiments with maize reported by Darwin
1899	Hopkins described the ear-to-row selection method of breeding in maize
1900	Mendel's law of heredity rediscovered independently by Correns of Germany, de Vries of Holland, and von Tschermak of Austria
1903	Pure-line theory of selection developed by W. L. Johannsen
1904-1905	Nilsson-Ehle proposed the multiple factor hypothesis for explanation of inheritance of grain colour in wheat
1908	Dominance hypothesis of heterosis proposed by Davenport
1908	Overdominance hypothesis of heterosis proposed independently by Shull and East
1908-1909	Hardy of England and Weinberg of Germany developed the law of equilibrium of populations
1908-1910	East published his work on inbreeding
1909	Shull conducted extensive research to develop inbreds to produce hybrids
1909	Bateson used the term *epistasis* to describe genes whose effects mask or cover the effect of other genes
1910	Dominance hypothesis of heterosis advocated by Bruce, and Keeble and Pellew
1917	Jones first developed commercial hybrid maize
1918	Jones suggested double cross hybrid in maize
1919	Hayes and Garber suggested possibility of commercial utilization of synthetic varieties
1925	Gametophytic incompatibility system originally called the oppositional-factor system was discovered by East and Mangelsdorf in *Nicotiana sanderae*
1926	Pioneer Hi-bred Corn Company established as first seed company
1927	Muller showed that X-rays are able to induce mutations in *Drosophila*
1928	Stadler showed that X-rays are able to induce mutations in corn and barley
1934	Dustin discovered colchicines
1935	Vavilov published *The Scientific Basis of Plant Breeding*
1935	Jenkins proposed early testing of inbreds for general combining ability

(Contd.)

Year	Milestone
1937	Harrington proposed the mass-pedigree method as a solution to one of the deficiencies in pedigree method of breeding
1940	Harlan used the bulk breeding selection method in plant breeding
1942	Tysdal, Kiesselbach and Westover suggested the term polycross to designate the progeny from seed of a line that was subject to out-crossing with selected lines growing in the same nursery
1944	Avery, MacLeod and McCarty discovered DNA is hereditary material
1944	Jones and Davis discovered cytoplasmic genetic male sterility (C type of male sterility) in onion
1944	Stadler proposed a modification of early testing , which he called ***gamete selection***
1945	Hull proposed recurrent selection for specific combining ability
1949	Comstock and Robinson proposed reciprocal recurrent selection simultaneously for both general combining ability and for specific combining ability
1950	Odland and Noll gave a detailed procedure to make use of incompatibility to produce hybrid cabbage varieties
1950	McClintock discovered the ***Ac-Ds*** system of transposable elements
1951	Seedless watermelon
1952	Chase discussed the use of monoploidy as a short-cut method in establishing homozygous lines in corn
1953	Watson, Crick, and Wilkins proposed a double helical model for DNA structure
1954	Lewis distinguished sporophytic system of self-incompatibility from gametophytic system of self-incompatibility
1956	Flor suggested gene for gene hypothesis in flax and flax rust
1970	Borlaug received Nobel Peace Prize for the Green Revolution
1970	Berg, Cohen and Boyer introduced the recombinant DNA technology
1994	FlavrSavr tomato developed as the first genetically modified food produced for the market
1995	Bt corn developed
1996	Roundup Ready GM soybean introduced
2002	Bt cotton commercialized in India
2004	Roundup Ready GM wheat developed
2004	*Golden Rice* (*GR*), a biotech/GM crop developed to contain beta-carotene (lycopene, a precursor of Vitamin A) in the seed, was first developed by co-inventors Ingo Potrykus and Peter Beyer
2014	Bt eggplant commercialized in Bangladesh
2017	Genetic Engineering and Appraisal Committee, of Ministry of Environment, Forest and Climate Change, government of India recommended environmental release of transgenic mustard hybrid, Dhara Mustard Hybrid 11 (DMH 11) having barnase, barstar, and bar genes developed by the Centre for Genetic Manipulation of Crop Plants (CGMCP) of the University of Delhi

Self- and Cross-Pollinated Crops

The crops which set self seed at least 95% are classified as self-pollinated crops. However, in some of these crops, out-crossing can go up to 10-50%, for example, in cotton and such crops are classified as often cross-pollinated crops. In contrast, cross-pollinated crops are those crops where normally out-crossing occurs at least 95%. For determining the mode of pollination, the first step is to examine the inflorescence and structure of flower. If the crop is dioecious, certainly it is cross-pollinated. Similarly, dichogamy, monoecism and other out-breeding devices are presumptive evidences that the crop is cross-pollinated. On the other hand, cleistogamy provides evidence of self-pollination.

Further, in order to establish, whether a crop is self-or cross-pollinated, the next step is to plant a single plant in space isolation. If there is no seed setting, the species is certainly cross-pollinated but the reverse may not be necessarily true. For example, maize is highly cross-pollinated but highly self-fruitful. Another good indicator of self-pollination is the effect of inbreeding. If there is no inbreeding depression, species is probably normally self-pollinated. *As a matter of fact based on no evidence of inbreeding depression in cucurbits which are monoecious are sometimes included in the list of predominantly self-pollinated crops.*

Self-Pollinated Crop Plants

Field Crops: Annual sweet clover, Broad bean*, Barley, Bur clover, Chick pea, Common bean, Cowpea, Cotton*, Flax, Foxtail millet, Hop clover, Jute, Lima bean, Mung bean, Oats, Peanut, Pea, Rice, Sorghum*, Soybean, Urd bean, Wheat, Velvet bean, Vetch

Horticultural Crops: Apricot, Citrus, Eggplant, Endive, Sweet pea, Lettuce, Nectarine, Okra, Parsnip, Peach, Pepper, Tobacco, Tomato

Cross-Pollinated Crop Plants

Field Crops: Alfalfa, Castor bean, Cassava, Crimson clover, Hemp, Hops, Maize, Pearl millet, Red clover, Rye, Safflower, Scarlet runner bean, Sunflower, Sweet clover

Horticultural Crops: Almond, Artichoke, Apples, Asparagus, Avocado, Banana, Asparagus, Beets, Blackberries, Bottle gourd, Bitter gourd, Broccoli,

*Frequently more than 10% out-crossed.

Brussels sprouts, Cabbage, Carrot, Cauliflower, Celery, Chard, Cherries, Chestnut, Chicory, Chinese cabbage, Collard, Cucumber, Date palm, Fig, Grapes, Kale, Mango, Muskmelon, Olive, Onion, Papaya, Parsley, Pear, Pecan, Pistachio, Plum, Potato, Pumpkin, Radish, Raspberries, Ridge gourd, Rutabaga, Smooth gourd, Spinach, Squash, Strawberries, Sweet potato, Turnip, Walnut, Watermelon

Methods Ensuring Cross-Pollination

Dioecy: The plant is either a male or female but not a hermaphrodite.

Monoecy: Male and female flowers located separately on the same plant.

Dichogamy: This happens in hermaphrodite flowers where cross-pollination is forced due to maturity of pollen grains before maturity and receptivity of pistil (**protandry**) or there is reverse condition (**protogyny**).

Self-incompatibility: Pollen from a flower is not tolerated by its own stigma.

Male-sterility: Pollen of the male is sterile compelling the plant to receive pollen from different flowers from different plants.

Heterostyly: A condition in which length of stamen and pistil differs significantly to make self-pollination less likely.

Genetic and Breeding Implications of Self-Pollination

Self-pollination is the most intense form of inbreeding and it promotes homozygosity of all the gene loci and the traits. The heterozygosity gets on eliminated with each generation of selfing. As a result, the population of a self-pollinated species comprises a mixture of homozygous lines after about 7 generations of selfing in a heterozygous plant. Once the lines become homozygous, they breed true that is they produce similar plants generations after generations, of course in absence of migration and mutation. Homozygosity is attained at the following rate with each generation of selfing:

$$\text{Proportion of homozygosity} = \left\{\frac{2^x - 1}{2^x}\right\}^n$$

Where x is the number of generations of selfing and n is the number of gene pairs for which the original plant is heterozygous. Let us assume one gene and three generations of selfing, then the proportion of homozygosity will be 7/8 (87.5%). After six generations of selfing with one gene pair, the proportion of homozygosity comes to 63/64 (98.4375%) and after seven generations, the value becomes 127/128 (99.21875%) implying thereby that by the 7th generation of selfing of an individual heterozygous for 1 gene pair, the homozygosity will be almost 100%.

Obviously, the number of individuals carrying heterozygous loci keeps on reducing on selfing and after only 7 generations less than one individual in 100 will still be heterozygous at any one locus. Even when 15 loci are considered together, the vast majority of the plants descended from original heterozygote will be completely homozygous after seven generations of selfing. The progress towards homozygosity in a population derived from a heterozygous individual can be generalized as follows:

Ratio of homozygotes to heterozygotes is given by $2^x - 1$: 1for any given locus where x is the number of generations of selfing. The percentage of heterozygotes for one locus is 100 $(1/2)^x$. For several independent loci, the composition of the population is given by expanding the following binomial:

$$\left\{(1/2)^x + (2^x - 1) / 2^x\right\}^n$$

where n is the number of factor pairs. For example, for two loci Aa and Bb and three generations of selfing, the composition of the population will be as follows:

{1/8 Aa + 7/8 (AA or aa)} {(1/8 Bb + 7/8 (BB or bb)} = 1/64 Aa Bb,

7/64 (AA Bb or aa Bb), 7/64 (Aa BB or Aa bb),

49/64 (AA BB, AA bb, aa Bb or aa bb).

Thus in this example, 49/64 of the derivatives after three generations of selfing are homozygous at both loci, 14/64 are homozygous at one locus and 1/64 are heterozygous at both loci (Williams, 1964).

From the above consideration, it is clear that inbred lines isolated after 7-10 generations of selfing will have achieved homozygosity at the majority of loci, and most of the inter-allelic variation which may have existed in the original population will have been lost through fixation. Variation between lines will, however, remain and number of genetically different lines possible is governed only by the number of heterozygous loci in the parent and by the viability of the inbred lines. The maximum number of genetically different lines that can be isolated by inbreeding from a single diploid plant is 2^n where n is the number of heterozygous loci. A total of 1024 different genotypes can be derived from a single parent possessing only 10 heterozygous loci.

Genetic and Breeding Implications of Cross-Pollination

Cross-pollinated crops are by and large heterozygous. Self-incompatibility occurs in such species. Unlike self-pollinated crops in which new gene combinations are prohibited, cross-pollinated crops share a wide gene pool from which new gene combinations are created to form the next generation. Further, cross-pollinated crops suffer inbreeding depression on selfing. Deleterious recessive alleles which are masked by the dominant alleles are allowed to express during inbreeding leading to inbreeding depression. However, this inbreeding depression is restored /reversed on cross-pollination in form of hybrid vigour. Thus, hybrid breeding to ensure heterozygosity in the population is quite common breeding methodology in these crops. Population improvement methods like mass selection, recurrent selection and development of synthetic cultivars are common methods of breeding in cross-pollinated crops. Several theories have been advanced to explain the basis of inbreeding depression in cross-pollinated species. These are:

- Directional dominance among alleles affecting metric characters.
- Homozygosis, either in respect of the organization of the gene milieu as a whole or in respect of specific loci.
- The elimination of homozygotes which transgress limits of physiological tolerance, thus lowering the average expression of characters among the descendants.

It should be now clear that a self-pollinated crop is a mixture of homozygous plants and that a commercial variety of such a crop is often one single genotype reproducing itself truly generation after generation. The breeder's job is to find or develop the most desirable single genotype or combination of genotypes.

A cross-pollinated crop may be compared to a pool of genes that combine in various ways to form the genotypes of each generation. A single plant produces many different kinds of gametes which combine mostly with the gametes of other plants. Here the breeder's goal is to develop the most desirable pool of genes. If hybrid varieties are intended to be developed, parental lines must be found that will combine to produce the most vigorous hybrid.

A comparison of self-and cross-pollinated crops is given in Table 1.1 with respect to their genetic and breeding characteristics and the breeding techniques used in their improvement (Briggs and Knowles, 1967).

Table 1.1 Comparison of self- and cross-pollinated crops based on breeding systems and breeding techniques.

Characteristic or breeding technique	Self-pollinated crops	Cross-pollinated crops
Breeding systems		
Genotype of sporophyte	Homozygous	Heterozygous
Genotypes of gametes of a single plant	All the same	All different

Characteristic or breeding technique	Self-pollinated crops	Cross-pollinated crops
Progeny of a single plant	Homogeneous	Heterogeneous
Self-incompatibility	None	Frequent
Inbreeding depression	None	Usual
Breeding technique		
Introduction	Yes	Yes
Pure-line selection	Often	Only to develop parental lines for hybrids
Mass selection	Occasionally	Frequently
Pedigree breeding	Yes	No
Bulk populations	Yes	Yes
Back-cross breeding	Yes	Yes
Hybrid breeding (varieties)	Occasionally specially in vegetable crops where seed requirement is less and there are more seeds/fruit	Yes
Recurrent selection	Occasionally	Yes
Synthetic varieties	Occasionally	Yes

Asexual Reproduction

Reproduction by asexual means like corms, bulbs, rhizomes, stolons or other vegetative organs is known as asexual reproduction. Budding and grafting are common modes of vegetative propagation in fruit crops. Asexual reproduction is common in fruit and nut trees, strawberries, black-berries, raspberries, grapes, pineapple, almost all ornamental shrubs and trees and field crops such as sugarcane, potato and sweet potato. The vegetatively propagated species are highly heterozygous and segregate widely upon sexual reproduction through flowers and fruits. Asexual reproduction leads to perpetuation of same genotype over cycles of reproduction/multiplication. This becomes of great advantage in plant breeding as large number of genetically identical genotypes are obtained year after years and the heterozygosity is maintained as in the original plant. The breeder can take advantage of such an outstanding genotype detected at any stage of breeding. Accordingly, breeding in such crops is easier than in sexually propagated crops. The new types are created by search for desirable mutants (sports) through natural or artificially induced mutations.

Breeding Implications of Vegetative Propagation

1. Vegetatively propagated crops can be subjected to selfing or hybridization in case these are producing seeds.

2. Clonal cultivars can be fixed immediately after hybridization and released provided a desirable genotype has been identified.
3. One single desirable clone is good enough for further multiplication without going for F5/F6 generations.
4. Heterosis if it occurs, is fixed in the hybrid product immediately after hybridization.
5. Species such as mango and citrus produce polyembryonic seedlings. This reproductive irregularity complicates breeding as clones of the parents are mixed with hybrid progeny.
6. Clonal crops are perennial out-crossers and are intolerant of inbreeding. They are highly heterozygous and are productive in the heterozygous state.

Apomixis

Ability to produce seed without fertilization is called as apomixis. Consequently, apomictically produced seeds produce clones of the mother plant. Such progenies are naturally devoid of recombination and variability. Plant families with highest apomixes are Gramineae (Poaceae), Compositae, Rosaceae and Asteraceae. Many species of citrus, mango and perennial forage grasses are apomictic. Some species can produce both sexual and apomictic seeds and these are called as facultative apomicts and the species that reproduce exclusively or nearly so by apomixes are called as obligate apomicts. When the progenies from a cross in a cross-pollinated species fail to segregate and appear uniform and resemble the mother plant, this can be taken as an indication of obligate apomixis.

Apomixis provides advantage in breeding programmes. It allows plant breeders to develop hybrids that can retain their original genetic properties indefinitely with repeated use of the seed produced over each cycle of seed production. Stated simply, hybrid can be produced from hybrid seed. However, facultative apomicts will pose greater challenge to implement this breeding procedure. The biggest advantage of apomictic cultivars would be to save seed from the field harvest of hybrid cultivar for planting the next season. There will be no need to reconstitute fresh hybrid seed each year.

2

Self-Incompatibility and Male Sterility

Self-Incompatibility

This is lack of self-fruitfulness and is a condition where the pollen from a flower is not receptive on the stigma of the same flower resulting into no seed setting. This happens despite the fact that pollen and ovule development both are normal and viable. This is caused by genetically controlled physiological hindrance to self-fertilization. This occurs widely among diverse families of flowering plants including the Leguminosae, Onagraceae, Rosaceae, Scrophulariaceae, Solanaceae, Compositae, Cruciferae, Papaveraceae and the Gramineae (now called as Poaceae). The incompatibility reaction is genetically governed by a single locus designated as S with multiple alleles whose number can go more than 100 in some species such as *Trifolium pretense*. East has estimated that this occurs in more than 3000 plant species distributed over 20 families. Lewis classified incompatibility systems into two main groups, namely, heteromorphic and homomorphic and homomorphic system has further been divided into gametophytic and sporophytic types as illustrated below:

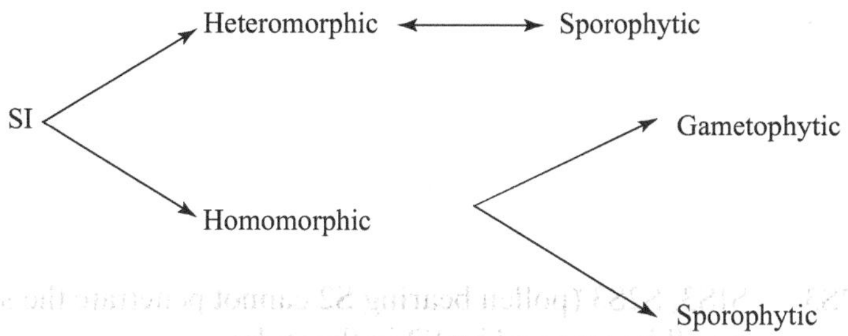

Fig. 2.1 Types of self-incompatibility.

Heteromorphic Self-Incompatibility

This is caused by difference in lengths of stamens and style (called heterostyly). Darwin has listed 38 genera having heterostyled species and described this in detail in *Primula sinensis*. In one flower type called the 'Pin', the styles are long and the anthers are short while in another type of flowers called

'Thrum', the reverse is true that is the styles are short and the anthers are long. Pin and Thrum flowers occur in ***Primula, Oxalis, Forsythia*** and ***Silia.*** This is genetically believed to be sporophytic in nature. Pin trait is conditioned by the genotype ss and the Thrum trait by the genotype Ss. The cross between the individuals of the same genotype is incompatible. Thus Pin x Pin (ss × ss) and Thrum x Thrum (Ss × Ss) are incompatible. The condition where styles have two different lengths is called as Distyly and in those cases where the styles have three different lengths, are described as tristyly.

The compatible crosses are:

Pin (ss) × Thrum (Ss) → 1 Pin (ss) : 1 Thrum (Ss)

Thrum (Ss) × Pin (ss) → 1 Pin (ss) : 1 Thrum (Ss)

As a matter of fact, the incompatibility reaction has nothing to do with the flower types but it depends on the reactions of the pollen tube and style. Pollen from the genotype ss will grow in the style of a plant of the genotype Ss and vice versa. In other words, the genotype of the plant producing the pollen determines the pollen behavior or reaction. This means that the pollen reaction is not under the control of its own genotype but the plant producing the pollen (sporophyte) and hence this system is under sporophytic control.

Gametophytic Self-Incompatibility

This originally called as 'oppositional factor system' was discovered by East and Mangelsdorf in *Nicotiana sanderae* in 1925. In this system, ability of pollen to function is determined by its own genotype (genotype of the gamete) and not the mother plant (sporophyte) that produces the pollen. It is more widespread than the sporophytic incompatibility. It is common in red clover, white clover and yellow sweet clover. The homomorphic gametophytic incompatibility (the inhibition of pollen tube growth in the style) occurs if there are identical alleles both in the pollen and style. The alleles show no dominance. Only heterozygotes for S allele are produced in this system. A few examples are as follows:

S1S2 × S2S3 → S1S3, S2S3 (pollen bearing S2 cannot penetrate the style as it will be opposed by S2 in the style)

S1S3 × S1S2 → S1S2, S2S3 (pollen bearing S1 cannot penetrate the style as it will be opposed by S1 in the style)

S1S2 × S1S2 →(fully incompatible as pollen bearing both S1 and S2 will be opposed by the S1 and S2 in the style

S1S2 × S3S4 → S1S3, S2S3, S1S4, S2S4 (fully compatible as both the alleles are not common in the pollen and the style)

Sporophytic Incompatibility

In this system of self-incompatibility, reaction of pollen is determined by the genotype of the plant (sporophyte) producing the pollen. It was first discovered in *Crepis foetida* by Hughes and Babcock and in *Parthenium argentatum* by Gerstel. It is most common in family Compositae and Cruciferae. The common crops having this incompatibility are broccoli, radish and kale. The S alleles exhibit dominance. However, it may have individual action in both pollen and style making this system complex. The incompatibility is expressed in form of decreased pollen germination and inhibition of pollen tube growth. Post fertilization failure which however is rare may also occur. Since, physiological hindrance to pollen germination (pollen and style having common alleles) occurs on the stigma, the self-incompatibility can be overcome by applying mature pollen grains on the immature stigma (bud pollination) for selfing in such crops to produce inbred lines. Bud selfing is quite common in tropical cauliflower, cabbage, broccoli and kale. Various mating systems and progenies obtained under gametophytic and sporophytic system are summarized in Table 2.1.

Table 2.1 Cross compatibility under the gametophytic and sporophytic systems of self-incompatibility.

System	Cross		Male gametes		
	Female	Male	Functional	Non-functional	Progeny
Gametophytic					
Both plants with the same genotype: no pollen functional	S1S2	S1S2	None	All	None
Plants differing in one allele: one half of pollen functional	S1S2	S1S3	S3	S1	S1S3, S2S3
Plants differing in both alleles: all pollen functional	S1S2	S3S4	S3, S4	None	S1S3, S2S3, S1S4, S2S4
Sporophytic assuming decreasing dominance in the pollen to be S1 > S2 > S3 > S4 and no individual action in style	S1S2	S1S2	None	All	None
Plants differing in one allele: all or none of pollen functional	S1S2	S2S3	None	S2, S3	None
Plants differing in both alleles: all pollen functional	S1S2	S3S4	S3, S4	None	S1S3, S2S3, S1S4, S2S4

Application of Self-Incompatibility in Plant Breeding

Sporophytic incompatibility has been utilized to some extent in producing seeds of hybrids on commercial scale in cabbage, kale and broccoli where inbred lines are developed through selfing in the base population by bud pollination and single cross, double cross and top cross hybrids are produced as follows:

S1S1 × S2S2
↓
S1S2 (Single cross hybrid)
↓

S1S1 × S2S2 → S1S2

S3S3 × S4S4 → S3S4

S1S2 × S3S4
↓
(S1S3, S2S3, S1S4, S2S4) (Double cross hybrid)
↓
S1S1 × Open pollinated cultivar
↓
Top cross hybrid

Fig. 2.2 Hybrid seed production using self-incompatibility.

Male Sterility

Male sterility is a condition in plants in which the anthers or pollen are non-functional manifested mostly by absence or extreme scarcity of pollen, severe malformation of flowers or stamens or failure of pollen to dehisce. Like self-incompatibility, male sterility enforces cross-pollination. Male sterility is of three kinds, namely, nuclear or genetic male sterility, cytoplasmic male sterility and cytoplasmic genetic male sterility.

Genetic Male Sterility

This is wide spread in barely, cotton, soybean, tomato, potato and lima bean. This is usually determined by a single recessive gene, often termed as *ms*. Accordingly, male sterile plant has the genotype *msms* and the fertile plant has the genotype *MsMs*. This male sterility is maintained by the procedure as outlined in Fig. 2.3. The male sterile plant *msms* is crossed as female with male fertile plant *MsMs*. The resultant F1 (*Msms*) is selfed and this produces an F2 having 1/4 *MsMs*, 2/4 *Msms* and 1/4 *msms*. The seed set on male sterile plants (*msms*) are harvested and planted as F3 which has 2/3 *Msms* (fertile) and 1/3 *msms* (sterile). Again, seed set on male sterile plants are harvested and the resultant progeny is planted as F4 in which *Msms* (male fertile) and *msms* (male sterile) plants segregate in 1: 1 ratio. This population is used to maintain the male sterility where seed set on male sterile plants are harvested and planted and the resultant progeny F4 onward segregates in a ratio of 1 male fertile to 1 sterile i.e. 50 % of the plants are always sterile. This plant population can be used to produce hybrid seed if fertile plants are rogued out before anthesis and the sterile plants are allowed to be crossed with a desired male parent.

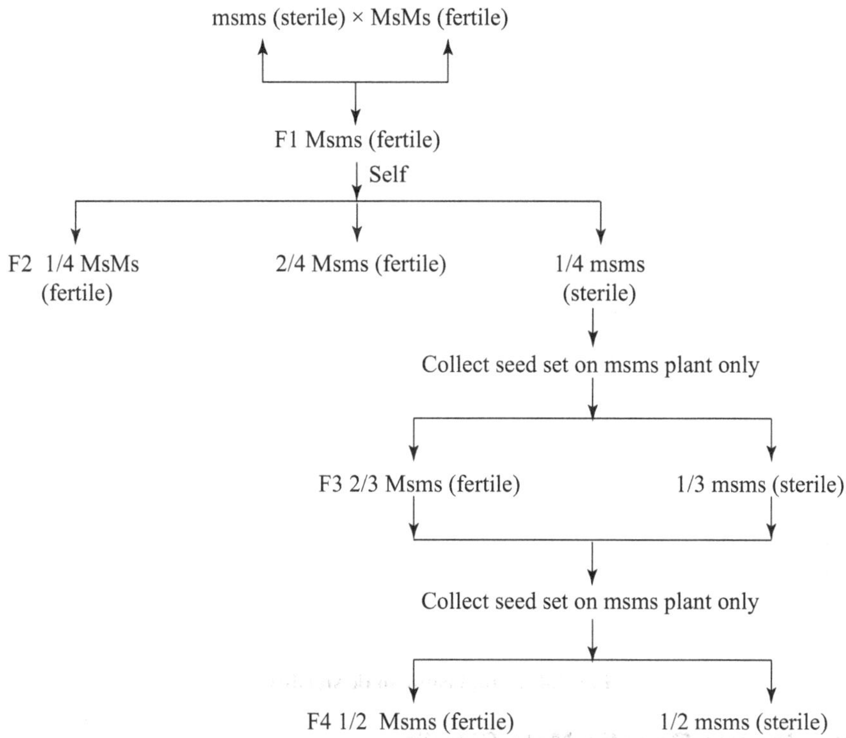

Fig. 2.3 Genetic male sterility and its maintenance.

F4 onward, this population will keep on segregating in 1: 1 ratio for fertile and sterile plants and will be the basic population to maintain *ms* gene. This population will be used as female parent to produce hybrid seed provided fertile plants are removed before anthesis and the left over sterile plants are crossed with another desired male parent. This has been used to produce hybrid seed on commercial scale in India in muskmelon by PAU and in pigeon pea at ICRISAT, Hyderabad by K. B. Saxena and his team. Hybrid pigeon pea could not become popular with seed producers and growers due to low level of cross-pollination and poor seed yield.

Cytoplasmic Male Sterility

The male sterility conditioned by the cytoplasm (mitochondrial gene) is known as cytoplasmic male sterility. This sterility may be influenced by nuclear gene also. A cytoplasm without sterility genes is described as normal (N) and the cytoplasm that causes male sterility is called as sterile cytoplasm (S).The cytoplasmic male sterility (CMS) is transmitted through egg only (maternal factor). The cross of a male sterile female with the fertile male having the normal cytoplasm (N) gives F1 which is male sterile as it has received only the sterile cytoplasm from the female parent. The male sterile plants tend to bloom longer than their seeded counterparts. Also the flowers remain fresh for a longer time. Thus, this male sterility is advantageous in ornamental plants and is unsuitable for production of hybrid seed in crops where the fruit or seed is commercial product. Its mode of inheritance is shown in Fig. 2.4.

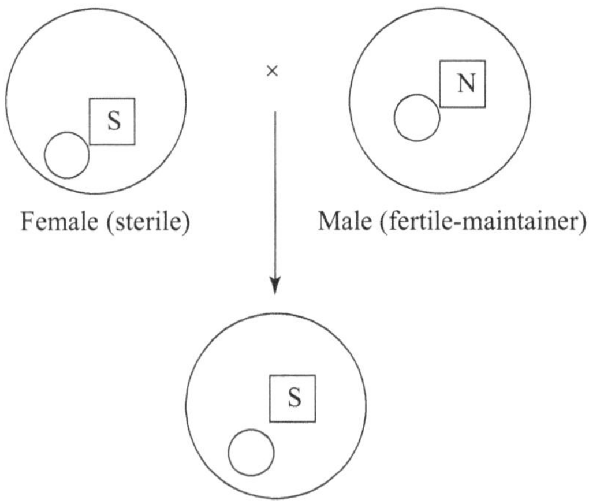

Fig. 2.4 Cytoplasmic male sterility.

Cytoplasmic Genetic Male Sterility

This is the third type of male sterility which differs from the cytoplasmic male sterility in the sense that the progenies of crosses of male sterile x male fertile plants can be male fertile in certain cases with certain male fertile parents. In this situation, the male fertile plants have been found to possess genes with power to restore the pollen producing ability of the male sterile cytoplasm. Thus the cytoplasmic male sterility is converted to cytoplasmic genetic male sterility merely by the discovery of 'restorer' genes (*Rf*) as they have been called. This has been initially discovered by Jones and Davis in 1944 in onion and since then it has been used for hybrid seed production on commercial scale in onion, corn, sorghum, pearl millet, sunflower, sugarbeet, carrot, rice and flax. The lines under hybrid seed production using this type of male sterility are known as female line also called as A line (S-*rfrf*) with sterile cytoplasm and fertility restorer gene in recessive condition, B line or the maintainer fertile line (N-*rfrf*) with normal cytoplasm and fertility restorer gene in recessive condition and the C or R line, the male line (N-*RfRf*) with normal cytoplasm and fertility restorer gene in dominant condition. The three factors involved in cytoplasmic genetic male sterility (CMS) are the normal cytoplasm, the male-sterile cytoplasm and the fertility restorer gene (*Rf*, *rf*). The essential features of the inheritance of this type of male sterility are summarized as follows:

System 1: *RfRf* or *Rfrf* or *rfrf* + normal cytoplasm = Male fertile

System 2: *RfRf* or *Rfrf* + sterile cytoplasm = Male fertile

System 3: *rfrf* +sterile cytoplasm = Male sterile

3

Plant Genetic Resources, Gene Pools and Centres of Origin

Plant genetic resources (PGR) are the genetic material of plants which are of value for present and future generations of humankind. Often used as a synonym to plant germplasm, it can be defined as a seed, a plant or plant part including cell cultures, genes and DNA sequences that are held in a repository or collected from wild as the case may be and that are useful in crop breeding, research or conservation because of genetic attributes. The term is used to describe a collection of genetic resources for an organism or genetic material which forms the physical basis of inherited qualities. In brief, germplasm is the sum total of the hereditary materials in a species. Germplasm is the foundation and lifeblood of plant breeding without which plant breeding activities cannot be conducted. Germplasm is the genetic material that can be used to perpetuate a species or a population. The base germplasm can be improved for better performance of the crop. In the initial stage of a breeding programme, breeder evaluates the available germplasm lines and makes a selection from them for further test and final release to the farmers. Later on, the germplasm serves as a source of the parental lines which are used to initiate another phase of the breeding programme. Besides, using the base germplasm, breeders generate new variability using a variety of breeding methods such as crossing parental lines and developing new lines, induced mutations and more recently gene transfer. This new variability is again subjected to appropriate selection methods leading to identification and further evaluation of new and promising genotypes for release as cultivars. Generally, germplasm includes following types:

- Commercial cultivars (current cultivars and old and obsolete cultivars)
- Breeding materials/lines available from previous breeding programme
- Landraces (farmers' developed and maintained cultivars, developed over a long period of time, adapted to the growing region and often heterogeneous)
- Plant introductions (imported, new, un-adapted genotype from outside the production region, usually from another country)
- Genetic stock (mutants and cytogenetic stock)

Spectrum of Germplasm and Its Source

Sneep and Hendriksen (1979) have summarized this matter through a simple diagramme as given in Fig. 3.1.

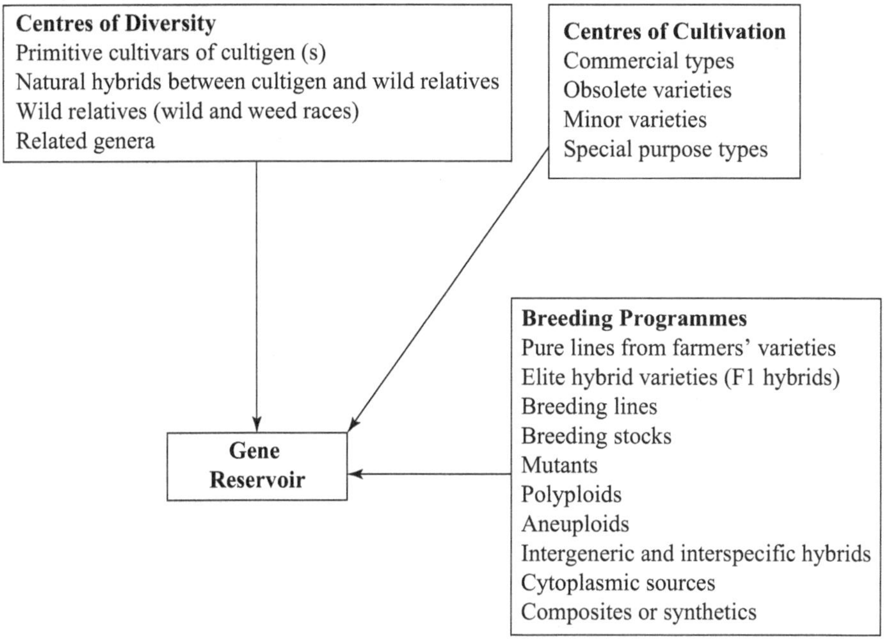

Fig. 3.1 Spectrum of germplasm in a crop species and its sources.

Conservation of Plant Genetic Resources

For effective use of germplasm continuously in on-going crop breeding programmes, effective conservation of plant genetic resources is absolutely necessary. Historically, this activity was given the due importance by Vavilov in USSR and by USDA in USA and subsequently several countries recognized its value and established genebanks. Great credit must be given to Sir Otto Frankel (CSIRO, Division of Plant Industry, Canberra) who since the 1950s has worked to devise effective measures against the loss of natural diversity via international cooperation (Frankel and Bennett, 1970). Presently, there are two types of broad PGR conservation methods.

In Situ Conservation: This means conserving biodiversity and plant genetic resources in the natural habitat applicable especially for wild relatives and forestry species on the locations where the landraces and the weeds related to the crops have evolved. In situ conservation is able to save larger parts of biological diversity.

Ex Situ Conservation: Ex-situ conservation entails planned conservation of targeted species (not all species), not in natural place of origin but under supervision of professionals, off site in locations called germplasm or gene banks. Plant material may be in form of seed or vegetative materials. Small samples of selected species are stored in a small space indoors or in a field outdoors under intensive management that facilitates their access to breeders.

Ex Situ conservation of PGR includes all conservation methods in which the plant species/varieties/all germpasm are taken out of their natural habitat and are kept in surroundings made by humans. This is done to minimize the erosion of plant biodiversity and to ensure that the required germplasm is conveniently and easily available to the plant breeders. The germplasms are conserved in scientific manner in seed storage structures called as "genebank". There could be "field genebanks" for the perennial plants/species propagated vegetatively, like cassava, sweet potato, mango, banana, citrus, cacao, sugarcane, etc. where the different types of germplasm could be planted in blocks and maintained properly with optimum crop management procedures. Field genebanks are also called as living collections. Ex Situ conservation of seeds in form of genebanks is quite common by all the CGIAR funded International Crop Research Centres, namely, CIMMYT, IRRI, CIAT, IITA, ICARDA, ICRISAT, etc.

Application of in vitro-culture is especially suited for vegetatively propagated species and recalcitrant seed storage type. Molecular techniques are being used to conserve genes and genomes (DNA libraries). Further, cells and tissues may be stored. It should be kept in mind that somaclonal variation occurs over the course of a long storage period. This phenomenon must be eliminated for in vitro-cultures to preserve the genetic integrity of stored material. For long term storage, undifferentiated tissue (callus or suspensions, in some species young embryos as well) can be stored under extremely low temperatures using liquid nitrogen known as cryopreservation.

Indian National Genebank at National Bureau of Plant Genetic Resources, New Delhi

The National Genebank conserves germplasm as per the genebank standards as base collections at –18 °C. The crop-wise details of the various accessions conserved in Indian National Genebank are listed in Table 3.1 (www.nbpgr.ernet.in). The current germplasm holdings in the National Genebank in the form of orthodox seed is 3,96,189 representing 1,584 species.

Table 3.1 Strength of plant genetic resources in the genebank at National Bureau of Plant Genetic Resources, New Delhi, India Conserved as Base Collection at – 18 Degree C.

Crop Group	Current status as in 2018	
	Species	Accessions
Cereals	90	1,56,526
Millets and forages	178	56,472
Pseudo cereals	30	6,825
Grain legumes	69	58,160
Oilseeds	58	57,479
Fibre crops	51	11,943
Vegetables	151	25,084
Fruits	35	530
Medicinal and aromatic plants and narcotics	661	6,771
Spices & condiments	17	3,721
Agro-forestry species	244	2,443
Duplicate safety samples		10,235
Total	1,584	3,96,189*

*Includes 4,141 released varieties and 2,221 genetic stocks.

In addition, a total of 28,735 exotic accessions of different crop groups, namely, cereals (15,018), grain legumes (1662), oilseeds (1329), vegetables (7982), fibers & forages (961), medicinal & aromatic (38), spices (894) and millets (19) have been kept as voucher specimen in the medium-term storage module.

US National Germplasm Conservation System

The US plant genetic conservation efforts are coordinated by the National Plant Germplasm System (NPGS). More than 400000 accessions exist in the inventory of the NPGS in the form of seed and vegetative material. Most of these materials are predominantly landraces and unimproved germplasm from overseas sources. This system has Plant Introduction Unit located in Beltsvile, Maryland and is part of the Plant Genetics and Germplasm Institute of the USDA-ARS (United States Department of Agriculture-Agricultural Research System). Each entry is given a plant introduction (PI) number. This unit does not maintain any plant material collection. The responsibilities to maintain, evaluate and release the germplasm accessions are assigned to four Regional Plant Introduction Stations (Western, North Central, North Eastern and Southern). The Plant Quarantine Facility of USDA operates from the Plant Introduction Station at Glenn Dale, Maryland. These regional centres determine germplasm needs within the region, assist with foreign explorations to fill regional needs; multiply, evaluate and maintain new plant and seed collections of crops adapted to the regions with minimal loss of genetic variability within the strains, and distribute the seed and plant accessions to plant scientists worldwide.

The base collections of the USDA are maintained at the National Seed Storage Laboratory at Fort Collins, Colorado. These base collections are seldom re-grown to avoid possible genetic changes. The laboratory provides long term backup storage for the NPGS. In addition to seed, there are National Clonal Repositories for maintaining clonal germplasm. These include Davis, California (grapes, nuts and stone fruits) and Miami, Florida (sub-tropical and tropical fruits, and sugarcane).

Types of Germplasm Collections

Base Collection: Base collections are generally not intended for distribution but maintained for long term storage. These represent most comprehensive collection of genetic variability of the designated species. Entries are maintained in original form. Storage temperature is -18 degree C and a seed moisture content of about 5 % (wet basis) using air-tight containers under low humidity or cryopreservation at -196 degree C. Under these conditions, viability of seeds will be preserved for many decades. All seeds of orthodox type are adapted to storage. This type is characterized by a logarithmic growth of viability when seed moisture content and storage temperature are reduced. The vast majority of species of temperate climate belongs to this type. The recalcitrant type of seed, however, does not follow this principle. These seeds cannot be dried without damage and they also display very limited natural viability, for example, coffee and cacao. A network of base seed collections has been established for major crop species. Genebanks holding base collections act as either global or regional depositories of specific crops. Detailed information on the physiological and technological aspects of seeds as well as on genebank management are given by Hanson (1985).

Backup Collection: This is to supplement base collection in case of disaster at base collection. In USA, the National Seed Storage Laboratory at Fort Collins, Colorado is a back-up collection for CIMMYT and IRRI germplasm. The Svalbard Global Seed Vault in Norway is an example of a secure facility for safety duplication of crop genetic resources. Located far beyond the Arctic Circle and 130 m deep inside a frozen mountain, permafrost provides an environmentally friendly solution to long-term secure conservation of crop diversity as a third safety duplicate that is only accessed in case of disaster or loss of the samples from the main safety backup. The vault can hold 4.5 million seed samples of crop diversity at storage temperatures of -18°C. Currently over 420 000 samples have been deposited there for safety.

Active Collection: Base and back-up collections are designed for long-term unperturbed storage. Active collections usually comprise the same material as in base collection. These are available for distribution to breeders and other users. Stored at 0 degree C, 8% moisture, these seeds are viable for 10-15 years. Multiple seed packets are maintained for easy distribution. These are frequently increased through field multiplication.

Breeders' or Working Collection: It consists of elite germplasm, enhanced germplasm and breeding stocks. Breeders' collection also includes products of rDNA research that can be used as parents in breeding programme.

Germplasm Enhancement

PGRs are of fundamental importance in improving agricultural productivity. These plant genetic resources stored in genebanks around the world include an assortment of alleles needed for resistance/tolerance to biotic and abiotic stresses including climate change induced harsh environments for which these are exposed in their natural habitats. However, introgression of these desirable traits into crop species has happened so far on a limited scale. Most breeders at times are reluctant to make use of germplasm of landraces and wild and weedy relatives available in active collections. As mentioned by Acquaah (2007), the valuable genetic resources are essentially "sitting on the shelf" in what have been dismissively called "gene morgues" (Hoisington *et al.*, 1999). Germplasm enhancement also termed as pre-breeding may be one of the keys to maximize the utilization of this germplasm existing in form of landraces/wild types/ weedy relatives. The germplasm enhancement/pre-breeding refers to the early component of sustainable plant breeding that deals with identifying a useful character, "capturing" its genetic diversity, and the transfer or introgression of these genes and gene combinations from non-adapted sources into breeding materials. The most common utilization of germplasm enhancement has been in resistance breeding using wild genetic resources as donors in crossing programme followed by backcrossing and selection.

Gene Pool

A gene pool is the total number and variety of genes and alleles in a sexually reproducing population that are available for transmission to the next generation. Gene pools as defined by Harlan and deWet (1971) have formed a valid scientific basis for the definition and utilization of plant genetic resources. Now plant transformations and genomics have added an additional dimension to the gene pool concept. The four identified gene pools are as follows:

Gene Pool 1: The biological species, including wild, weedy, and cultivated races.

Gene Pool 2: All species that can be crossed with gene pool 1, with some fertility in the individuals of F1 generation, gene transfer is possible but may be difficult.

Gene Pool 3: Hybrids with GP1 do not occur in nature, they are anomalous, lethal or completely sterile. Gene transfer is not possible without applying radical techniques.

Gene Pool 4: This includes any synthetic strain with DNA/RNA sequences that do not occur in natur.

A simple diagramme (Fig. 3.2) illustrates relationship among these four gene pools.

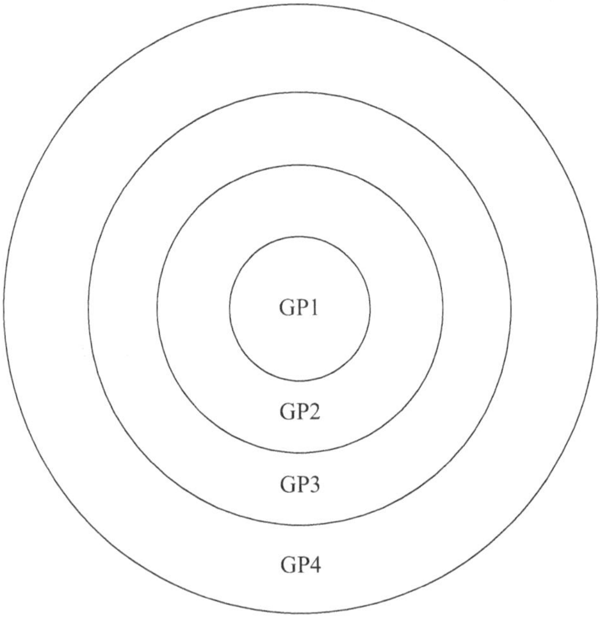

Fig. 3.2. Gene pool concept for plant genetic resources.

Management of Plant Genetic Resources

The key activities of a germplasm curator include regeneration of accessions, characterization as per standard germplasm descriptors, evaluation in association with other scientists, namely, breeders, pathologists, entomologists, biochemists, etc. monitoring seed viability and genetic integrity during seed storage and maintaining redundancy among collections. Germplasm banks receive new germplasm lines on regular basis. However, these germplasm lines must be properly distributed to plant breeders for their proper utilization.

Issue of Redundancy and Concept of Core Collection

In few crops like wheat and corn, the number of germplasm collections is very large and hence some of the accessions are bound to be duplicates. Since the cost of maintaining and regenerating the germplasm is high, it is desirable to devise an effective and efficient method of germplasm conservation and utilization. Therefore, redundancy should be minimized in the collections. However, eliminating duplicates is also as expensive as perhaps maintaining them. Therefore, concept of core collections has come into practice. Core collection is a reduced sub-set of entire germplasm which represents the diversity of the entire collection of a given species preserved in the genebank. The procedure uses passport and characterization data and statistical tools and in general 10% of entire collections are adequate to make a core collection and also to represent the adequate diversity of the entire germplasm. The entire collection is reduced to a manageable sample or "core set" that would represent the genetic diversity of the crop and its relatives with minimum repetition. The core set would serve as a working collection and enable easy retrieval of germplasm and related information needed in various research programmes, thereby facilitating better assembly, management, and use of genetic resources by the gene bank curator. This would also allow plant breeders to extensively examine the crop genome for not only specific alleles but also desirable gene blocks. In addition, the core set can provide an indication of the group(s) that should be examined in detail for particular descriptors. The accessions not included in the core set are designated as a reserve collection. Core-collection concept has been advocated by Harlan, Frankel, Brown and others. The procedure for identifying a core set uses information on the origin and characteristics of the accessions, the size of the core set, the grouping of accessions in the entire collection, the number of accessions to be selected from a group, and the sampling theory. It has been suggested to use sampling theory of selectively neutral alleles and that if about 10% of germplasm accessions were drawn randomly from the entire collection, about 70% of the allelic variation of the collection would be retained. A good core set should capture maximum genetic diversity, with a minimal number of genotypically redundant entries representing the entire collection, and the core set should be small enough to manage.

The concept of core set has been successfully used in several crop species such as soybean, lentil, peanut, chickpea, alfalfa, cassava, common bean, coffee, barley, okra, mungbean, quinoa, sesame, sweet potato, chili, pigeon pea, safflower, finger millet , and brinjal.

Crop Plant Domestication and Centres of Origin

Domestication: Domestication in most simple form is defined as the process of bringing wild species under the management of man. Under this process, wild plants are subjected to changes through the selection process imposed by humans. This is a evolutionary process in which both natural and artificial selection operate on the wild and primitive forms leading to morphological, physiological and other changes so that the derived products are adapted to the supervised agricultural conditions and possess the characteristics that are preferred by the producers and consumers. In a layman language, this amounts to taming a wild animal into a pet. Like evolution, domestication is a process of genetic change in which a population of plants can experience a shift in its genetic composition in the direction of selection as imposed by the domesticator. Under domestication, new plant types are continually selected for as domesticates as new demands are imposed leading to gradually moving the selected plants farther away (genetically, morphologically and physiologically) from their wild progenitors. Domestication has been under way for over 10,000 years since beginning of the agriculture.

Centres of Plant Domestication/Centres of Origin/Centres of Diversity

N. I. Vavilov contributed tremendously to the knowledge of global dispersal of crop plants and their wild relatives. Vavilov (1951) has shown that certain areas of the world are centres of much of the diversity in crop plants. Vavilov (1887-1943) investigated regions of great diversity in a systematic, global way. By means of extensive expeditions, Vavilov explored large areas of world from 1920-1943 and made an inventory of cultivated plant species, their primitive and wild relatives. After comprehensive evaluation of collections, he identified a concentrated diversity of certain crop species in some regions of the world. Vavilov proposed the concept of centres of diversity to summarize his observations. He defined the centre of origin of crop plant as geographic area(s) where it exhibits maximum diversity (i.e. where the greatest number of races and botanical varieties occur). Vavilov identified eight such centres some of which were sub-divided. These centres of origin or centres of diversity with examples of few associated crops are as follows:

1. Chinese (lettuce, rhubarb, soybean, turnip, buckwheat)
2. Indian/Hindustan (cucumber, eggplant, pigeon pea, mango, rice, oriental cotton)
   (a) Indo-Malayan (banana, citrus, yam, mango, sugarcane, tea, coconut, rice)

3. Central Asiatic/Central Asia (north India, Afghanistan, Turkmenistan)- (almond, flax, lentil, wheat, barley, onion, pea, fig, date, pear)
4. Near-Eastern/Asia Minor (alfalfa, apple, cabbage, rye, sugar beet,)
5. Mediterranean (celery, chickpea, durum wheat)
6. Abyssinian/Ethiopia (coffee, grain sorghum, pearl millet, African rice, yam, watermelon, cowpea, coffee)
7. South Mexican and Central American (lima bean, maize, papaya, upland cotton)
8. South American (Peruvian-Ecuadorian-Bolivian)- (Egyptian cotton, potato, tomato)
   (a) Chiloe (potato)
   (b) Brazilian-Paraguayan (yam, pineapple)

Vavilov called the centres with maximum diversity as the primary centres and the places where types migrate to as the secondary centres. Lately North America has been recognized as the centre of origin for plum, blueberry and sunflower. Different centres of diversity/origin of crop plants are depicted in Fig. 3.3.

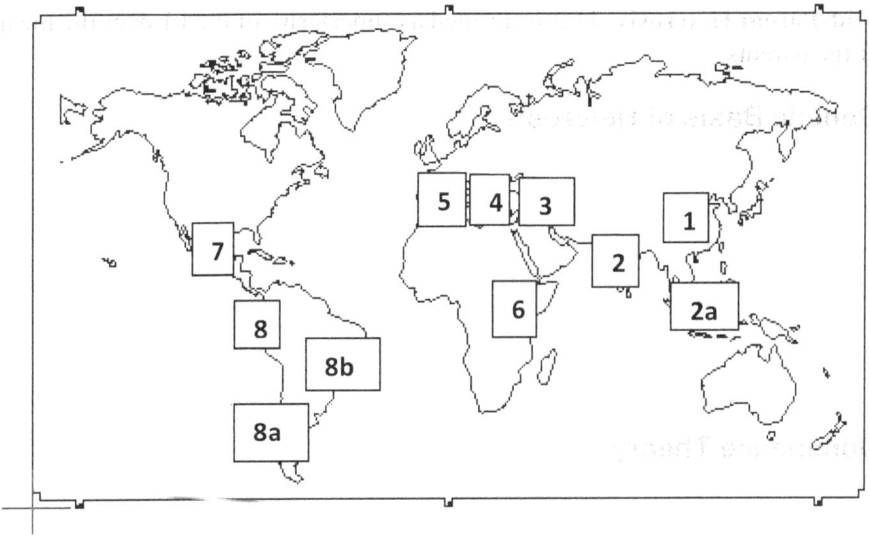

Fig. 3.3 Centres of origin of crop plants.

4

Heterosis and Inbreeding

Heterosis or hybrid vigour is the condition where an F1 hybrid falls outside the range of the parents with respect to some character or characters. It is also defined as a manifestation of heterozygosity, expressed as increased vigour, size, fruitfulness and resistance to diseases, insects or climate extremes. It has been observed in nearly every crop including both self and cross-pollinated crops. It is converse of inbreeding depression. However, even the species that show little or no inbreeding depression, often benefit from crossing and resultant hybrids. Therefore, heterosis has been more emphasized than the inbreeding depression. It is defined in two basic ways:

Better-Parent Heterosis: This is calculated as the degree by which the F1 mean exceeds the better parent in the cross. It is also known as heterobeltiosis.

Mid-Parent Heterosis: This is defined as superiority of the F1 over the mean of the parents.

Genetic Basis of Heterosis

Two theories have been advanced to explain the expression of heterosis. Dominance theory first proposed by Davenport in 1908 was later elaborated by Lerner. The second theory known as over-dominance theory was first proposed by Shull in 1908 and later by Mather and Jinks. A third theory, the mechanism of epistasis (non-allelic gene interactions) has also been proposed. Any viable theory should account for both inbreeding depression in cross-pollinated crops upon selfing and increased vigour in F1 upon hybridization.

Dominance Theory

This theory assumes that hybrid vigour in plants is conditioned by dominant alleles and recessive alleles being deleterious or neutral in effect. It follows then that a genotype with more number of dominant alleles will be more vigorous than the one with lesser number of dominant alleles. Crossing two parents with complementary dominant alleles will lead to concentration of more number of dominant favourable alleles in the F1 hybrid than the either parent. This is more favoured than the over-dominance theory by most of the scientists, although neither is completely satisfactory. Based on this theory, it should be possible to isolate true breeding inbred lines having all the desirable dominant genes together in homozygous condition and this line should be as vigorous as the F1 hybrid. However, this does not occur and it is hypothesized that linkage and large number of genes are the major hurdles in developing inbred lines that contain all homozygous dominant alleles. Inbreeding depression is observed as the deleterious recessive alleles that are suppressed in the heterozygous condition, are expressed in the homozygous condition. However, lately highly productive inbred lines in maize have been developed paving the way for developing and commercializing single cross hybrids in maize. The dominance theory of heterosis is illustrated with the following example where each dominant genotype contributes two units and a recessive genotype contributes one unit to the phenotype. The example shows that F1 is better than the better parent and shows heterosis.

P1	×	P2
AAbbCCddEE	×	aaBBccDDee
2 + 1 + 2 + 1 + 2 = 8	×	1 + 2 + 1 + 2 + 1 = 7

↓

AaBbCcDdEe

2 + 2 + 2 +2 + 2 = 10

D. S. Falconer developed a mathematical relationship between heterosis, degree of dominance and difference in gene frequency in the parents of the cross. The equation is as follows.

HF1 = $\sum dy^2$ where

HF1 = Deviation of hybrid performance from the mid-parent value.

d = Degree of dominance.

y = Difference in the gene frequency in the parents involved in the cross.

Maximum heterosis will occur when the values of the two variables (*d* and *y*) are each unity. This means that the parents to be crossed are fixed for opposite alleles ($y = 1.0$) and there is complete dominance ($d = 1.0$).

Over-dominance Theory

The first theory of heterosis (heterozygosity theory of heterosis) was proposed by Shull and East independently in 1908. It was restated later by Shull in 1948 as "Hybridity itself - the union of unlike elements, the state of being heterozygous – has……..a stimulating effect upon the physiological activities of the organisms whose effect disappears as rapidly as continuous inbreeding redues the progenies to the homozygous types". For this, he proposed the term "heterosis". According to this theory, heterozygote is superior to either homozygote i.e Aa is always superior to AA or aa. Initially this theory was applicable to single locus but later on it was extended to a series of alleles where the increased heterosis was in the order as a1a2 < a1a3 < a1a4, etc.

Numerically, over-dominance theory of heterosis can be demonstrated as follows, if we assume that the recessive, heterozygote, and homozygote dominants contribute one, three and two units to the phenotypic value, respectively. Heterozygosity per se is most superior of the three genotypes.

	P1	×	P2
	(aabbCCDD)		(AABBccdd)
Phenotypic value	1 + 1 + 2 + 2 = 6	↓	2 + 2 + 1 + 1 = 6
		AaBbCcDd	
		3 + 3 + 3 + 3 = 12	

Biometrical Concepts of Heterosis

Virmani *et al.* (2004) have reviewed the literature on this subject and summarized the information. The simplest genetical model for estimating heterosis is based on the difference between the hybrid and parental means which is called mid-parent heterosis and is often expressed as a percentage of mid-parent. Heterosis may also be defined as the amount by which the mean of F1 family exceeds its better parent. Quantitative genetic theory of heterosis is often developed on the basis of single locus theory. D. S. Falconer in 1960 demonstrated that mid-parent heterosis also called as panmictic mid-parent heterosis could be expressed as $H_{mp} = y^2d$, where d is the level of dominance and y is the difference in allele frequency of the parents. Therefore, the level of heterosis expressed in a hybrid depended largely on allelic frequency differences in two parents and presence of certain level of dominance. If two inbred lines instead of populations as in the case above are crossed, y can only be 0 or 1. Therefore, heterosis in a cross of two inbreds is a function of level of dominance at loci with different alleles only. Heterosis shown by F2 population derived from inbreeding of F1 (also called inbred mid-parent heterosis) is expressed as $H_{mp} = 1/2y^2d$. The difference between panmictic and inbred mid-parent heterosis reflects the vigour lost due to inbreeding. Genetic divergence and dominance are involved in both inbred and panmictic mid-parent heterosis.

Single locus theory has directly led to two prominent theories of heterosis called dominance and over-dominance hypothesis. Heterosis under dominance hypothesis is due to sheltering of deleterious recessive genes by their counterpart dominant alleles. Heterosis under over-dominance assumes that heterozygosity *per se* is important for hybrid vigour. Both the theoretical expectations and the experimental evidences suggest absence of genuine over-dominance. Non-allelic interaction and linkage disequilibrium appear to be misinterpreted as over-dominance. Beyond single locus theory, one can easily recognize that a complex trait like yield is actually due to several underlying multiplicative acting traits which are controlled by a number of genetic loci. Therefore, epistasis may actually contribute a great deal to the heterosis expressed in a cross. Several experimental evidences have conclusively proved that epistasis contributes a part of total hereditary variance in case of quantitative characters such as grain yield in maize.

For self-pollinated crops, if there were mid-parent heterosis but no inbreeding depression, then there would be good evidence for existence of additive by additive epistasis. Studies in rice, a self-pollinated crop, indicate wide-spread epistatic loci. Despite tremendous efforts made during several decades on nature of gene action to explain heterosis, not much is known about the exact number of loci involved in heterotic expression or their distributions for the effect or the interactions. It appears rather difficult to clarify the issues involved only using classical methods of quantitative genetics. Developments in biochemical and molecular techniques have opened new vistas to study the phenomenon of heterosis more clearly.

Physiological and Biochemical Concepts of Heterosis

Grain yield is a complex trait resulting from the action and interaction of several essential metabolic processes over a period of time. The increase in yield for many crop species is generally attributed to changes in allocation of assimilates rather than changes in total production. This attribute commonly known as harvest index has been significantly improved in rice and wheat and several other crops. However, this desirable feature causing increased yield must eventually be traceable to certain key metabolic reactions. Metabolic studies in sorghum have revealed that these differences certainly contribute to yield differences. There are views that metabolic efficiency may substantially relate to heterozygosity and heterosis. Accordingly, it is believed that advantage in metabolic establishment of heterosis should be a major reason for heterosis. R. H. Hageman and colleagues in 1967 proposed the metabolic balance concept for determining heterosis. According to this, (i) inbreds tend to have unbalanced metabolic systems with enzymes controlled at variable levels of activities from low to high, (ii) highly homozygous lines may have some important enzyme reactions that are severely limiting metabolism , and (iii) the specific limitations probably differ among individual lines which may be overcome in heterozygote by a proper choice of complementary inbred parents giving rise to better balanced metabolic system in the resulting hybrid. H. K. Srivastava in 1981 proposed the concept of mitochondrial complementation and suggested that heterotic hybrids were endowed with superior systems of chloroplasts and mitochondria and concluded that such superiority was provided by genomic and inter-genomic interactions.

There are reports that maize DNA is highly methylated and the methylation was found to be genotype, tissue and developmental stage specific as well as affected by the environment. Hybrids were found to be less methylated than inbreds. Inbreds were reported to vary in their methylation status and improved lines were less methylated than older and low yielding ones.

Barnawal *et al.* (2012) have summarized the latest position on heterosis as follows after extensive literature survey and analysis of the results and the concepts. Perceived by Charles Darwin in many vegetable plants and rediscovered by George H Shull and Edward M East in maize, heterosis or hybrid vigour is one of the most widely utilized phenomena, not only in agriculture but also in animal breeding. Although, numerous studies have been carried out to understand its genetic and/or molecular basis in the past 100 years, our knowledge of the underlying molecular processes that result in hybrid vigour can best be defined as superficial. Even after century long deliberations, there is no consensus on the relative/individual contribution of the genetic/epigenetic factors in the manifestation of heterosis. However, with the recent advancements in functional genomics, transcriptomics, proteomics, and metabolomics-related technologies, the riddle of heterosis is being reinvestigated by adopting systems-level approaches to understand the underlying molecular mechanisms. A number of intriguing hypotheses are converging towards the idea of a cumulative positive effect of the differential expression of a variety of genes, on one or several yield-affecting metabolic pathways or overall energy-use efficiency, as the underlying mechanism for the manifestation of heterosis.

Concept of Heterotic Pool/Group

A heterotic group has been defined as a group of related or unrelated genotypes from the same or different populations which display a similar combining ability when crossed with genotypes from other germplasm groups. A heterotic pattern on the other hand is a specific pair of heterotic groups which may be populations or lines that express in their crosses high heterosis resulting into high hybrid performance. Such information is required especially in commercial hybrid breeding programmes where the hybrid seeds are marketed to the farmers. This information helpd plant breeders in identifying complementary parental lines for maximizing hybrid vigour. New heterotic groups can be created by introgression of new traits from diverse sources. This concept has its origin in researches of maize breeders who observed that the inbred lines derived from certain populations tended to produce superior commercial hybrids. The heterotic groups may have come into existence because of unique allelic diversity among groups on account of genetic drift, accumulation of unique diversity by mutation or selection. Inter-allelic interaction (over-dominance) or repulsion phase linkage among loci manifesting dominance (pseudo-over-dominance) could explain the occurrence of high heterosis following a cross between lines derived from genetically divergent populations. In such cases, inter-group hybrids perform better than the intra-group hybrids.

The procedures to identify heterotic groups may include pedigree analysis, geographic isolation inference, measurement of heterosis and combining ability, and molecular marker analysis. A. E. Melchinger, a German maize breeder has listed following criteria to identify heterotic groups.

- High mean performance and genetic variance in the hybrid population.
- High per se performance and good adaptation of the parent population to the target region.
- Low inbreeding depression during inbreeding and inbred line development.

Inbreeding

Inbreeding is defined as any system of mating that leads to an increase in homozygosity. This is also defined as the mating of individuals more closely related than individuals mating at random. Selfing is the most intense form of inbreeding which provides the most rapid approach to homozygosity. When selfing is not possible, full-sib and half-sib mating are used to do inbreeding. The progeny of selfed plant is symbolized by S_1 and the selfed progeny of the S_1 by S_2 and so on. Occasionally symbol (x in a circle⊗) is used to designate selfed seed. The effects of inbreeding are as follows:

1. Production of chlorophyll deficient plants.
2. Defective endosperm.
3. Increase of homozygosity and development of uniform inbred lines.
4. Decline in vigour and reduction in productivity.

Coefficient of Inbreeding

Level of inbreeding is measured by coefficient of inbreeding (F) which is the probability of identity of alleles in the progeny by descent. It ranges from 0 (no inbreeding) to 1 (prolonged selfing and 100 % homozygosity). Increase in percentage homozygosity under selfing can be demonstrated as follows (Fig. 4.1):

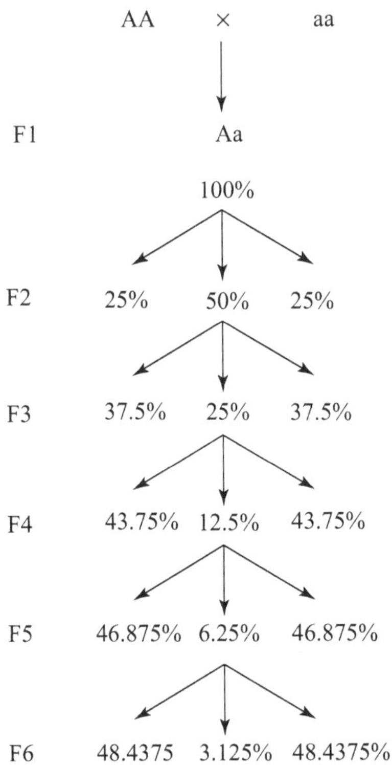

Fig. 4.1 Increase in percentage homozygosity under selfing.

Further, coefficient of inbreeding (F) devised by Wright is based primarily on the number and closeness of ancestral connections between male and female parents. The formula for coefficient of inbreeding is as follows:

$$F_x = \sum_A (1/2)^{n1+n2+1} (1 + F_A), \text{ where}$$

A is an ancestor common to the pedigree of both male and female parents of individual X, $n1$ is the number of generations between the male parent and A, $n2$ is the number of generations between the female parent and A, and F_A is the inbreeding coefficient of ancestor A. For self fertilization, $n1 = n2 = 0$ and the formula is reduced to as follows:

$F = 1/2\ (1 + F')$ where prime indicates the coefficient of the preceding generation.

The coefficient of inbreeding of an individual may be calculated by counting the number of arrows (*n*) which connect the individual through one parent back to the common ancestor and back again to its other parent and applying the formula as follows:

$$F_x = \sum(1/2)^n\ (1 + F_A).$$

For illustration, see the following example (Fig. 4.2):

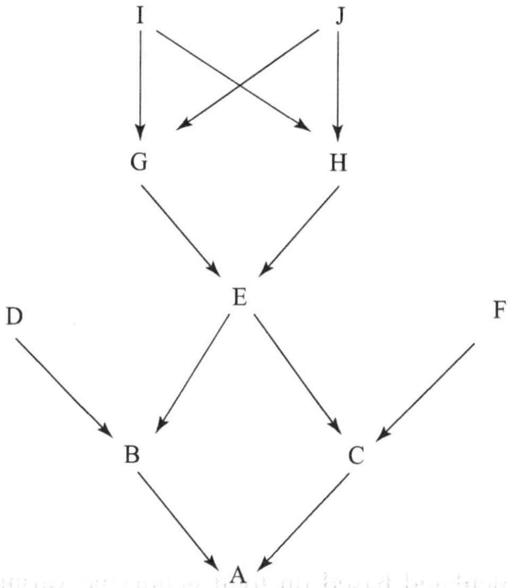

Fig. 4.2 Pedigree diagramme of an inbred.

There is only on pathway from *B* to *C* and that goes through ancestor *E*. However, ancestor *E* is itself inbred. Parents of *E* are full sibs.

F_E = Through route EGIH, there are three arrows and hence $(1/2)^3 = 1/8$

Another route is EHJG involving three arrows and hence will give rise to $(1/2)^3 = 1/8$

Sum of both these (1/8 + 1/8) comes to 1/4 = 0.25

Now, $F_A = \sum(1/2)^n\ (1 + F_A) = (1/2)^3\ (1 + 0.25)$

$= 1/8 \times 1.25 = 0.156$

(*n* is number of arrows connecting the individual '*A*' through one parent *B* back to common ancestor '*E*' and back again to its other parent '*C*').

5

Heritability, Genetic Advance and Gene Action

Any phenotype is the result of interaction of genotype and the environment. Any genotype cannot express itself unless proper environment for its expression is made available. Likewise, environment cannot create any phenotype unless a genotype is there to be acted upon by the environment. However, the effect of genotype and the environment is not fixed, its keeps on changing depending upon the relative role of the genotype. Heritability therefore specifies the proportion of the total phenotypic variability that is due to genetic causes. In other words, heritability is defined as the ratio of genetic variance to the total phenotypic variance. Heritability is the degree to which the characteristics of a plant are repeated and expressed in the progeny. This can be expressed quantitatively as follows:

Heritability (h^2) = Genotypic variance/Phenotypic variance

= Genotypic variance/Genotypic variance + environmental variance

= Variance G/Variance G + Variance E

The heritability calculated based on total genotypic variance in relation to total phenotypic variance is termed as "heritability in broad sense". This can be illustrated with the following simple numerical calculation.

Let us say we have an F2 population of wheat segregating for plant height where the height of 100 F2 plants has been measured and the calculated variance based on 100 numerical figures hypothetically comes to 300 which can be considered totally phenotypic variance as it includes genotypic variance based on 100 plants segregating for plant height and simultaneously it includes environmental variance also. Environmental variance is easily calculated by taking plant height measurements on non-segregating populations, like parental lines. Let us say the variance observed for plant height in parent 1 is 50 and the variance observed for plant height in parent 2 is 70. It should be clear that the variance observed in these two non-segregating lines is totally environmental as all the plants have same genotype. The average of these two variances is (50 + 70)/2 = 60. From these values heritability in broad sense can be easily calculated as follows:

Total phenotypic variance in F2 = 300

Total environmental variance = 60

Total genotypic variance = 300 – 60 = 240

Heritability in broad sense = 240/300 = 0.80 = 80%

Further, it is well recognized that genetic variance is further sib-divided into three components, namely, additive genetic variance, dominance variance and epistasis. The proportion of additive genetic variance to the total phenotypic variance is termed as "heritability in narrow sense". Different methods of estimating heritability include variance component method and parent off-spring regression method. Heritability in narrow sense is more useful to the plant breeders as additive component of genetic variance determines the response to selection. However, when breeding for clonally propagated crops like sugarcane and banana, etc. in which both additive and non-additive gene actions are fixed and transferred from parent to offspring, broad sense heritability is also useful. The magnitude of narrow sense heritability cannot exceed, and is generally less than the corresponding broad sense heritability estimate.

Variance Component Method of Estimating Heritability

A most common method to estimate heritability is using F2 and back-cross population variances as given below:

$$V_{F2} = V_A + V_D + V_E$$

$$V_{B1} + V_{B2} = V_A + 2V_D + 2V_E$$

$$V_E = (V_{P1} + V_{P2} + V_{F1})/3$$

A hypothetical data is given as follows:

	*P*1	*P*2	*F*1	*F*2	*BC*1	*BC*2
Mean	20	40	30	35	25	32
Variance	10	15	8	55	45	60

From above data, one can calculate as follows:

$V_E = (V_{P1} + V_{P2} + V_{F1})/3 = (10 + 15 + 8)/3 = 11$

$V_A = 2V_{F2} - (V_{BC1} + V_{BC2}) = 2 \times 55 - (45 + 60) = 110 - 105 = 5$

$V_D = (V_{BC1} + V_{BC2}) - V_{F2} - V_E = 45 + 60 - 55 - 11 = 39$

Heritability in broad sense = $(V_A + V_D)/V_A + V_D + V_E$

$= (5 + 39)/5 + 39 + 11 = 44/55 = 0.8 = 80\%.$

Heritability in narrow sense = $V_A/ V_A + V_D + V_E = 5/55 = 0.09 = 9\%.$

V_A = Additive genetic variance, V_D = Dominance variance,
V_E = Environmental variance.

Parent-Offspring Regression Method of Estimating Heritability

This method is simple and straightforward. The means of the parent and the offspring are obtained and regression of offspring on mid-parental values is calculated. Heritability in narrow sense is as follows:

Heritability in narrow sense (h^2): $b_{op} = V_A/V_P$ (b = regression coefficient, V_A and V_P, additive and phenotypic variance, respectively).

This estimate of heritability is based on certain assumptions like (i) trait of interest has diploid inheritance, (ii) population from which parents originated is randomly mated, (iii) the population is in linkage equilibrium implying no linkage among loci controlling the trait, (iv) parents are non-inbred and (v) there is no environmental correlation between the performance of parent and offspring.

Estimation of Heritability Through Evaluation of Parental Germplasm in a Randomized Complete Block Design

The heritability is obtained from analysis of variance table (ANOVA) as shown below:

Analysis of Variance Table

Sources of variation	Degrees of freedom	Expected mean sum of squares (EMS)	Hypothetical numerical mean sum of squares
Replication	$r - 1 = 4 - 1 = 3$	-	
Genotypes	$g - 1 = 20 - 1 = 19$	$6_e^2 + r6_g^2$	100
Error	$(r - 1)(g - 1) = 57$	6_e^2	60

$$\text{Genotypic variance } (6_g^2) = \frac{\text{(Genotypic mean sum of squares – error mean sum of squares)}}{\text{Number of replications}}$$

$$= \frac{(100 - 60)}{4} = 10$$

Phenotypic variance (6_p^2) = Genotypic variance + Error variance = 10 + 60 = 70

Heritability in broad sense = Genotypic variance/Phenotypic variance = 10/70 = 0.14 = 14%.

Applications of Heritability

1. Heritability estimate gives an idea if plant breeding methods can be applied successfully in the material under improvement.
2. Heritability estimates provide an indication about the most effective breeding methods.
3. It helps in predicting response to selection.

Genetic Advance or Response to Selection

It is quite common to select certain high yielding or better performing individuals for a specific trait from a variable population available to the breeder. Then, the selected individuals are grown and a new population better than the original base population for the target trait is created. In this process, we encounter three mean values for the target trait. Let us say the mean of the original population is $X1$, the mean of the selected plants is $X2$ and the mean of the population grown from the selected plants is $X3$. The differences between two means are defined as follows:

$$\text{Genetic advance or response to selection } (R) = X3 - X1$$

Response to selection (*R*) is the difference between the mean phenotypic value of the offspring of the selected parents and the whole of the parental generation before selection. In other words, response to selection is simply the change of population mean between generations following selection. Selection differential (*S*) is the mean phenotypic value of the individuals selected as the parents expressed as deviation from the population mean (i.e. from the mean phenotypic value of all individuals in the parental generation before selection). As mentioned by Allard (1960), it can be shown by method of the calculus that the expected genetic advance is equal to:

$$R = i6_p h^2$$

where

i = Selection intensity whose value can be seen from a table. It depends on the fraction of current population retained to be used as parents for the next generation. At 1%, $i = 2.668$, at 5%, $i = 2.06$ and at 10%, $i = 1.755$. More the plants selected, less stringent is selection and accordingly selection intensity is also reduced when more plants are selected.

6_p = Sigma *P* is phenotypic standard deviation of the mean yields of the original lines in the base population in which selection was initiated.

h^2 = Heritability coefficient estimated as the ratio of genotypic variance to the phenotypic variance.

Response to selection is related to heritability in the following manner also. $R = Sh^2$ (S = Selection differential, h^2 = Heritability).

The diagrammatic illustration of genetic advance taking certain hypothetical numerical values is given in Fig. 5.1.

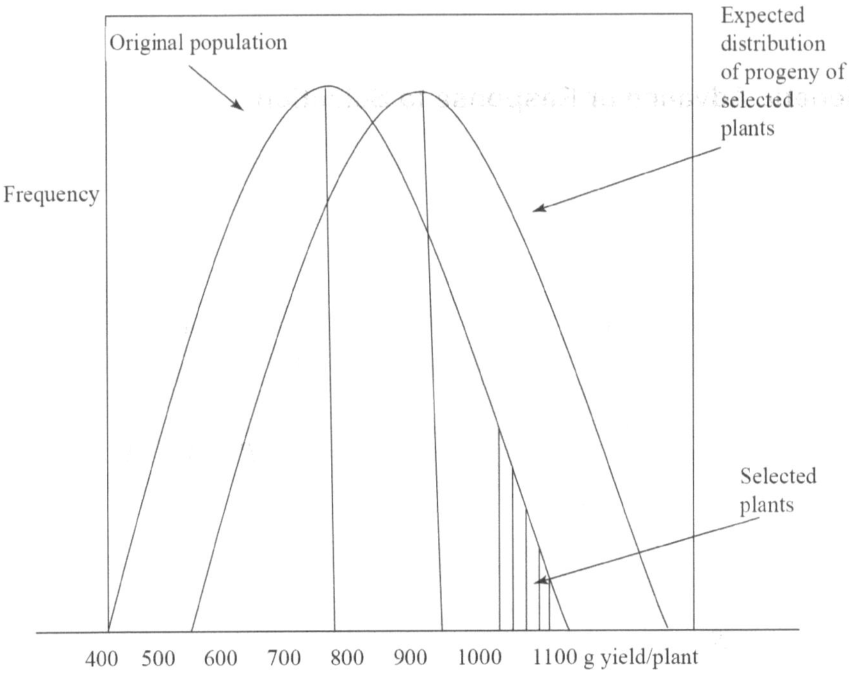

Fig. 5.1. Diagrammatic illustration of response to selection.

Let us say the original population has a mean $X1$ equalling 700g yield/plant and a standard deviation estimated to be 70 g/plant. The hypothesized heritability is 0.40. The expected mean of 5% of highest yielding plants is $X1 + i6_p = 700 + (2.06) \times (70) = 844.2$ g/plant. The progeny of these plants are expected to have a mean of $X1 + i6_p h^2 = 700 + (2.06 \times 70 \times 0.40) = 700 + 56.68 = 757.68$ g/plant. Therefore, the genetic advance under selection in this case is expected to be $757.68 - 700 = 57.68$ g/plant.

Correlated Response to Selection

Correlation is the degree of association between two related traits. In terms of response to selection genetic correlation is useful. In presence of strong genetic correlation between two traits, selection for one trait will cause a corresponding change in other related trait. This response to change by genetic association is known as correlated response. This could be caused by pleiotropism or linkage disequilibrium. While calculating correlated response to selection, normally genetic correlation is used, however, phenotypic correlation estimated as average over several environments may also be used. The correlated response (CR_y) to selection in primary trait (y) for a secondary trait (x) is given by the following equation (Falconer, 1960).

$CR_y = ih_x h_y r_A 6_{PY}$ (i = Selection intensity, h_x and h_y = Square root of heritabilities of respective traits x and y, r_A = Genetic correlation between traits and 6_{PY} = Standard deviation of Y).

Correlated response has important breeding application in homozygous, self-fertilizing species and apomicts. Additive genetic correlation is important in selection in plant breeding programmes. Additive breeding value is what is transferred to offspring and can be changed by selection. Therefore in those situations where two traits are additively genetically correlated, selection for one trait will produce a correlated response in another.

Selection for Multiple Traits

For simultaneous selection for several traits, the plant breeders adopt the following types of selections:

Tandem Selection: In this mode of selection, selection is focused on one trait at a time. Once this is achieved, selection for another trait is exercised and so on. Thus, this selection is also known as sequential selection or serial improvement. This takes a long time and is effective when genetic correlation does not exist between the target traits.

Independent Culling: This is also called as truncation selection and here selection is practiced for multiple traits in one generation. Here one might select for all the characters at the same time but independently, rejecting all individuals that fail to come to a certain level or standard for each character regardless of their values for any other of the characters.

Selection Index: It has been shown that the most rapid improvement of economic value is expected from selection applied simultaneously to all the component characters together where appropriate weights are given to each character according to its relative economic importance, its heritability and genotypic and phenotypic correlations between different characters as conceived and elaborated by Hazel and Lush in 1942. The practice of selection for economic value is thus a matter of some complexity. The component characters have to be combined together into a score or index in such a way that selection applied to the index, as if index were a single character, will yield the most rapid possible improvement of economic value. If the characters are uncorrelated, each character is weighed by the product of its relative economic value and its heritability. But if the genetic correlations are known, efficiency of selection index can be improved.

In selecting a multivariate phenotype, the breeder explicitly or implicitly assigns a weighting scheme to each trait resulting in the creation of a univariate trait also known as index based on which individuals are selected. The index is the best linear prediction of an individual's breeding value. It takes the form of a multiple regression of breeding values on all the sources of information available for the population. The methods used for constructing a selection index usually include information on heritability estimates, the relative economic value of each trait, and genetic and phenotypic correlation between the traits. The classic selection index is expressed as follows:

$$I = b_1x_1 + b_2x_2 + b_3x_3 \ldots\ldots$$

Where

x_1, x_2, x_3 are the phenotypic performance of the traits of interest, and b_1, b_2 and b_3 are relative weights attached to the respective traits. The index is meaningless unless it is used in comparing several individuals on a relative basis. Further, in comparing different traits, the breeder is faced with the fact that the mean and variability of each trait is different and frequently the traits are measured in different units. However, standardization of variables resolves this problem. It is hardly used in applied plant breeding in the field while carrying out selection of desirable plants/lines.

Gene Action

Four types of gene action are involved in expression of any quantitative trait. These are additive, dominance, epistatic and over-dominance. Conceptually, gene action is same for major genes and minor genes, however, difference being that the action of a minor gene is small and significantly influenced by the environment.

Additive Gene Action: The effect of a gene is said to be additive when each additional gene enhances the expression of the trait by equal increments. Assuming that one gene adds one unit, the genotypes aabb, Aabb, AABb and AABB will have phenotypic values as 0, 1, 3 and 4 respectively. The heterozygote will be exactly between the parents (AA = 2, Aa = 1, aa = 0). The performance of an allele will remain same irrespective of other allele at the same locus. In absence of environmental effect, the phenotype is a true reflection of genotype in additive gene action. Furthermore, a superior phenotype will breed true in the next generation making selection more effective. Additive gene action can be fixed through development of homozygous cultivars.

The average effects of genes of the parents determine the mean genotypic value of the progeny. The value of an individual judged by the mean value of its progeny is known as 'breeding value' of the individual. This is the value that is transferred from an individual to its progeny. This can be measured unlike the average effect of a gene. If an individual is mated to a number of individuals taken at random from the population, then the breeding value is twice the mean deviation of the progeny from the population mean. This deviation has to be doubled as the parent in question provides only half the genes in the progeny while the other half comes at random from the population.

Dominance Gene Action: This describes the relationship of alleles at the same locus. Dominance variance has two components, namely, variance due to homozygous alleles (additive variance) and variance due to deviations from additivity that make the heterozygote resemble one parent more than the other. When AA = Aa, the dominance is complete. The extent of dominance (d_A) is calculated as the deviation of the heterozygote (Aa) from the mean of the two homozygotes (AA and aa). In case there is no dominance, $d_A = 0$. If dominance is complete, $d_A = a_A$, whereas $d_A < a_A$ for incomplete or partial dominance, and $d_A > a_A$ for over-dominance. Here 'a' is the change which happens by replacing 'a' by 'A' in the genotype 'aa'.

Over-dominance Gene Action: This type of gene action happens when each allele at a given locus produces a separate effect on the phenotype and the joint effect of both the alleles exceeds the independent effect of both the alleles. For example, aa = 1 unit, AA = 2 units and Aa = 3 units. From breeding standpoint, over-dominance effects can be fixed only in the F1 hybrid cultivars through apomixis or chromosome doubling of the product from a wide cross.

Epistasis Gene Action: Epistasis is a term originally used by Bateson in 1909 to describe genes whose effects mask or cover the effect of other genes. Epistasis is often defined as the masking effect of expression of a gene by another locus. This is also described as non-allelic gene interaction in contrast to dominance which refers to non-additivity of alleles at the same locus that is intra-allelic interaction. The simplest epistatic di-hybrid F2 ratios for qualitative traits are 9 : 7, 13 : 3, 15 : 1, 9 : 3 : 4, and 12 : 3 : 1. With three gene interactions, the ratios such as 37 : 27, 55 : 9, and 27 : 9 : 9 : 19 can occur. For quantitative traits, the typical example of epistatic gene action could be as Aabb = 0, aaBB = 0 and A-B- = 4 which implies that individually A and B genes do not produce any effect but when they come together, there is an obvious effect.

6

Hardy-Weinberg Equilibrium

In a large random-mating population both gene frequencies and genotype frequencies remain constant from generation to generation in absence of migration, mutation and selection and the genotype frequencies are determined by the gene frequencies. These properties of a random mating population were first explained by G. H. Hardy and W. Weinberg independently in the year 1908 and are generally known as Hardy-Weinberg law. The operational understanding of this law involves three steps as follows:

1. From parents to the gametes they produce.
2. From union of the gametes to the genotypes in zygotes produced.
3. From genotypes of the zygotes to the gene frequencies in the progeny generation.

The gene and genotype frequencies in the original parent generation are assumed as follows:

Genes		Genotypes		
A	a	AA	Aa	aa
p	q	D	H	R

$P + q = 1$ and $D + H + R = N$

From the above, one can deduce the following:

Frequency of A (p) = (D + 1/2 H)/N

Frequency of a (q) = (1/2 H + R)/N

For example in a group of 80 individuals (4, 24, 52 respectively for AA, Aa, aa), following gene frequencies are calculated:

Proportion of 'A' genes (p) = (4 + 12)/80 = 0.2 and proportion of 'a' gene (q) = (12 + 52)/80 = 0.8.

When the proportion of zygotes is given in proportions, then p and q are straight away as follows:

$$p = D + 1/2\ \mathrm{H} \text{ and } q = 1/2\ \mathrm{H} + R.$$

Accordingly a population of 4, 24 and 52 individuals for AA, Aa and aa becomes 0.05, 0.30 and 0.65 in terms of proportions. From these proportions, the p and q will be as follows:

$$p = 0.05 + 0.15 = 0.20 \text{ and } q = 0.15 + 0.65 = 0.80 \text{ and } p + q = 1.00.$$

From the random mating among AA, Aa and aa, genotypes, the resultant zygotes will be as follows (Table 6.1):

Table 6.1 Composition of zygotic population after random mating among AA, Aa and aa genotypes.

Type of mating	Frequency of mating	Offspring (zygote)		
		AA	Aa	aa
AA × AA	D^2	D^2	-	-
AA × Aa	2 DH	DH	DH	-
Aa × Aa	H^2	1/4 H^2	1/2 H^2	1/4 H^2
AA × aa	2 DR	-	2 DR	-
Aa × aa	2 HR		HR	HR
aa × aa	R^2	-	-	R^2
Total	1.00	$(D + 1/2\ H)^2 = p^2$	$2\ (D + 1/2\ H)\ (1/2\ H + R) = 2\ pq$	$(1/2\ H + R)^2 = q^2$

The gene frequencies (p) and (q) in the population in Table 6.1 will be as follows:

$$p = p^2 + 1/2\ (2\ pq) = p^2 + pq = p(p + q) = p \text{ as } p + q = 1.$$

and $$q = q^2 + 1/2\ (2\ pq) = q^2 + pq = q(q + p) = q \text{ as } p + q = 1.$$

Thus, any population which has gene frequencies of A and a as p and q, will reach to an equilibrium for the different zygotes, namely, AA, Aa and aa in the ratio of p^2, 2 pq, q^2. This state will reach immediately after one generation of random mating disregarding the different genotype frequencies but with the same gene frequencies in the original population. This is numerically demonstrated as follows:

Original population

AA	*Aa*	*aa*
.05	.30	.65
.01	.38	.61
.18	.04	.78
0	.40	.60
.20	0	.80

$= p = .20$, $q = .80$ and the composition of the population obtained after random mating will be p^2, $2pq$, q^2 = .04, .32, .64.

7

Stability Parameters

Phenotype is the product of genotype and environment. In presence of genotype x environment interaction (differential performance of genotypes over the environments), phenotype is the product of genotype, environment and genotype x environment interaction. A stable genotype is one which interacts less with the environment or shows a minimum of g x e interaction. When g x e interaction is significant, one must investigate the specific types of g x e interactions such as genotype x location, genotype x year and genotype x location x year interactions by growing selected genotypes over multiple locations and years. Normally, the genotypes are evaluated for three years. A typical analysis of variance for g x e interaction is given in Table 7.1 (Allard, 1960).

Table 7.1. Analysis of variance based on evaluation of certain cultivars (genotypes) at two or more number of years and locations.

Source of variation	Degrees of freedom	Expected mean squares
Varieties	$V-1$	$\sigma_e^2 + R\sigma_{aly}^2 + RY\sigma_{al}^2 +$ $RL\sigma_{ay}^2 + RLY\sigma_a^2$
Varieties x years	$(V-1)\times(Y-1)$	$\sigma_e^2 + R\sigma_{aly}^2 + RL\sigma_{ay}^2$
Varieties x locations	$(V-1)\times(L-1)$	$\sigma_e^2 + R\sigma_{aly}^2 + RY\sigma_{al}^2$
Varieties x years x locations	$(V-1)\times(Y-1)\times(L-1)$	$\sigma_e^2 + R\sigma_{aly}^2$
Error	$(R-1)\times(\mathrm{F}YL-1)$	σ_e^2

V, R, L, Y refer to the number of varieties, replications, locations, and years, respectively.

From the above table, one can estimate the following components of variance:

σ_a^2 = A genetic component arising from genetic differences among varieties. = 2400

σ_{al}^2 = A component arising from interaction of varieties and locations. = 150

σ^2_{ay} = Variety-by-year component of variance. = 200

σ^2_{aly} = Variety-by- location-by- year component. = 3000

σ^2_e = The error variance. = 12000

The numerical values are the hypothetical values from which, total phenotypic variance can be calculated as follows:

$$\sigma^2_A = \sigma^2_a + \sigma^2_{al}/L + \sigma^2_{ay}/Y + \sigma^2_{aly}/LY + \sigma^2_e/RLY$$

$$= 2400 + 150/2 + 200/2 + 3000/4 + 12000/8 = 4825$$

Heritability = Genotypic variance/Phenotypic variance

= 2400/4825 = 0.49 = 49%.

Measures of Stability

Various parameters to measure stability/adaptability have been proposed by various authors like Wricke (1962), Finlay and Wilkinson (1963), Eberhart and Russell (1966), , Breese (1969), Hanson (1970), Tai (1971) and Shukla (1972) but the most commonly used is that of Eberhart and Russell (1966) according to which a desirable stable variety is one that has high mean yield, higher than the average yield of all genotypes under test, regression coefficient (*b*) close to unity and small deviation (s^2_d) from regression possibly close to zero. In this widely used model, growth response index (regression coefficient) measures response to increments in an improving environment. Genotypes with regression coefficient greater than one would be adapted to more favourable environment, while the one with regression coefficient less than unity would be adapted to less favourable growing environment. This approach is diagrammatically illustrated in Fig. 7.1.

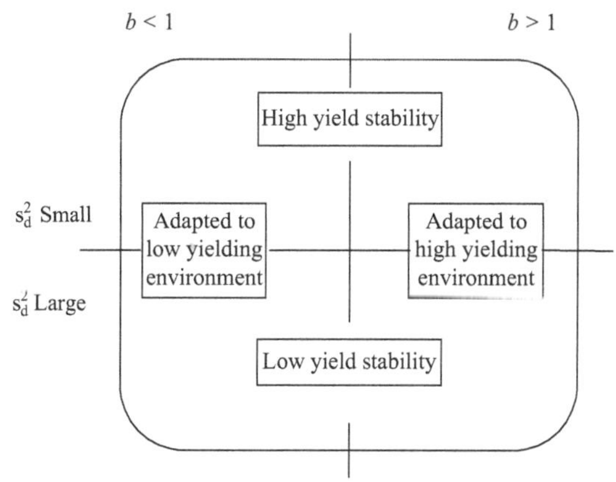

Fig. 7.1. Schematic illustration of stability parameters as defined by Eberhart and Russell (1966).

Breeding for Wider Adaptability

In self-pollinated crops, a parent of proven stability is selected and used in the crossing programme using as complementary parents and the standard pedigree method of breeding is practiced for isolation of recombinant pure-breeding lines equipped with high yield and greater stability. The segregating generations are grown in a number of contrasting target environments. This method of breeding is based on the principle of disruptive selection where selection is practiced for extreme phenotypes. International Wheat and Maize Improvement Centre (CIMMYT), Mexico practiced this method on a large scale and termed it as "Shuttle Breeding". Here the individuals/families having plasticity in different traits resulting into stability of yield over different environments are selected. Finally the individuals are selected on the basis of mean yield performance in such environments and are advanced.

In cross-pollinated crops while developing single cross hybrids, parental inbreds should be thoroughly evaluated and only those inbreds having higher adaptability should be used to produce hybrids. Again as in self-pollinated crops, while developing inbreds through pedigree method, individuals/ families should be tested in contrasting environments and selection should be based on the means of inbreds in different environments. In nutshell, while breeding for wider adaptability, there is obvious need of multi-location testing.

8

Mutation Breeding and Wide Crosses

The discovery of mutagenic effects of X-rays on the fruit fly (*Drosophila*) by H. Muller in 1927 and by L. J. Stadler in barley and maize in 1928 paved the way for researchers to experiment with its effects on various organisms. In 1928, H. Stubbe showed the use of mutagenesis in producing mutants in crop plants like tomato, soybean and others. The first commercial mutant was produced in tobacco in 1934. Reports by B. Sigurbjornsson and A. Micke mentioned 77 cultivars that were developed through mutagenesis prior to 1995. In 1995, this number was 484. This number has since been significantly increased covering a large number of crops, namely, corn, wheat, pea, chrysanthemum, poinsettia, dahlia, citrus, apple, peach and the modified traits include plant maturity, winter hardiness, lodging resistance, protein and lysine content and numerous ornamental mutants. By the end of 20th century, the number of mutant varieties rose to 2252 covering a range of crops like cereals, oilseeds, pulses, vegetables, fruits, fibres, and ornamentals. Majority of these mutants were released as direct mutants and the rest as a result of cross breeding with the mutants. Most of the mutant varieties have come as a result of using physical mutagens (X-rays, gamma rays, thermal and fast neutrons) with gamma rays alone accounting for more than 60% of the mutant varieties. By 2017, nearly 3250 mutant varieties belonging to 175 plant species have been developed and released where China, Japan, India, Russia, the Netherlands, Germany and USA are the leading countries. India with 345 commercialized mutants stands third after China (810 mutants) and Japan (481 mutants). Thus mutation breeding is a significant contributor to the development and release of commercial cultivars in several crops and for several traits.

Mutation breeding is considered to be more important in breeding vegetatively propagated species including field crops, ornamentals, and fruit and forest species. It is especially useful in ornamentals where novelty is often advantageous and of commercial significance.

Types of Mutations

Spontaneous or Natural Mutations: These are the natural mutations that occur spontaneously in nature at a very low rate of one in 100,000 to 1,000,000 gametes.

Induced Mutations: These are higher rate mutations caused by physical and chemical mutagens. Ionizing radiations including X-rays, neutrons and protons as well as the alpha, beta and gamma emissions of radioactive chemicals, ultra-violet light (a non-ionizing radiation) and a number of chemical substances (mustard gas, ethyl methane sulfonate –EMS, etc) cause induce mutations.

Mutations are also classified according to the type of structural changes produced: These are as follows:

Genomic Mutations: These are changes in chromosome number (gain or loss in complete set of chromosomes or parts of a set).

Structural Mutations: These mutations involve changes in chromosome structure (duplication of segments, translocation of segments).

Gene Mutations: These mutations arise as a result of changes in the nucleotide constitution of DNA.

Mutations may occur in nuclear DNA or in extra-nuclear genetic system. Cytoplasmic genetic male sterility is a good example of mutation which occurs in the chloroplasts.

In terms of gene action, mutations may be of following types:

Recessive Mutations: This occurs due to change of a dominant allele to a recessive allele for example change of 'A' to 'a'.

Dominant Mutations: This is change of a recessive allele to a dominant allele, for example, change of 'a' to 'A'.

The most frequent and common phenotypes in natural condition for any crop plant are called as wild types and the rare ones are called as mutant types. Mutation from wild to mutant form is known as forward mutation and the mutation from mutant to wild type is known as reverse mutation.

Based on changes in the nucleotide constitution of the DNA sequence, the mutations are classified as follows:

Transitions: DNA consists of four bases, A, T, G and C. A pairs with T and G pairs with C. Mutation resulting from conversion of one purine base to another purine or conversion of one pyrimidine base to another pyrimidine is called as transition.

Transversions: A transversion involves the substitution of a purine by a pyrimidine base and the vice-versa.

Mutation Breeding Process

1. **Material to be Treated and the Mutagen:** Seeds are the most common material to be treated with appropriate mutagenic agent. Usually dry

seeds are used for exposure to physical mutagens and for chemical mutagenic treatment, the seeds are soaked for about 24 hrs in the solution prepared from the chemical. The Treatments dosages that kill 50–70% of the seeds or the developing seedlings are preferred for mutagenesis. A few common dosages based on literature surveys are as follows:

Corn:	0.5 krad	for gamma irradiation
Peas:	5–15 krad	for gamma irradiation
Barley	10–15 krad	for gamma irradiation
Wheat	15–20 krad	for gamma irradiation
Mustard	> 100 krad	for gamma irradiation
Most crops	10–25 krad for treatment with X-rays	
Most crops	< 1%–4% for EMS	

While treating the seeds, it should be kept in mind that the M2 population should consist of 20,000-50,000 plants. While deciding the appropriate size of smaple to be treated, it should be understood that about 50% plants will fail to germinate and there might be some sterility in the M1.

2. **Growing of M1:** M1 especially in cereals are grown in dense stand so that only one tiller grows from each seed. This is suggested as first tiller in cereals has been found to show maximum number of mutations. It should also be kept in mind that the plant density should not be so dense that many of the plants may fail to develop since plants carrying mutations frequently are less competitive. Here experience will matter. The common physiological defects caused by the mutagenic treatment and observed in M1 are reduction in seedling height, chromosomal abnormalities and sterility. M1 plants rarely exhibit gene mutations as mutations are mostly recessive and the recessive mutations are detected in the M2 generation. Of course dominant mutations are detected in M1. Usually M1 plants with reduced fertility are discarded because sterility almost always results due to changes in chromosomes. Such plants do not show a higher rate of gene mutation in succeeding generations. M1 plants are bagged to prevent out-crossing. In self-incompatible species, several M1 plants should be bagged together or the entire M1 should be grown under isolation.

   Treated seed (M1) are planted in small plot to produce M2 seed that is harvested for planting an M2 spaced population. It is always desirable to plant plots of the untreated genotype (M0) in the mutation breeding project for comparison and identification of mutants.

   The seed is multi-cellular and therefore a mutation in a single cell will produce a chimeric plant. Thus, an M1 plant is subject to both diplontic selection (competition between mutated and normal tissue during the vegetative phase of development) and haplontic selection (competition between mutated and normal pollen during fertilization) in order to be included in the tissues forming seed on the plants.

It is worth mentioning that M1 plant population may have different mutations and hence it is logical to handle M1 population as segregating population like an F2.

3. **Handling of M2 Generation:** The common procedures to handle M2 plant generations are single seed descent and typical pedigree method. In SSD, one or few M2 seeds are selected from each M1 plant and bulked for planting M2 plants. Desirable M2 plants are harvested and progeny rowed. In pedigree method, each M1 plant is harvested separately. M2 progenies are grown from M1 plants. Desirable M2 plants are harvested and progeny rowed. Desirable rows are harvested and bulked separately. The $M_{2.4}$ lines are evaluated in replicated trials. Bulk selection method may also be applied to handle M2 generations. A weakness in bulk method is likelihood of exclusion of desired mutant as only a sample of the M2 is planted and quite often mutants have low productivity and hence may be excluded under bulk selection method.

   Normally selection begins in the M2 generation. Therefore, M2 should be grown under conditions that will permit the best expression of the target character. For example, if resistance to stem rust in wheat is desired, nursery should be inoculated with rust. Further, in M2 generation, every 10th or 20th row should be sown to untreated seed. Plants with desirable characteristics should be taken from as many rows as possible. Two-three plants should be selected from each selected row. Selected plants should be selfed especially if the crop is cross-pollinated. In only rare cases, will normal plants be harvested and carried forward into M3, even though mutations are not expressed sometimes until M3.

   In case the treated variety is an adapted type and it is desired to change one particular gene, then it will be necessary to grow very large M2 plant population with about 10,000 plants or more as the probability of occurrence of a desirable mutation without other undesirable characteristics is very rare event.

   In M3, it will be possible to have many rows breeding true for the character selected in the M2. The M3 and succeeding generations in case of self-pollinated crops are normally handled as per pedigree method. However, in case of cross-pollinated crops, M3 materials homozygous for the desired character could be bulked and improved by mass or recurrent selection.

Mutation Breeding in Vegetatively Propagated Species

The general radiation dose for vegetatively propagated crops ranges from 2000 *r* to 4000 *r* which is further reduced by half for sensitive herbaceous materials and is generally doubled for lignified or dormant tissues. Physical mutagens are preferred over chemical mutagens. Vegetatively propagated species tend to be highly heterozygous and therefore inbreeding is accompanied by inbreeding depression. In such situation mutation breeding has been found to be efficient since the genetic structure of the material remains largely undisturbed except the mutant gene. The plant parts used for mutagen treatment are normally those that can produce a bud from which a new plant can develop. These parts include tubers, rhizomes, shoot apex, cuttings and bulbs. The exposure to mutagen should occur as early as possible in the development of the bud and target the meristematic cells (the tissue in most plants containing undifferentiated cells). A mutation in meristematic cell is of critical importance to avoid chimeras which can delay selection because breeder may not be able to identify the mutant without propagating all material a number of times.

Chimeras are desirable in breeding of certain ornamental species such as African violet. Induced mutations in clones are dominant mutations unless the starting material has been heterozygous which could yield recessive mutants. Clonally induced mutations are primarily chimerical and start as sectorial mutants and later turn into periclinal types. Mutation being single cell event results in part of whole layers exhibiting chimeras. A periclinal chimera is one that involves the entire layer (the whole layer is mutated) whereas a mericlinal chimera involves only part of one layer. Periclinal mutations are stable. The number of initial cells in the layers in the growing tip determines the sectorial patterns. Sectorial patterns are mostly unstable, eventually producing periclinal structures. Because of this behavior, the breeder has to grow several clonal generations successively and select for the desired mutant as well as vegetative stability which is attained when uniformly periclinal lines of sub-clones are isolated. In a vegetative system, the treated plant material meristem is known as M1V1 and the next vegetative plant as M1V2, etc.

Clonally propagated woody species of crop plants have been considered particularly suitable for improvement through mutation breeding because beneficial mutants can be readily fixed by vegetative propagation and also the long juvenile period from germination to flowering in many tree crops prevents the full exploitation of the standard breeding methods in these crops. Many technical problems are encountered in studies on mutation production and in spite of initial optimism, it now seems unlikely that artificially induced mutations will replace the standard methods of breeding even in the most difficult breeding subjects like fruit trees.

Wide Crosses

Crosses between species of the same or different genera are commonly known as wide crosses. In other words, crosses involving material outside cultivated species are collectively described as wide crosses. When the wide cross involves another species, it is called as inter-specific and when it involves plants from another genus, it is called as inter-generic. Such crosses have contributed to plant improvement, gene and genome mapping, studies of chromosomal behavior and evolution as reviewed by Sharma (1995). These crosses have attracted significant attention as source of desirable characteristics for genetic improvement of crops.

Natural Wide Crosses as Example of Success in Crop Improvement

Natural wide crosses have been determined by the breeders and geneticists to be the origin of several modern day plants of economic importance. These include ornamentals such as irises, cannas, dahlia, roses, violets, tree crops such as apples, cherries and grapes and field crops such as wheat, tobacco, cotton and sweet potatoes. G. L. Stebbins observed that wide crosses may be more valuable in vegetatively propagated species than seed propagated species.

Synthetic (Artificial) Wide Crosses

Gene Transfer Between Species With The Same Chromosome Number: wide crosses between two tomato species, namely, *Lycopersicon pimpinellifolium* x *Lycopersicon esculentum* have been attempted to transfer genes related to diseases such as leaf mold, Fusarium wilt and tomato yellow leaf curl virus (TYLCV). Gene transfer in which both the parents have same chromosome numbers is usually without complication barring a few minor ones like reduction in pollen fertility. It is estimated that nearly all commercial tomatoes resistant to Fusarium wilt have derived their resistance from a wild source.

Gene Transfer Between Species With Different Numbers of Chromosomes: Common wheat is an allo-hexaploid with genomic formula of AABBDD having 21 pairs of chromosomes ($2n = 6x = 42$). There is a diploid species einkorn (*Triticum monococcum*) with 7 pairs of chromosomes and genomic formula of AA. There is a tetraploid wheat (*T. dicococcum*, AABB, $2n = 4x = 28$). Transfer of genes from species of lower ploidy to common hexaploid wheat is possible but reverse is not always possible. Stem rust resistance gene transfer is one such example in wheat.

Gene Transfer Between Two Genera: Common what comprises three genomes out of which genome DD has been contributed by *Aegilops*. Consequently, leaf rust resistance has been transferred from Aegilops to common wheat (*Triticum aestivum*) which is an example of gene transfer between two genera.

Overcoming Challenges of Crossing Barriers in Wide Crosses

There are reproductive barriers for gene transfer in wide crosses. The primary difficulty relates to obtaining fertile F1 hybrids due to mechanisms that promote especially gametic incompatibility. The gametic incompatibility leads to prevention of pollen from reaching to the stigma of other species, prevention of pollen germination and pollen tube growth down the style/ prevention of fertilization, and prevention of development of the zygote into a seed and into a mature plant. Even if there is no gametic incompatibility, there are additional barriers related to hybrid breakdown. Several techniques have been devised to increase the chances of recovering viable seeds and plants from a wide cross. These techniques are as follows:

1. Conducting reciprocal crosses.
2. Shortening the length of style as practiced easily in corn.
3. Applying growth regulators like naphthalene acetic acid and gibberellic acid to promote rapid pollen tube growth.
4. Modifying ploidy level where both the species are brought to the same level of ploidy.
5. Use of mixed pollen from a compatible species with pollen from an incompatible species.
6. Removing stigma as applicable to potato.
7. Grafting of female parent to the male parent has been reported to promote pollen tube growth.
8. Protoplast fusion.
9. Embryo rescue.

9

Polyploidy in Plant Breeding

Polyploid is any organism with other than two basic sets of chromosomes, that is, monoploid, triploid, tetraploid, and various aneuploids. Species that have their chromosomes as multiples of a basic number (x) are polyploids. Many of the species which have been developed most recently are polyploids. Autoploidy is not known to have significantly impacted the evolution of species. Autoploids of commercial importance include banana, a triploid which is seedless (diploid bananas have hard seeds not desirable for consumption). Other important autoploids are tetraploid crops, namely, alfalfa, peanut, potato and coffee. Spontaneous autoploids are important in horticultural crops where gigas feature has produced superior varieties of flowering ornamentals of narcissus, tulip, hyacinth, gladiolus, and dahlia among others. Triticale is the most common example of man-made cereal through allopolyploidy. A number of economically important crops are natural alloploids. These include wheat, oat, tobacco, cotton, sugarcane, strawberry and blueberry, wild mustard, black mustard, rutabaga/rape, etc.

Induction of Polyploidy

The use of ccolchicine facilitated the production of the vast number of polyploids that have been experimentally developed in the past 50 years. Colchicine is an alkaloid obtained from crocus (*Colchicum autumnale*) . Colchicine has been most successful when it has been applied to rapidly developing somatic tissues like germinating seeds, young seedlings and developing buds. For seed and seedling treatments, the concentration of colchicine in aqueous solution has been 0.1 to 0.4% for the time varying from 30 minutes to 3 hrs. After treatment, seeds or seedlings are thoroughly washed. Extreme care should be exercised while handling colchicines. It should be kept out of eyes and hands should be properly washed after any contact to avoid skin irritation. Colchicine disrupts spindle formation and prevents the migration of the undivided chromosomes to the poles at anaphase. In other words, a restitution nucleus is formed with twice the normal number of chromosomes. Without chromosome migration to poles, cell division does not take place.

Effects of Autoploidy

- Increase in stomatal cell size.
- Larger cell size.
- Greater leaf thickness.
- Often shorter and broader leaves.
- Increase in size of plant organs (*gigas* character).
- Increase in vitamin A content in tetraploid corn than in diploid.
- More vitamin C reported in many vegetables and fruits following chromosome doubling.
- In general, slower growth rate in tetraploids than in diploids due to lower auxin contents.
- Delayed and prolonged flowring.
- Reduced branching and tillering in grasses.
- Poor pollen production and sterility in autoploids limiting their utilization in agriculture.
- Break-down of self-incompatibility in tetraploids over diploids.

Terminology in Polyploidy

Ploidy refers to number of copies of entire chromosome set in a cell of an individual. The complete chromosome set is characteristic of, or basic to, a species. The number of chromosomes in ancestral diploid ancestors of polyploids is represented by 'x'. Furthermore, basic set is monoploid set. The haploid number (n) is the number of chromosomes that are found in gametes. This (n) number represents half the chromosome numbers found in somatic cells which is designated by '$2n$'. A diploid species such as corn has $n = 10$ and $2n = 20$. In diploid species like this $2n = 2x$. In case of autotetraploid, $2n = 4x$ and $n = 2x$. Sometimes a species may have more than two genomes from different sources. In such cases, each genome is represented by different letters. For example, wheat a hexa-allopolyploid has chromosome sets from three different origins and accordingly its genomic formula is given by AABBDD.

Two types of chromosome numbers have been observed. In one type called euploidy, the individuals contain multiples of the complete set of chromosomes that is characteristic of the species (the basic number, x). In another called aneuploidy, individuals contain incomplete set of chromosomes that may be equivalent to the euploid number plus or minus one or more specific chromosomes. When euploids contain multiples of genomes from the same species, they are called autoploids and when a combination of genomes from different species is involved, then that is called as alloploid. A polyploid whose somatic chromosome number is made up of the entire somatic complements of two species, is termed as amphidiploid. Various terms are summarized in Table 9.1.

Table 9.1. Terminology of polyploidy.

Name	Genome/Formula	Description (somatic chromosome complement where A, B and C are non-homologous chromosomes
Diploid	$2x = (ABC)\ (ABC)$	Contains two of a basic chromosome set
Euploid		
Monoploid	$x = (ABC)$	Only one basic set of chromosomes is present
Triploid	$3x = (ABC)\ (ABC)\ (ABC)$	Three sets of basic chromosomes are present
Autotetraplod	$4x = (ABC)\ (ABC)\ (ABC)\ (ABC)$	Four sets of basic chromosomes are present where each of the repeated sets of chromosomes is considered to be identical
Allotetraploid	$2x + 2x' = (ABC)\ (ABC)\ (A'B'C')\ (A'B'C')$ or $(ABC)\ (ABC)\ (DEF)\ (DEF)$	The hybrid is made between two basic diploid species whose chromosome sets are completely differentiated
Aneuploid		
Nullisomic	$2x - 2 = (AB)\ (AB)$	Both members of one pair of chromosomes are missing
Monosomic	$2x - 1 = (ABC)\ (AB)$	A monosomic lacks one chromosome of the normal complement
Double monosomic	$2x - 1 - 1 = (AB)\ (AC)$	A double monosomic lacks two different chromosomes of the normal complement
Trisomic	$2x + 1 = (ABC)\ (ABC)\ (C)$	A trisomic has two complete sets of chromosomes plus a single extra chromosome that is one particular chromosome is present in triplicate
Double trisomic	$2x + 1 + 1 = (ABC)\ (ABC)\ (A)\ (B)$	An extra chromosome each of two different pairs is present
Tetrasomic	$2x + 2\ (ABC)\ (ABC)\ (A)\ (A)$	Two extra chromosomes of one particular chromosome are present

Application of Polyploidy in Breeding of Seedless Watermelon

In this case, a superior line of diploid watermelon is converted into tetraploid through treatment with colchicines. It is better to use tetraploids with gray colour rind and the diploids with striped rinds to identify and discard selfed tetraploids (gray rind) in the progeny. Colchicine (1-2%) is applied to the shoot apex of the diploid plant seedlings as soon as the cotyledons emerge from the soil. This causes chromosome doubling and development of tetraploid seedlings. Normally one drop of the colchicines solution is applied to smaller seedlings and two drops to the larger seedlings. The success rate of chromosome doubling is about to 1% only. Tetraploids are detected by chromosome count or by count of number of chloroplasts in each side of the guard cells. Diploids have 5-6 chloroplasts/side in comparison to 10-14/side in tetraploids. Morphologically, tetraploid watermelons tend to have thicker leaves, shorter stems and slower growth than the diploids.

The putative tetraploids (T1) are evaluated in the next generation (T2) to confirm the tetraploidy. Selected plants are selfed/sibbed in isolation to stabilize the new genotypes and to increase their fertility to have more seeds of tetraploid lines for evaluation. Tetraploid lines are evaluated for performance as triploid hybrids by making crosses of tetraploids with the diploids. Successful parents produce triploid and seedless (lacking empty seed coats) hybrids with high yield and gray rind colour. Triploids are produced by crossing tetraploids with diploids using hand pollination or insect pollination in isolation. When developing triploids by insect pollination in isolation, the two parents should have different rind colours for easy identification of hybrid than the selfed ones. As mentioned earlier, usually tetraploid female parent has gray rind and the diploid male parent has striped fruits. The triploid produced will have striped fruits. The procedure is outlined in Fig. 9.1 (Acquaah, 2007).

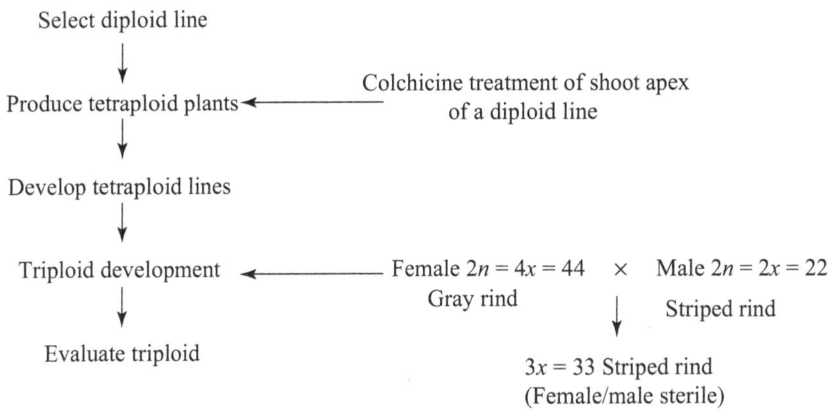

Fig. 9.1. Generalized steps in the breeding of seedless watermelons.

Genetics of Polyploids

Allopolyplloids or Alloploids: Alloploids arise from the combination and subsequent doubling of different genomes. The genomes that are combined differ in degree of homology, some are close enough to allow pairing with each other whereas others are too divergent to pair. Sometimes only segments of the chromosomes are different and the resulting alloploid is called as segmental alloploid. Chromosomes from two genomes are said to be homeologous when they are similar but not identical (homologous). Most alloploids have evolved certain genetic systems that ensure that pairing occurs between the chromosomes of the same genome. A classical example is available in wheat ($2n = 6x = 42$) where a gene on chromosome 5B designated as *Ph* enforces diploid like pairing within genomes of wheat and wheat behaves like a typical diploid. In absence of the gene *Ph*, there is pairing between homeologous chromosomes of the three genomes (A, B and D) resulting in formation of multivalents at meiosis I.

Alloploids exhibit several meiotic features. Sometimes, chromosomes pair as bivalent and produce disomic ratios. Tetrasomic ratios are expected for loci in case multivalent associations are found in allotetraploids. Duplicate factor ratio is a general indication for alloploid origin of species. This ratio when observed in forms having normal diploid inheritance, may be taken as a slender evidence that a species has undergone some increase of ploidy at least in respect of particular chromosomes or chromosome segments.

Autopolyploids: For adherence to simple concept and understanding, this section is limited to autotetraploids. In an autoterploid, five genotypes are possible for a given locus as shown below:

1. AAAA (quadruplex)
2. AAAa (triplex)
3. AAaa (duplex)
4. Aaaa (simplex)
5. Aaaa (nulliplex)

These zygotes arise from the fusion of three different types of gametes, AA, Aa and aa whose relative frequency will depend upon cytological events at meiosis. In a newly formed autotetraploid, each chromosome of the original haploid set will be represented by four completely homologous chromosomes and naturally there will be quadrivalent formation in place of bivalents observed in the diploids. The quadrivalents may pass through diakinesis as a ring or chain of four chromosomes, as chain trivalent and one univalent or as two bivalents depending upon number and location of chiasmata at diplotene. If pairing between four chromosomes is perfectly random (each chromosome has an equal chance of pairing with any one of the other three) and if random orientation of centromeres occurs at first metaphase of meiosis, the chromosomes will separate randomely to the poles and gametes will have random pairs of chromosomes drawn from the set of four chromosomes. Depending upon the mode of distribution of sister chromatids (chromatids derived from the same chromosome), two types of chromosome segregations occur. These are described as follows:

Random Chromosome Segregation

This occurs under the situation when only bivalents are formed. As a result two chromatids carrying a particular allele (let us say two dominant alleles) are always joined to the same kinetochore and always move to the same pole in first anaphase. However, they separate in the second anaphase. Thus, two dominant alleles will never end up in the same gamete. This type of partitioning of chromatids where the sister chromatids never end up in the same gamete is called as 'random chromosome segregation or assortment'. Random chromosome segregation is simplest and one of the extreme forms of segregation in autotetraplod. It can occur only under special circumstances of random pairing and absence of chiasmata between centromere and locus. Random chromosome segregation by twos can be more readily understood if the four chromosomes bearing a given locus are numbered 1-4 and three segregating genotypes as A1A2A3a4, A1A2a3a4 and A1a2a3a4. It should be clear that the remaining two genotypes (A1A2A3A4 and a1a2a3a4) will not segregate. The gametic output of the three genotypes based on random distribution of entire chromosome can be illustrated as follows:

Genotype of autotetraploid	Genotype of gametes	Gametic ratio
A1A2A3a4	A1A2, A1A3, A1a4 A2A3, A2a4 A3a4	1 AA: 1Aa
A1A2a3a4	A1A2, A1a3, A1a4 A2a3, A2a4 A3a4	1 AA: 4Aa: 1aa
A1a2a3a4	A1a2, A1a3, A1a4 a2a3, a2a4 a3a4	1 Aa: 1 aa

Gametic output under random chromosome segregation for a triplex can be shown as follows also where eight chromatids have been identified by numerals from 1 to 8. While deriving gametes, care has been taken that no two sister chromatids end up in the same game i.e. A1 and A2, the two sister chromatids are not allowed to go together and so on.

	A1	A2	A3	A4	A5	A6	a7	a8
A1			A1A3	A1A4	A1A5	A1A6	A1a7	A1a8
A2			A2A3	A2A4	A2A5	A2A6	A2a7	A2a8
A3					A3A5	A3A6	A3a7	A3a8
A4					A4A5	A4A6	A4a7	A4a8
A5							A5a7	A5a8
A6							A6a7	A6a8
a7								
a8								

The gametic ratio = 12 AA: 12Aa = 1 AA: 1 Aa. Various phenotypic ratios expected from random chromosome segregation are shown in Table 9.2.

Table 9.2. Phenotypic ratios in autotetraploids under random chromosome segregation and assuming complete expression of dominant allele in single dose.

Parents	Gametic ratio	Phenotypic ratio in the zygotes
A1A2 a3a4 × A1A2a3a4	(1 AA: 4 Aa: 1 aa) × (1 AA: 4 Aa: 1 aa)	35 A: 1a
A1A2a3a4 × A1a2a3a4	(1 AA: 4 Aa: 1 aa) × (1 Aa: 1 aa)	11 A: 1 a
A1A2a3a4 × a1a2a3a4	(1 AA: 4 Aa: 1 aa) × (all aa)	5 A: 1 a
A1a2a3a4 × A1a2a3a4	(1Aa: 1 aa) × (1Aa: 1 aa)	3 A: 1 a
A1a2a3a4 × a1a2a3a4	(1 Aa: 1 aa) × (all aa)	1 A: 1 a
A1A2A3A4 × A1A2A3A4	(All AA) × (all AA)	All A
A1A2A3a4 × A1A2A3a4	(1 AA: 1 Aa) × (1 AA: 1 Aa)	All A
a1a2a3a4 × a1a2a3a4	(All aa) × (all aa)	All a
A1A2A3a4 × A1A2a3a4	(1 AA: 1Aa) × (1 AA: 4 Aa: 1 aa)	All A

The ratios given in Table 9.2 are expected under the extreme situation that there is always bivalent formation which means no quadrivalent formation and there is no crossing over between the centromere and the locus. Thus, obviously the ratios given in the table cannot be expected in every case and departures from the expectation are highly probable.

Random Chromatid Segregation

When there is formation of quadrivalents (pairing among four homologous chromosomes) and 50% crossing cover between the locus under consideration and the centromere, sister chromatids at this locus can end up attached to different kinetochores giving rise to a situation where sister alleles may be included in the same gamete or in different gametes depending upon the distribution of the chromatids at the two meiotic anaphases. Under this situation, the partitioning of eight chromatids to gametes will be random and accordingly this process is called as random chromatid segregation. The gametic out-put in a triplex under random chromatid segregation is shown as follows (Fig. 9.3) where each chromatid of a chromosome has been identified, for example, the two sister chromatids of chromosome A1 are identified as A1.1 and A1.2.

	A1.1	A1.2	A2.1	A2.2	A3.1	A3.2	a4.1	a4.2
A1.1		A1.1A1.2	A1.1A2.1	A1.1A2.2	A1.1A3.1	A1.1A3.2	A1.1a4.1	A1.1a4.2
A1.2			A1.2A2.1	A1.2A2.2	A1.2A3.1	A1.2A3.2	A1.2a4.1	A1.2a4.2
A2.1				A2.1A2.2	A2.1A3.1	A2.1A3.2	A2.1a4.1	A2.1a4.2
A2.2					A2.2A3.1	A2.2A3.2	A2.2a4.1	A2.2a4.2
A3.1						A3.1A3.2	A3.1a4.1	A3.1a4.2
A3.2							A3.2a4.1	A3.2a4.2
a4.1								a4.1a4.2
a4.2								

Fig. 9.2. Random chromatid segregation in a triplex (AAAa = A1.1A1.2 A2.1A2.2 A3.1A3.2 a4.1a4.2).

The gametic ratios from a triplex as shown in Fig. 9.3 are 15 AA: 12 Aa: 1 aa. The repetition of the same exercise as shown in Fig. 9.3 will yield a gametic ratio of 1 AA: 12 Aa: 15 aa from a simplex (Aaaa) and a ratio of 3 AA: 8 Aa: 3 aa from a duplex (AAaa). Various gametic and phenotypic ratios under random chromatid segregation are given in Table 9.3 (Williams, 1964).

Table 9.3. Gametic and phenotypic ratios arising out of random chromatid segregation in an autoteraploid assuming full expression of dominant allele in single dose.

Parents	Gametic ratio	Phenotypic ratio
AAAa × AAAa	(15AA: 12Aa: 1aa) × (15AA: 12Aa: 1aa)	783A: 1a
AAAa ×AAaa	(15AA: 12Aa: 1aa) × (3AA: 8Aa: 3aa)	389A: 3a
AAAa × Aaaa	(15AA: 12Aa: 1aa) × (1AA: 12Aa: 15aa)	769A: 15a
AAAa × aaaa	(15AA: 12Aa: 1aa) × (all aa)	27A: 1a
AAaa × AAaa	(3AA: 8Aa: 3aa) × (3AA: 8Aa: 3aa)	187A: 9a
AAaa × Aaaa	(3AA: 8Aa: 3aa) × (1AA: 12Aa: 15aa)	347A: 45a
AAaa × aaaa	(3AA: 8Aa: 3aa) × (all aa)	11A: 3a
Aaaa × Aaaa	(1AA: 12Aa: 15aa) × (1AA: 12Aa: 15aa)	559A: 225a
Aaaa × aaaa	(1AA: 12Aa: 15aa) × (all aa)	13A: 15a

Breeding Implications of Autotetraploidy

It can be seen that random chromatid segregation in AAAa (triplex) and Aaaa (duplex) genotypes will produce gametic ratios of 15 AA: 12 Aa: 1aa and 3

AA: 8 Aa: 3 aa respectively yielding a small proportion of aa or AA gametes respectively. It is thus possible to recover a small proportion of fully recessive phenotypes from selfing of a triplex and a similar number of homozygous dominant phenotypes from selfing a simplex genotype in an autotetraploid. Thus, we need to raise a population having large number of plants while dealing with the autotetraploids.

Another aspect of autoploid genetics with a plant breeding implication is difficulty of distinguishing between a triplex (AAAa) and a quadruplex (AAAA) on the basis of a progeny test assuming random chromosome segregation. Both genotypes (AAAa and AAAA) will breed true for the dominant allele. To identify a triplex plant, the breeder will have to advance the progeny one more generation to identify the duplex plants of S1. Attaining genetic purity in autotetraploids is difficult not only because it is challenging to identify triplex plants but also because deleterious genes may persist in an autotetraploid manifesting themselves only rarely in the homozygous genotype. The breeder would need an additional two generations in order to identify the homozygous dominant genotype unequivocally (Aquaah, 2007).

D. R. Dewey has summarized the properties of plant species suited for induction of polyploidy as follows (Aquaah, 2007):

1. The species has low chromosome number.
2. The economic part of the plant is vegetative matter (for example, forage species).
3. The plant is cross-pollinated.
4. The plant is perennial in habit.
5. The plant has the ability to reproduce vegetatively.

10

Pure-Line Selection in Self-Pollinated Crops

A line homozygous at all loci, ordinarily obtained by successive self-fertilization in plant breeding is known as pure-line. W. L. Johannsen (1903), a Danish biologist provided a sound scientific basis for selection in self-pollinated crops when he defined 'pure-lines' and described the genetic mechanism by which they are established through a series of experiments for seed weight selection in bean, variety Princess. Pure-lines in principle are supposed to be 100% homozygous but that hardly happens in practical plant breeding programmes and usually the so called pure-lines are 'near pure-lines. Pure-lines tend to be uniform in the traits of interest. Genetic mutations, admixtures and limited out-crossing are the main causes of variation within pure-lines. This is a rapid and most in-expensive method. Pure-line theory may be summarized as follows:

- Lines that are genetically different may be successfully isolated from within a population of mixed genetic types.
- Any variation that occurs within a pure-line is due to environmental factors and is not heritable and consequently further selection within a pure-line is not effective.

Pure-lines thus developed are used as cultivar after due evaluation and following release procedures. They are also used as parents in crosses in development of new cultivars/breeding lines.

Pure-Line Breeding Procedure

The steps involved, in general, are as follows:

1. **Year 1:** Have a variable base population (new introduction, segregating population from crosses, landrace) and space plant the same under appropriate growing conditions. Harvest better looking plants individually and put the clean seed into individual bags/packets without mixing the seed of any two plants.

2. **Year 2:** Grow individual plant-to-progeny rows from the seed of plants harvested in first year. Harvest better and uniform looking progenies individually and bulk the seed from such selected and harvested progenies separately into individual bags. Ensure no mixing of seed from any two progenies harvested. Each bulk is termed as a new strain/breeding line.
3. **Year 3-5:** Conduct station trial of newly created strains/breeding lines along with the standard checks in replicated trial and record all desirable traits.
4. **Year 6-9:** Conduct coordinated multiplication trials for the lines selected based on station trial for three years at multi-locations for detailed observations and identification of a few outstanding lines for further release and notification provided the selected line(s) are found to be better than the currently grown best cultivar.

11

Pedigree Selection in Self-Pollinated Crops

This method has been first described by H. H. Lowe in 1927. This is the most widely used method to handle segregating generations following crosses between two parents to develop new cultivars combining desirable traits from both the parents. In this method, superior types are selected in successive segregating generations and a record is maintained of all parent-progeny relationships. The records help the breeder to advance only progeny lines with plants that exhibit genes for the desired traits. Selection begins in F2 generation where individual plants are selected which in judgment of breeder will produce the best progeny in F5/F6 when the selection is finally concluded. In F3 and F4 generations, many loci will have become homozygous and family characteristics begin to appear. By F5/F6 generation, most families are expected to be homozygous and uniform and hence selection within such families is no longer effective. At this stage selection emphasis shifts to the selection among families rather than within families.

Selection of parents is of great significance for the success of this method. Normally, this method is used to replace already established variety. Therefore, one parent should be selected on the basis of proven performance in the particular area where the breeding is in progress. The second parent should be such that it complements the first parent for the character under improvement. Sometimes, three or four parents may also be used to supply the desired character by way of three-way or double crosses.

Common Plant Breeding Notations Used in Pedigrees/Crosses

Symbol for Basic Crosses and Selfing Generations: The symbol F (filial) denotes the progeny of a cross between two parents. The subscript (x) represents the specific generation (F_x). The progeny from crossing of two parents is written as F_1 and selfing of F_1 ($F_1 \times F_1$) yields F_2 plant. Planting seed from F_1 plant will give rise to F_2 plant population, the most diverse generation following a cross in which plant breeders generally practice selection. Selfing an F_2 plant produces F_3 plants and so on. It should be understood that the seed is one generation ahead of the plant, that is, an F_2 plant bears and produces F_3

seed. Selfed generation products, that is, inbred lines are denoted as S_1 and S_2 where subscript 1 and 2 indicate the number of selfing generations. Selfing is denoted by writing letter 'x' inside a circle.

Inbred lines are described by two systems. System I describes an inbred line based on the generation of plants that are being currently grown. System II describes both the generations of the plant from which the line originated as well as the generation of plants being currently grown. For example:

Case 1: The base population is F_2. The breeder selects an F_2 plant from the population and grows the F_3 seeds in the next season.

System I: The planted seed produces an F_3 line.

System II: The planted seed produces an F_2 derived line in F_3 and this line will be designated as $F_{2:3}$ line.

If seed from the F_3 plants is harvested after selfing and bulked, and the breeder samples F_4 seed in th next generation, the designation of inbred lines will be as follows:

System I: The planted seed produces an F_4 line.

System II: The planted seed produces an F_2 derived line in F_4 to be designated by $F_{2:4}$ line.

Notations for Pedigrees: Pedigree is an ancestral record of any derived line and it helps the breeder in retracing the steps in a breeding programme to reconstitute a cultivar. Some common notations in writing pedigrees are as follows:

1. A slash, /, indicates a cross.
2. A figure between slashes, /2/, indicates the sequence or order of crossing. A /2 / is equivalent to / / and indicates the second cross. Likewise, / , is the first cross, and / / /, the third cross.
3. A back-cross is designated by *, and for example, *3 indicates the genotypes was back-crossed three times to another parent.

The following example will be used to explain the concept/notations (Acquaah, 2007).

Pedigree 1: MSU48-10/3/Pontiac/Laker/2/MS-64.

Interpretation

(i) The first cross was made Pontiac (as female) x Laker (as male).

(ii) The second cross was [Pontiac/Laker (as female) × MS-64 (as male).

(iii) The third cross was MSU48-10 (as female) × [Pontiac/Laker (as female) × MS-64 (as male).

Pedigree 2: MK2-57*3/SV-2 equivalent to MK2-57 /3/MK2-57 /2/MK2-57/ SV-2.

Interpretation

The genotype MK2-57 was back-crossed three times to genotype SV-2.

Year-wise, General Breeding Procedures under Pedigree Selection

The procedures are as follows:

Year 1 (Crossing): Identification of desired parents and making sufficient crosses to have about 50-100 crossed seeds/parental combination.

Year 2 (F1): Growing 50 F1 plants including parents to authenticate hybridity and harvesting of F2 seeds in bulk from F1 plants.

Year 3 (F2): Approximately 5000 F2 plants in cereals are space planted to ensure full expression of the characters and evaluation of individual F2 plants. F2 population provides first opportunity to start selection. Obviously, undesirable plants are often removed as soon as they are identified. It is here that beginning plant breeder will need to develop a ruthless-ness toward unpromising plants. About 300 desirable F2 plants are selected and harvested separately keeping records of their identities. From these 300 plants, about 50 plants are discarded after laboratory examination for seed size, shape, colour, etc. Thus, there are 250 F3 seed bags/packets each having come from separate individual F2 plants. Seed from any two packets should not be mixed. Parental/ check cultivars should also be planted at appropriate intervals.

Year 4 (F3): Seed from selected individual F2 plants are progeny rowed. There are approximately 250 individual progeny rows/cross. F3 generation is the beginning of line formation. It is desirable to plant check cultivars for comparison and to select the superior rows and superior plants within superior rows. The progeny rows are space planted for easy visual selection and record keeping. F3 family row should contain large number of plants (25-30) to permit the true family features to be evident so that most desirable plant(s) can be selected. Selection at this stage is both within and between rows by first identifying superior rows and selecting 3-5 plants from each selected progeny row to plant the next generation. In total 125 plants are selected from 50 superior rows.

Year 5 (F4): Selected F3 plants are planted in plant-to-row fashion as in F3 generation in 125 rows. The progenies become more homogeneous (homozygosity is 87.50%). Distinct lines start appearing in F4 generation. Therefore, selection is exercised more between rows rather than within rows. In total about 90 superior individual plants are selected from selected

40 progeny rows. Families that are distinctly inferior should be discarded while more than one plant may be selected from exceptional and outstanding families. However, generally, number of plants selected and advanced does not exceed the number of F3 families.

Year 6 (F5): Plants selected in F4 numbering 90 are grown in 90 rows in F5. These 90 rows are subjected to rigorous selection keeping the end objectives in mind and about 35 best performing progeny rows are selected. In total about 80 individual plants are selected from the short-listed 35 progenies. Depending upon seed requirement for trial in the next generation, selected plants within progeny rows can be bulked to have sufficient seeds for trial.

Year 7 (F6): 80 breeding lines as constituted from F5 are planted and evaluated in initial station trial along with other breeding lines pooled from other crosses and checks in a randomized block design with two replications. Each trial normally has 20 genotypes (18 new breeding lines and two checks). From each trial, breeding lines found to be superior to the checks are selected for further evaluation in next year in station advanced varietal trial.

Year 8 (F7): Superior lines from each initial station trial are pooled together for evaluation in station advanced varietal trial where each trial normally has 20 new breeding lines (16 new lines plus 4 checks). This trial is planted in randomized block design with four replications. Superior new breeding lines (about 4)/cross are short-listed for further evaluation under coordinated multi-location trials.

Year 9-11 (F8-F10): The selected lines are entered into national coordinated varietal trials where there are yield evaluation for three years and then the most outstanding line is identified for release and ultimately the release proposal is submitted to Central Sub-Committee on Crop Standards, Notification and Release of Varieties (Field/Horticultural Crops) for their consideration and release and notification so that the product enters into formal seed production and distribution chain.

The generalized steps in breeding procedure by pedigree selection as applicable to cereals are given in Fig. 11.1 after modification from Briggs and Knowles (1967). The number of plants/lines planted and selected will vary from crop to crop and will be influenced by the availability of land and other resources at the breeding station.

Generation	Procedure	Number grown		Number selected	
		Plants	Lines	Plants	Lines
F_1	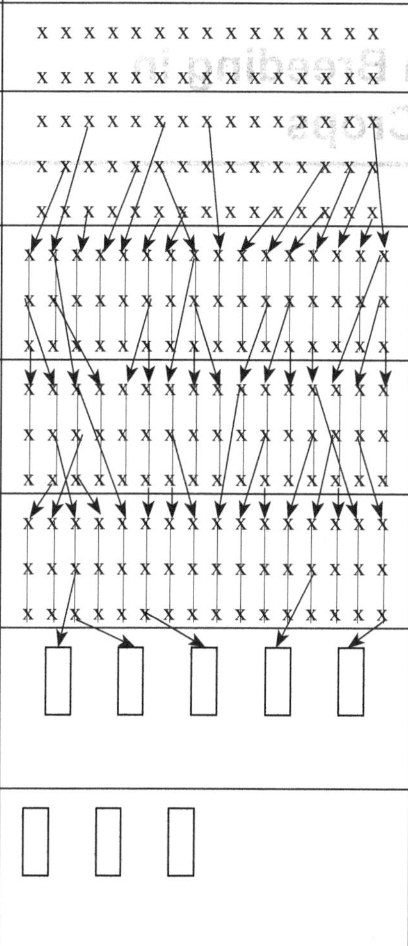	50	-	50	-
F_2		5,000	-	250	-
F_3		-	250	125	50
F_4			125	90	40
F_5			90	80	35
F_6		Replicated station initial varietal trial	80	-	15
F_7		Replicated station advance varietal trial	15	-	4
F_8-F_{10}	Multiloction trial of the most outstanding breeding line for three years along with other lines and checks and commercial release of the product subject to its superior performance over the checks.		4	-	1

Fig. 11.1. Pedigree method of breeding of handling hybrid population. 'x' indicates single plants. 'xs' joined by lines represent lines/families examined. Rectangles represent selections grown and evaluated in bulk. Arrows indicate the route of material from one year to the next.

12

Bulk Population Breeding in Self-Pollinated Crops

Bulk population breeding developed by the Swede H. Nilsson Ehle in 1908 and later on supported and practiced by H. V. Harlan in barley breeding in 1940s is a breeding strategy in crop improvement in which natural selection is allowed to play its full role where the F2 generation is planted in a plot large enough to accommodate several hundreds of plants. Planting density and cultural practices are usually the normal ones followed for commercial crop planting. At maturity, the plot is harvested in bulk and the seeds are used to plant a similar plot in the next season. This process is repeated till at least F5 where the plants are space planted to allow individual plant evaluation for effective selection. During period of bulk propagation natural selection plays a key role in shifting gene frequencies in the bulk population. This role is of crucial importance as bulk handling is continued over a number of generations. Bulk population handling can be aided by artificial selection at any time during the period of bulk propagation. Aggressive and highly competitive but undesirable plants may be physically rogued out of the population to avoid increasing the frequency of undesirable genes. For example, determinate bush types of beans do not compete well against indeterminate, vine types. And if the breeder is interested only in determinate types, artificial selection against the indeterminate types is certainly called for. There is also good reason to apply artificial selection against seed colour, pubescence, awns or other characters that are undesirable in the variety to be developed but may be more or less neutral in survival. Bulk population handling often provides opportunities to make use of artificial aids to selection. For example, bulk harvesting when only part of plants are mature is an effective and inexpensive way to select for earliness. If particular sizes of seeds are required, screens or mechanical devices provide a rapid way of eliminating inferior types. Naturally occurring or artificially induced disease or insect infestation, of course, provide opportunities to rogue susceptible individuals from the bulk plot. In F6, the plants selected in F5 are progeny rowed and superior progenies are selected from where again few plants can be selected and seed bulked within a selected row. Depending upon seed requirement for yield trial in the following years, even the seed from selected entire row can be bulked.

The follow up yield evaluation trials are as given in case of pedigree selection. This method is operationally best suited for seed crops such as small grains, beans and soybean and is almost totally unsuited for fruit crops and for most vegetable crops.

When two types compete in a bulk population, the theoretical survival value of the poorer competitor is given by the following formula (Allard, 1960).

$A_n = as^{n-1}$ where A_n is the proportion of the poorer competitor in the nth generation, a is the initial proportion of the poorer competitor and s is the selective index. As an example, consider that two types have been mixed in equal proportion ($A_1 = 0.5$) and survival values for the better and poorer competitors are 1 and 0.9 respectively ($s = 0.9$). Then the expected proportion of the poorer types in the fifth generation will be as follows:

$$A_5 = (0.5)\ (0.9)^4 = 0.3645 = 36.45 \text{ per cent}$$

The proportion of the better competitor is 1 – 0.3645 = 0. 6355 = 63. 55 per cent.

The operational procedure of bulk population breeding is given in Fig. 12.1 (Briggs and Knowles, 1967).

Generation	Operations	Numbers grown		Numbers selected	
		Plants	Lines	Plants	Lines
F_1	x x x x x x x x x x x x x x x x x x x x x x x x x x x x x x x x x x x x x x x	50	-	50	-
F_2	Bulk planting	5000	-	5000	-
F_3	Bulk planting	5000	-	5000	-
F_4	Bulk planting	5000	-	5000	-
F_5	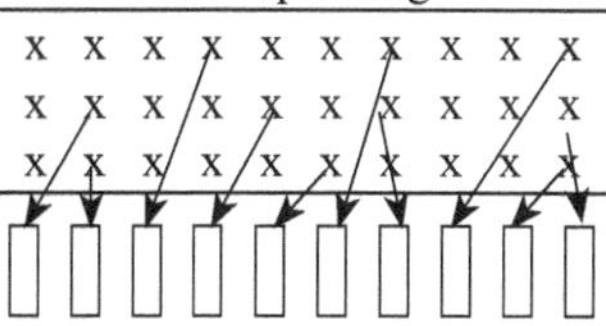	5000	-	250	-
F_6		Initial yield trial	250	-	15
F_7		Advanced yield trial	15	-	5
F_8-F_{10}	Multi-location testing in replicated trials along with relevant checks for 3 years and finally selecting 1-2 the most outstanding breeding lines for formal release as given in pedigree method of breeding.				

Fig.12.1 Bulk population breeding in self-pollinated crop.

The bulk and pedigree methods are not mutually exclusive. J. B. Harrington in 1937 proposed and used the *mass-pedigree method* in which hybrid populations are grown in bulk till circumstances are favourable for expression of important characters when single plant selections are made and population carried forward by the pedigree method. Some scientists have preferred to use the pedigree method in early generations to eliminate undesirable types quickly and then to bulk the better plants into one population and follow bulk method thereafter.

Composite Cross

Composite cross bulk population breeding or evolutionary method of breeding was developed by Love in 1927 when he stated that several crosses could be composited. H. V. Harlan and his team (M. L. Martini and H. Stevens) in 1928 and 1940 gave prominence to this method and suggested that F2 of several crosses could be bulked into one composite population, frequently termed as "composite cross" which is handled as per bulk/pedigree method of breeding. A composite cross may be derived in a variety of ways. Harlan and colleagues crossed 28 barley varieties in all combinations and composited the same amount of F2 seed from each cross. This procedure permits the germplasm of each variety to be combined with that of every other variety but still means if cross-pollination is excluded, that all hybrid plants can have germplasm from no more than two parents. Thus, there is no opportunity to combine germplasm from three or more parental lines. Another method, assuming 16 parental varieties would involve making 8 crosses including all 16 varieties followed by crossing 8 F_{1s} into four crosses and in turn F1 of these crosses can be combined into a final single cross as shown in Fig. 12.2.

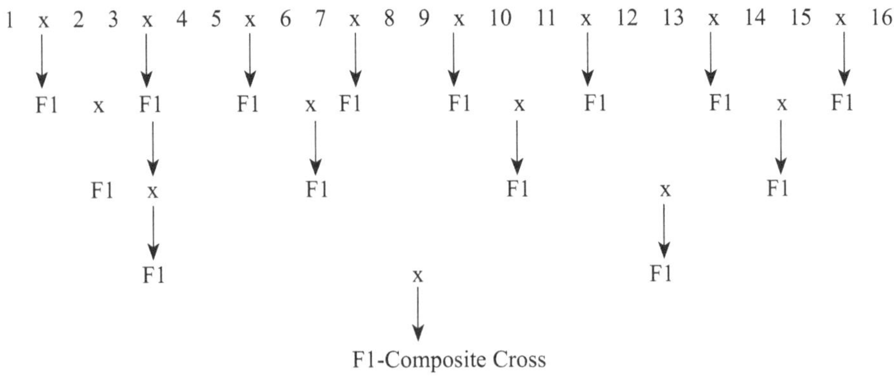

Fig. 12.2 Composite cross.

Single Seed Descent Method in Self-Pollinated Crops

The method of single seed descent (SSD) was born out of a need to speed up the breeding programme by rapidly inbreeding a population prior to initiating individual plant selection and evaluation while reducing a loss of genotypes during the segregating generations. The concept as per Acquaah (2007) was first proposed by C. H. Goulden in 1941 when he obtained the F6 generation in two years by reducing the number of generations grown from a plant to one or two while conducting multiple plantings per year using greenhouse and off-season planting. H. W. Johnson and R. L. Bernard described the procedure of harvesting a single seed per plant for soybean in 1962. However, Brim (1966) suggested modified pedigree method i.e. single seed descent (SSD) as an economical way to develop homozygous lines and simultaneously preserve the genetic variation in the population. In SSD, only one seed is chosen at random from each plant for further use. Since the number of plants to be grown is considerably reduced by such a procedure, the individual generations can be grown in greenhouses where two to three generations per year can be achieved.

This method allows the breeder to advance the maximum number of F2 plants through F5 generation. This is achieved by advancing one randomly selected seed per plant through early segregating generations. The focus on the early stages of the procedure is on attaining homozygosity as fast as possible and without selection. Discriminating/selection among plants starts after attainment of homozygosity. This method is best suited for small grain crops as well as legumes that can tolerate close planting and produce some seeds per plant. The species that can be forced to mature rapidly are more suited for breeding by SSD method. The handling of plants/lines starting from F6 and onward is as given in case of pedigree selection. The classical procedure of SSD has sometimes been modified and practiced as single pod/multiple seed method where 2-4 seeds per plant are taken and only half of the bulk seed thus produced is used to advance the generations.

13

Back-Cross Breeding in Self-Pollinated Crops

The application of back-cross breeding first proposed by H. V. Harlan and M. N. Pope in 1922 provides a precise way of improving varieties that excel in a large number of attributes but are deficient in a few key characteristics. F. N. Briggs too started an extensive back-crossing programme to develop bunt resistant varieties of wheat in 1922. As the name implies, the back-cross method of breeding makes use of a series of back-crosses to the variety to be improved during which the character or characters in which improvement is sought is maintained by selection. At the end of back-crossing, the gene (or genes) being transferred will be heterozygous. Selfing at the end of last back-cross produces homozygosity for this gene pair and coupled with selection will result in a variety similar to the well adapted and recurrent parent but superior for the particular characteristic for which improvement programme was initiated.

In back-cross method of breeding, the adapted and highly desirable parent is called the recurrent parent and the source of desirable character being transferred is called as donor parent. It is most effective and easy to apply when the missing trait is qualitatively inherited, dominant and produces a phenotype that is readily observed in a hybrid plant. The procedure for transferring a recessive trait is similar to that for dominant trait but involves an additional step of selfing. Back-crossing is also used to transfer entire sets of chromosomes in a foreign cytoplasm to create a cytoplasmic male sterile (CMS) genotype as has been practiced in maize, onion and wheat, etc. This is done by crossing the donor (of chromosomes) as male until all donor chromosome are recovered in the cytoplasm of the recurrent parent. Back-cross breeding has also been used to develop isogenic lines (lines that differ only in alleles at a specific locus) for traits like disease resistance and plant height, etc.

Transfer of Dominant Gene by Back-Crossing

The generalized steps in back-cross breeding for a dominant trait are given in Fig. 13.1 (Acquaah, 2007). With the availability of markers, the foreground and background selection can be made accurate and now it is possible to reduce the number of years of back-crossing substantially.

Year	Generation	Crosses	Action
1	P	A (susceptible) x B (resistant) rr x RR	Select parents, make 10-20 crosses, harvest F_1 seed
2	F_1	Rr (resistant) x rr	Grow F_1 plants and back-cross F_1 to recurrent susceptible parent
3	BC_1F_1	rr, Rr x rr	Discard susceptible plants (rr); back-cross Rr (about 30-50 heterozygous plants resembling with the recurrent) with rr
4	BC_2F_1	rr, Rr x rr	Discard susceptible plants (rr); back-cross Rr (about 30-50 heterozygous plants resembling the recurrent parent) with rr
5	BC_3F1	rr, Rr	Grow BC_3F_1
6	BC_3F_2	rr, rr, Rr, RR	Discard susceptible plants (rr); progeny row
7	BC_3F_3	rr, Rr, RR RR	Select BC_3F_3 progenies with resistance and high yield
8		RR x rr	Back-cross superior lines to rr
9.	BC_4F_1	Rr x rr	
10.	BC_5F_1	rr Rr x rr	Discard susceptible plants; back-cross resistant plants to rr, BC_5 should closely resemble the recurrent parent
11.	BC_6F_1	Rr rr	Self BC_6, select 300-400 desirable plants and harvest them individually
12.	BC_6F_2	rr, Rr, RR	Discard susceptible (rr) and grow progeny rows
13.	BC_6F_3	rr, Rr, RR, RR	Discard susceptible (rr) progenies and select true breeding resistant progenies (RR), conduct yield test of the back-cross derived line with the recurrent parent to determine equivalence before release.

Fig. 13.1. Generalized steps in transferring a dominant gene in back-cross breeding.

In general, the steps for transferring a dominant gene are straightforward. Following the first cross between the parents, phenotypic selection is adequate for selecting plants that exhibit the target trait. Recessive genotypes are discarded. The recurrent parent traits are not selected at this stage. The next cross is between selected F1 and recurrent parent. This step is repeated for several cycles. After satisfactory recovery of the recurrent parent, the selected plant in BC_5/BC_6 will be homozygous for other alleles but heterozygous for the desired traits. The last back-cross is followed by selfing to stabilize the desired gene in the homozygous state. All homozygous recessive segregates are discarded.

The number of times that a recurrent parent is used in a back-cross programme is indicated either by a superscript or subscript. For example, the population after the third back-cross would be indicated by the symbol A^4 x B or A_4 x B. This indicates the original cross of A to B and three back-crosses to A. With every cross or back-cross to the recurrent parent, the average proportion of germplasm from the donor parent is reduced by one half. Accordingly, any F_1 from A x B will have one half of its germplasm from both the parents. The cross of this F_1 with A will be designated as BC_1F_1 and proportion of donor parent in this back-cross will now be one quarter. If the number of crosses and back-crosses to the recurrent parent is n, then the proportion of germplasm from the donor parent is given by $(1/2)^n$. Thus, after five back-crosses, it would be $(1/2)^6 = 1/64$. The proportion of homozygosity is given by the same formula as applicable to selfing series which is as follows:

$$(2^m - 1)/2^m$$

Where m is the number of back-crosses. Thus, after five backcrosses, the proportion would be $(2^5 - 1)/2^5 = 31/32 = 96.875\%$. The distinction from selfing series is that under back-crossing the homozygous loci are those from the recurrent parent and in selfing series, the homozygous loci are from different combinations of loci from both the parents (Briggs and Knowles, 1967).

The proportion of homozygous genotypes is also given by the following formula of the selfing series:

$$\left[\frac{2^m - 1}{2^m}\right]^n$$

Where m is the number of back-crosses and n is the number of heterozygous loci in the F_1 of the original cross. Using the example of selfing series of ten heterozygous loci and assuming that five back-crosses have been made, 72.79 = 72.8% of BC^5 would be homozygous for the ten loci of the recurrent parent. All homozygous genotypes will be those of recurrent parent.

The numbers of back-crosses that are used in breeding programmes vary mostly from three to ten. When ten back-crosses are used, almost complete identity with the recurrent parent is achieved. Most common number of back-crosses followed is five to six. When there is scope to select strongly for the recurrent type in the early generations using markers as background selection, three back-cross generations are sufficient. After third back-cross, the material is handled by pedigree or bulk methods of breeding.

Transfer of Recessive Gene by Back-Crossing

The generalized steps as given by Acquaah (2007) are as follows:

Years 1: Select the donor and recurrent parents and make crosses.

Year 2: Grow F1. All F1s will be susceptible. Make back-cross with the susceptible recurrent parent. Harvest BC1F1 seed.

Year 3: Grow BC1F1 plants and self. Harvest the BC1F2 seed and bulk, All BC1s will be susceptible.

Year 4: Grow BC1F2 plants. Identify disease resistant plants and back-cross 10-20 such plants to the recurrent parent to get BC2F2 seed.

Year 5: Grow BC2 plants. Select 10-20 plants that resemble the recurrent parent and cross with recurrent parent (RR).

Year 6: Grow BC3 plants, harvest and bulk BC3F2 seed.

Year 7: Grow BC3F2 plants. Screen and select desirable resistant plants. Back-cross 10-20 resistant (rr) plants with susceptible recurrent parent (RR).

Year 8: Grow BC4 plants, harvest and bulk BC4F2 seed.

Year 9: Grow BC4F2 plants, screen for disease resistance, select desirable resistant (rr) plants. Back-cross 10-20 resistant plants (rr) with susceptible recurrent parent (RR).

Year 10: Grow BC5 plants, harvest and bulk the BC5F2 seed.

Year 11: Grow BC5F2plants, screen and back-cross desirable resistant plants (rr) with susceptible recurrent parent (RR).

Year 12: Grow BC6 plants, harvest and bulk BC6F2 seed.

Year 13: Grow BC6F2 plants, screen for resistance to disease, select 400-500 resistant (rr) plants and harvest separately for growing progeny rows.

Year 14: Grow progenies of selected plants, screen and select about 100-200 uniform progenies, harvest and bulk seed.

Year 15-16: Follow the procedure for yield trial and equivalence test as in transfer of a dominant gene.

The key difference between transfer of dominant and recessive gene in back-cross breeding is that in the latter case, phenotype identification of desirable plants is not possible after a cross. Each cross needs to be followed by selfing so that the progeny with the homozygous recessive genotype can be identified and back-crossed to the recurrent parent. Extensive advance testing is not necessary in a back-cross breeding programme because the new cultivar already resembles the adapted old cultivar except for the newly introduced gene. It is possible to transfer two or more genes in back-crossing programme by simultaneous selection among the progeny. However, this will require a larger population than if two genes are transferred independently. Introgression of genes from weedy, adapted, exotic or wild germplasm is possible by back-crossing. However, such transfers often take longer than typical transfer because of the time needed to remove the undesirable agronomic traits brought in by these distantly related sources.

Multiline Breeding and Making Cultivar Blends in Self-Pollinated Crops

Multi-lines are developed for self-pollinated crops and consist of a mixture of specifically developed isolines or near isogenic lines which differ only in a single gene or defined set of genes. Isolines are developed primarily for disease resistance. Isolines are developed using back-crossing procedure in which F1 is repeatedly crossed to one of the recurrent parents that lacked the gene of interest, for example, disease resistance. In principle, Jensen (1952) proposed multi-line varieties as a means of stabilizing the race situation in stem rust and of preventing the rapid increase in new races. As per this original proposition, multi-line varieties would be a composite of selections or varieties with similar height, maturity, yield capacity and other agronomic features, with each constituent selection of multi-line variety carrying a different gene for resistance. No matter what race was present, there would be something resistant to it in the multi-line variety. Because a race could attack only one or a few constituents of the multi-line variety, the disease spread will be slow and less damaging even to susceptible types.

Borlaug (1958) proposed a similar approach except that the components lines would be developed by the back-cross method and consequently would be phenotypically similar but would carry different genes for resistance. The composition of the variety can be rapidly modified as per changing population of rust races. Later on Borlaug advocated the development of multi-lineal hybrid wheat varieties which are basically supposed to be produced by three-way crosses. The first cross incorporates several genes for resistance to rust into a male-sterile variety and the second imparts heterosis and restores male fertility. However, this should be noted that these types of varieties could not be accepted on commercial scale in recent times as now the focus is more on uniformity and better options are available to breed cultivars having resistance to several rust races.

14

Development of Hybrid Cultivars in Self-Pollinated Crops

Initially, exploitation of heteosis was confined to corn and a few other open-pollinated crops but presently heterosis is being exploited practically in all species cutting across mode of reproduction. More and more evidences are emerging that heterotic expression in a cross is a function of genetic backgrounds and genes. Heterotic grouping would remain a key strategy in management of diverse genetic backgrounds and genes for full exploitation of heterosis. The commonly used methods in developing hybrids in self-pollinated crops are listed and explained as follows:

Manual Emasculation and Pollination Method

In this method, female parent flower buds are emasculated manually and the emasculated buds are pollinated by the pollen grain collected from protected flowers of the male parent. This method is effective and successful where the crossed fruits produce plenty of seeds and the seed requirement for planting the commercial hybrid seed is much less. Consequently, this method is being used in a routine manner on a commercial scale in vegetable crops mostly by private sector seed companies in India and south-east Asian countries. The crops covered under this method of hybrid seed production are as follows:

Tomato, hot-pepper, sweet pepper, eggplant, okra, bottle gourd, bitter gourd, ridge gourd, smooth gourd, cucumber, pumpkin, water melon and muskmelon. Cucurbits are included here as although they are cross-pollinated but behave like self-pollinated crops in terms of almost no inbreeding depression.

Manipulation of Sex Expression Using Growth Regulators

Cucumber plants are typically monoecious. However, now gynoecious lines are available through the initial efforts of L. M. Pike, G. E. Tolla and C. E. Peterson. The gynoecious lines produce predominantly female flowers and can be easily used as female parent in hybrid seed production of cucumber using hand pollination or even the hybrid seed can be produced in isolation by planting gynoecious and male lines side by side and allowing insect pollination. The gynoecious lines are maintained by induction of male flowers by spraying of gibberellic acid i.e. GA 4/7 (2000 ppm) or silver nitrate (200-300 ppm) on seedling leaves.

Monoecious lines of squash can be converted into female by applying ethephon (two sprays, 600 ppm at 2- and 4-leaf stage) which results into complete male flower suppression during fruiting stage. This has now become a commercial reality.

Use of Genetically Controlled Male Sterility

Male sterility is defined as the inability of a plant to produce or release functional pollen. The occurence of male sterility has been documented as early as 1930 (Kaul, 1998). It was first deployed in hybrid onion breeding and now is being used in a number of crops, like, chili, muskmelon, sorghum, pearl millet, pigeon pea, rice, etc. It is of different types, namely, cytoplasmic male sterility (CMS) which is due to cytoplasmic factors; nuclear or genetic male sterility (NMS or GMS) which is controlled by nuclear gene(s); cytoplasmic nuclear male sterility (CNMS or CGMS) which results from nuclear-cytoplasmic interaction; and environment-sensitive male sterility (EGMS) which is induced by the interaction of genetic and environment factors. Different systems are briefly described as follows.

Cytoplasmic Male Sterility (CMS): This is used in those crops where vegetative parts rather than seed is of economic importance such as a few vegetables and ornamentals. Male sterile female parent is crossed with male fertile male parent which carries fertile cytoplasm. The F1 thus produced has high heterosis for vegetative growth and vigour but is male sterile.

Nuclear Male Sterility: In case of NMS, female parent has recessive gene for sterility (*msms*) and the male parent has dominant fertile gene (*MsMs*). Male sterile female parent is maintained and multiplied as a population having male sterile (*msms*) and male fertile heterozygote (*Msms*) genotypes in a ratio of 1:1. For multiplication, this population is grown in isolation, *msms* plants are identified and seed set on these male sterile plants as a result of pollination from heterozygous male fertile (*Msms*) plants are harvested. These seeds when grown again, segregate in a ratio of 1 sterile: 1 fertile and thus the sterile gene is maintained generation after generation. For hybrid seed production, this seed of female parent having *msms* and *Msms* in equal proportion is planted as female and normal fertile parent (*MsMs*) as male. Male fertile plants (*Msms*) in female rows are removed before anthesis and the seed set on remaining 50% sterile plants (*msms*) is harvested as hybrid seed. This system has been used on commercial scale in pigeon pea, safflower, cotton, muskmelon and chili.

Hybrid seed production using NMS/GMS is cumbersome and costly. However, the seed production could be greatly facilitated and aided if appropriate male-sterility/fertility allele is closely linked to a morphological marker gene that expresses at seedling stage or at least before anthesis and helps in the identification and removal of male fertile plants from female parent row.

Cytoplasmic-Nuclear Male Sterility/Cytoplasmic Genetic Male Sterility: In this case, the female parent carries cytoplasm as well as genes for male sterility and is commonly known as CMS or 'A' line. 'A' line has a counterpart maintainer commonly known as maintainer or 'B' line which has same genotype as 'A' line except that it carries a normal fertile cytoplasm. 'A' line is maintained by crossing it as female with 'B' line whereas 'B' line is maintained in isolation. Hybrid seed is produced by crossing 'A' line with a restorer 'R' line which has genes for fertility restoration. The 'R' line is also maintained in isolation. The hybrid between 'A' line and 'R' line is male fertile and shows heterosis. This system is also called as three line system designated as A-B-C or A-B-R lines system. This is being commercially used in rice and chili among self-pollinated crops but is quite common in cross-pollinated crops like sorghum, pearl millet, maize and sunflower.

Environment-Sensitive Genetic Male Sterility (EGMS): In EGMS, male sterility is induced when the plant, at a sensitive stage of its development is exposed to a particular environmental factor(s) whereas fertility is restored at some other level of environmental factor(s). During the sterile phase, EGMS line can be used as female parent to produce hybrid seed. During the fertile phase, the line is maintained by self-fertilization. Thus only two lines are required and therefore this system is also called as two-line system of hybrid seed production.

If the EGMS is induced in response to variation in the photoperiod, the resulting male sterility is known as photoperiod sensitive genetic male sterility (PGMS) as has been observed in *japonica* rice cultivars as back as in 1973. The plant turned out to be male sterile under long day condition and male fertile under short day condition. Since then the PGMS trait has been transferred to several *indica* and *japonica* rice cultivars. Likewise, EGMS induced in response to variation in temperature is known as temperature–sensitive genetic male sterility (TGMS). The first TGMS line of rice has been isolated as spontaneous mutant by Z. C. Tan and colleagues in 1990. Following this, Maruyama *et al.* (1991) induced TGMS mutant in rice variety Remei which exhibited no seed set under 31/24 ° C temperature, partial fertility under 28/21° C and complete fertility under 25/15° C. A reverse trend of TGMS where sterility occurs at low temperature (24° C) and fertility at higher temperture (27° C) has also been reported in rice by Chinese scientists (Dhillon *et al.*, 2004). This system offers the advantage in the sense that several parental lines can be used in hybrid development and seed production. The field area ratio between male-sterile multiplication, hybrid seed production and commercial cultivation of hybrid becomes more economical.

Gametocide Induced Male Sterility: Gametocides also known as chemical hybridizing agents (CHAs) offer the advantages like that of EGMS in terms of widening the choice of parental lines and simplifying the hybrid seed production. Any two parents which are heterotic in a hybrid combination can be directly used for hybrid seed production. An ideal CHA should result into complete male sterility without affecting female fertility, should have wide spectrum to induce male sterility in successively emerging tillers, should be least phytotoxic , no carcinogenicity and no residual toxicity to human being and animals. Application of CHA should be easy and economical. Several CHAs such as etherel, oxinilates, arsenates, GA 3, GA 4/7, maleic hydrazide and 2, 3-dichloroisobutyrate have been reported to have gametocidal properties but none of these has reached to commercial application level.

Hybrid Breeding in Selected Self-Pollinated Crops Using Cytoplasmic Genetic Male Sterility

Rice: The commercial exploitation of hybrid rice by L. P. Yuan in China in 1977 using CNMS system motivated the breeders of self-pollinated crops to embark on hybrid breeding in rice, pigeon pea, Indian mustard and chili in India and elsewhere. Hybrid technology efforts in rice remained mostly academic till the discovery of stable CMS source in the wild species *O. sativa, spontanea* designated as 'wild abortive' or 'WA type' that made hybrid technology in rice a commercial feasibility. This development is considered a landmark in history of rice breeding. Using WA-based CMS lines such as V20A, Chinese breeders started exploiting hybrid vigour successfully on commercial scale since late seventies. This success story prompted IRRI and countries like India to develop appropriate WA-based CMS lines well adapted to typical tropical conditions. Two CMS lines developed at IRRI, namely, IR58025A and IR62829A provided the base for India to launch hybrid breeding on a network mode in 1989-90. This network resulted into half a dozen of first generation hybrids of rice (APRH1, APRH2, TNRH1, KRH1, CNRH1, DRRH1) with reasonably good production package of hybrid seed. Convinced with nearly 1 ton/ha yield advantage of these hybrids over the best variety, several other hybrids, namely, KRH2, Pant Sankar Dhan1, CORH2, Sahyadri, Narendra Sankar Dhan2, PHB71 were added subsequently. Now Private sector seed companies like Bayer Crop Science and Syngenta are pursuing this technology in rice aggressively and marketing hybrid seed of their own hybrids. Medium duration CMS lines having high amylose content such as PMS3A, PMS10A, PMS12A, APMS5A, CRMS31A, CRMS32A, IR68888A and IR68897 are now being used widely. PRH 10 is the world's first hybrid of Basmati rice developed at IARI, New Delhi. It matures in 110 days and out-yields Pusa Basmati 1 by over 30 per cent (Siddiq *et al.*, 2004).

As far as the maintenance of genetic purity of seed is concerned, an integrated approach involving traditional nucleus seed production manually or by paired crossing, maintenance of proper isolation distance and grow-out test for production of breeder/foundation seed and adoption of molecular marker technology for ensuring purity of CMS lines, maintainers and restorers is required.

Enhancement of yield heterosis is of greater importance for sustenance of hybrid technology in rice. Among the breeding efforts, evolution of progressively more heterotic hybrids, structured parental line improvement and inter-varietal (*indica*/*japonica*) combinations exploiting wide compatibility gene through conventional 3-line as well as temperature sensitive male sterility-based 2 line approach are considered to be important.

Cotton: India is credited as the first country in the world which developed and produced hybrid seed of cotton hybrid 'Hybrid 4' manually under the leadership of C. T. Patel from Surat centre in 1971. Since then, hybrid seed production in cotton is being done manually on a large scale by the large number of private seed companies especially with launch of Bt. cotton Bollgard I in 2002 and Bollgard II in 2006. The conventional open-pollinated seed in India has now been replaced by hybrid seed to the extent of about 90% for 11 million hectares area. Of the total hybrid seed required, about 90% is being produced through manual method of emasculation and pollination employing family members of hybrid cotton seed producers under guidance of the private sector seed company personnel.

Use of male sterility system, either genetic or cytoplasmic-genetic is one option to reduce the cost of hybrid seed production in cotton. Although both the sources of genetic and cytoplasmic-genetic male sterility were known to exist, the former was exploited first. Double recessive male sterile source (*ms5ms5*, *ms6ms6*) developed by Weaver has been used first to reconstitute successfully the popular Indian hybrid H 4 and Varlaxmi followed by CPH 2, MECH 1, Ankur 15 and a few others. Subsequently, cytoplasmic-genetic male sterility system based on *G. haarknessii* cytoplasmic sterility source was used taking advantage of restorer sources available in exotic genotypes of *hirsutum-barbadense* background. MECH 4 released for central zone, CAHH 468 and a few more from private sector are CMS-based hybrids.

Indian Mustard (*Brassica juncea*): Heterosis has been exploited in *B. napus* using CMS-based male sterility. Napus hybrids based on *polima* cytoplasm have been released in China, Europe, Canada and Australia while *tournefortii* CMS-based PGSH 51 is the first to be released in India. Hybrids in Indian mustard are running into difficulties due to lack of strong sources of fertility restoration to many of the CMS sources. There were encouraging reports on the identification of restorer sources for CMS lines based on *moricandia, lyratus* and *canariense* cytoplasm as investigated by S. Prakash, S. S. Banga and S. K. Banga from 1997 to 2001.

Recombinant DNA technology has introduced a novel means of inducing male sterility across plant species by selectively disturbing one or the other stages of pollen development. Mariani *et al.* (1990) were the first to demonstrate in tobacco that plants spliced with a chimeric dominant gene *barnase* (bacterial RNAse from *Bacillus amyloliquefaciens*) driven by the tapetum specific promoter TA 29 from tobacco induce male sterility. They also demonstrated the possibility of restoration of fertility by a dominant gene *barstar* extracted from the same bacterium. The protein encoded by *barstar* suppresses the cytotoxic effect of *barnase* and restores thereby pollen fertility. Maintenance of male sterility requires as in GMS system, a strategy to selectively remove fertile plants when the plant population segregates in the ratio of 1 fertile: 1 sterile. Linking of a dominant herbicide resistance gene *bar* with *barnase* (*Ms*) under constitutive promoter (CaMV 35S) helps to eliminate fertile plants when sprayed with the herbicide. Indian mustard hybrids using *barnase-barstar* system developed by Pro-Agro in India could reach to field evaluation but could not be commercialized.

Status of Hybrid GM Mustard in India: 2016 was the turning point for biotech in India as it transcended from the shadows of the moratorium on IR-Bt. brinjal (eggplant) imposed in 2010 to the ultimate step of commercial release of biotech mustard in 2016. India has successfully completed the process of inviting public comments on bio-safety dossier of biotech mustard seeking permission for environmental release of transgenic mustard hybrid DMH-11 and parental lines containing events bn 3.6 and modbs 2.99 expressing barnase, barstar and bar genes. Biotech mustard is the first genetically modified crop developed indigenously by the Centre for Genetic Manipulation of Crop Plants (CGMCP) of the University of Delhi. In January 2016, the GEAC constituted a Sub-Committee of scientific experts to thoroughly address each aspect of the dossier of biotech mustard, which was under consideration for permission for environmental release. Over the next 8 months, the Sub-Committee and the Department of Biotechnology's (DBT) Bio-safety Support Unit (BSU) deliberated on the bio-safety dossier of biotech mustard, and prepared a comprehensive document "Assessment of Food and Environmental Safety (AFES)". This was presented to the Genetic Engineering Appraisal Committee (GEAC) for consideration in its meeting held on 20 June 2016. The Report submitted to GEAC contained a thorough assessment of the bio-safety data generated by the applicant, its comparison with such international assessment by well-known regulatory agencies such as the European Food Safety Authority (EFSA), Office of the Gene Technology Regulator (OGTR), and Canadian Regulatory Authorities, and existing scientific literature on the subject in the peer reviewed journals, as well as addressing the specific uses of mustard in Indian context.

The Sub-Committee was of the opinion that the biotech mustard is "as safe as conventional mustard" and "does not raise any public health or safety concerns for human beings and animal health." The Sub-Committee concluded that the environmental release of parental lines for hybrid production DMH-11 "may not pose any risk to biodiversity and the agro-system". In order to address the concern raised by anti-biotech activists, the GEAC also held a special meeting on 18 July 2016 to record their objection regarding the environmental release of transgenic mustard hybrid DMH-11 and parental lines. Subsequently, MOEF&CC, on the recommendation of the Genetic Engineering Appraisal Committee (GEAC), published the "Assessment of Food and Environmental Safety (AFE S)" report on the MOEF&CC website and made available a biotech bio-safety dossier for review at MOEF&CC office for public comment from 5 September to 5 October 2016. The MOEF&CC received over 750 comments from various stakeholders including students, farmers, researchers and other key stakeholders on the report by 5 October 2016. These public comments were sent to the Sub-Committee, which submitted their report on the public comments to MOEF&CC by the end of 2016. The GEAC of MOEF&CC is likely to review the final assessment of the biotech mustard dossier, including public comments and expected to take a final view on the permission for environmental release of transgenic mustard hybrid DMH-11 and parental lines some time soon. However, due to pending PIL in Supreme Court and watch and wait position adopted by Government of India, The environmental release of this transgenic mustard DMH-11 (Dhara Mustard Hybrid-11) developed by Dr. Pental and his team remain uncertain.

The assessment of food and environmental safety reports noted that biotech mustard hybrid DMH-11, a cross between Varuna bn 3.6 and EH-2 modbs 2.99 is superior compared to the parents, showing proof-of-concept of the technology, exhibiting heterosis and hybrid vigor. The assessment of food and environmental safety report concluded that the presence of transgenes including bar, barnase and barstar genes in the hybrid DMH-11, does not lead to any unintended effect on the agronomic parameters, and further evaluation provided evidence of hybrid vigor in the hybrid DHM-11. The barnase-barstar technology of GM mustard will accelerate mustard breeding programme by both public and private sector, and result in the introduction of high-yielding and superior mustard hybrids capable of revolutionizing mustard farming and edible oil production in the country. The development of biotech mustard is a classic example of India's scientific ability to harness the science of biotechnology and farm innovation in agriculture.

Pigeon Pea: Exploitation of heterosis as a means to raise genetic yield in pigeon pea started with the isolation of genetic male sterile lines *viz.,Ms3A* and *Ms4A* from the germplasm accessions ICP 1555 and ICP 1696, respectively at ICRISAT, Hyderabad which has pigeon pea as one of its mandate crops (Reddy *et al.*, 1978). Later on ICRISAT achieved a major breakthrough in hybrid breeding technology in pigeon pea under the leadership of K. B. Saxena and his team (Saxena *et al.*, 1986). In the absence of CMS, available GMs sources were used to develop hybrids and their seed production technology. This extensive exercise led to release of world's first pigeon pea hybrid ICPH 8 in 1991 using GMS.

The male sterile line for producing hybrid seed of pigeon pea is Prabhat. It is determinate and flowers in 60-70 days. Plants are compact and erect and attain a height of about 1 m. The male sterile line is characterized by translucent (white) anthers, devoid of pollen grain. The seed produced from sib-mated *msms* x *Msms* will always segregate into 1 sterile: 1 fertile. For hybrid seed production this seed is used as female and ICPL 161 as male. Fifty per cent male fertile plants from the male sterile rows are removed before anthesis. This hybrid was taken up by private sector seed company Mahyco for large scale production and marketing but the hybrid could not make a big success primarily due to high cost of hybrid seed production. Since release of ICPH 8, several genetic male sterility based pigeon pea hybrids have been commercialized. However, their genetic base being quite narrow no further progress in yield could be achieved.

Research efforts to identify cytoplasmic-genetic male sterility have been continued. Wild pigeon pea species like *C. sericeus*, *C. acutifolius*, *C. cajanifolius* and *C. albicans* have been used for developing alloplasmic lines of *C. cajan*. CMS lines have been developed by two methods, namely, (i) wide hybridization involving *C. sericeus* and *C. cajan* and substitution back-crossing (Van der Maesen, 1990) and (ii) multiple cross genome transfer. The male sterility is maintainable by pigeon pea genotypes ICPL 85030 and ICPL 90035 of ICRISAT and GT 288A. The first CGMS stable line was developed at GAU, Gujarat using *C. scarabaeoides* . However, no hybrid based on CGMS could be commercialized in pigeon pea.

15

Mass Selection in Cross-Pollinated Crops

Mass selection is the simplest, oldest and least expensive method of plant breeding as applicable to cross-pollinated crops. The method involves planting balanced bulk seed of the population to be improved, in isolation, inferior plants rogued out before pollination wherever possible and selecting and bulking superior genotypes that already exist in the population. A balanced bulk of selected ears, for example, in corn is made for planting the next cycle. Selection units are individual plants. Selection is surely on phenotypic performance. This method exploits additive gene effects plus additive x additive interactions. The efficiency of this method depends largely on the heritability of the trait under selection. Selection results using mass selection have been quite effective for less complex traits. Mass selection could be very effective for those traits that can be identified or selected prior to flowering, for example, curd/head shape, size, colour in cauliflower and cabbage, root shape, size, colour and quality traits in radish and carrot and bulb shape, size, colour and weight in onion. A few examples of such traits in maize could be flowering, leaf angle, prolificacy, resistance to thrips, fall armyworm, reduced plant and ear height, reduced anthesis-silking interval and barrenness and tolerance to pre-flowering diseases.

Lack of pollination control is major limitation of this method as both desirable and undesirable pollen will be involved in pollination of the selected plants. Further, if selection intensity is very high (small plant population size advanced), the possibility of inbreeding depression is increased. This selection is not very effective if the heritability of the trait under selection is low (less than 10% on individual plant basis). Mass selection is likely to be more effective for individual or a combination of few traits rather than a whole series of characters. Despite some limitations, mass selection can be very useful and relevant procedure in breeding of cross-pollinated crops in many situations characterized by resource limitations, non-availability of trained scientific manpower and additional test sites, lack of seed storage facilities even for short term and lack of continuity of technical personnel. Lack of control on soil variability also makes this method less effective.

To overcome soil heterogeneity problems, a modification of simple mass selection known as 'stratified mass selection' has been suggested by Gardner in 1961. The basic steps of this procedure are same as in simple mass selection except that the field research area in which the crop is grown for mass selection is divided into a number of grids and that the superior plants are selected from each grid to reduce environmental variation. Simple or modified mass selection can also be practiced in small grain cereals to increase the frequency of desirable alleles in the segregating generations.

Despite some limitations, mass selection is effective in cross-pollinated crop improvement programmes especially for those traits that are governed by simple genetic system and are highly heritable. Even for complex traits, this method is effective provided some secondary trait can be identified which has high heritability and is correlated with the target complex character. It can be successfully deployed to select for photo-period insensitive genes, to permit use of exotic germplasm and to broaden the germplasm base in a breeding programme. Under resource constrained conditions, mass selection will continue to be a preferred method (Vasal *et al.*, 2004).

16

Recurrent Selection in Cross Pollinated Crops

The concept of recurrent selection grew out of need to increase the number of superior genotypes that could be obtained from breeding stock of corn. This could have been possible by increasing the frequency of superior genes in the gene pool and by increasing the chances for genetic recombination to occur. Both can be achieved by recurrent selection which is defined as "a method of breeding designed to concentrate favourable genes scattered among a number of individuals by selecting in each generation among the progeny produced by mating ***inter se*** of the selected individuals (or their selfed progeny) of the previous generation." The original idea behind recurrent selection was proposed first by Hayes and Gerber in 1919 and independently by East and Jones in 1920. Jenkins (1940) was the first to describe the method in detail and Hull (1945) proposed the name 'recurrent selection'. Each cycle of recurrent selection requires (i) evaluation and selfing of selections and (ii) crossing of the progenies of superior selfed plants in all combinations and bulking of equal amount of seed from each cross. The selection of plants before selfing is possible for maturity, plant height, ear/head characters, kernel characteristics and resistance to diseases and insects. In general, these are the characters which have high heritability. It is clear that the selection of plants before selfing is more for morphological traits and these evaluations do not include yield. Yield evaluations are done on the top cross progenies (crossing of selected plants to a tester). Such top crosses will be grown in the second year of the cycle during which time the selfed seed is held over until the evaluations for yield are over. Or the selfed seed may be grown out as S_1 lines which in turn are selfed. After yield evaluation is over, the selfed seed of superior plants is grown out for crossing. Progenies of selfed plants are crossed in all possible combinations. This is done manually or by growing progenies together in an isolated block. In isolation planting, random mating should be ensured. After manual crossing or random mating in isolation, equal amounts of seed of each are bulked and this bulk seed is used to start the second cycle of selection, selfing and crossing.

The base population to start recurrent selection may be an open-pollinated variety, a single or double cross hybrid or a synthetic variety. Cycles may be continued till there is no gain after selection. The outcome of the recurrent selection is usually called as synthetic although other terms like composites and populations have also been used. There are four types of recurrent selection as listed below:

- Recurrent selection for phenotype or simple recurrent selection
- Recurrent selection for general combining ability
- Recurrent selection for specific combining ability
- Reciprocal recurrent selection

Recurrent Selection for Phenotype/Simple Recurrent Selection

This is the simplest form of recurrent selection where no tests for combining ability are required and each cycle may be as short as one year. Since selection is based on phenotype, it is effective primarily for the traits with high heritability. Thus, in principle amounts to an extension of mass selection with the refinement that the selected plants are selfed and the selected plants are crossed before another round of selection. In those cases where desired plants can be identified before flowering, for example, in cauliflower, cabbage, carrot, radish, etc., the inter-crosses of selections can be done in the first year itself and second year's progeny test is eliminated from each cycle of selection. Typical phenotypic recurrent selection is illustrated in Fig. 16.1.

Recurrent Selection for General Combining Ability

In the simplest form, this is a three year cycle recurrent selection scheme in which selected plants are self-pollinated and crossed to a broad-based heterozygous tester in the first year. Self-seed from first year are kept in reserve. In the second year, the test-crosses are evaluated and superior ones are identified. In the third year, the remnant selfed seed which gave rise to superior test-cross progenies are planted as progeny rows. All possible crosses are made among these rows or the progeny rows are planted in isolation in such a way that random pollination takes place. Equal quantity of seed from each cross is bulked and this bulk which is an improvement over the original base population is used to initiate the next cycle of recurrent selection. This procedure is illustrated in Fig. 16.2.

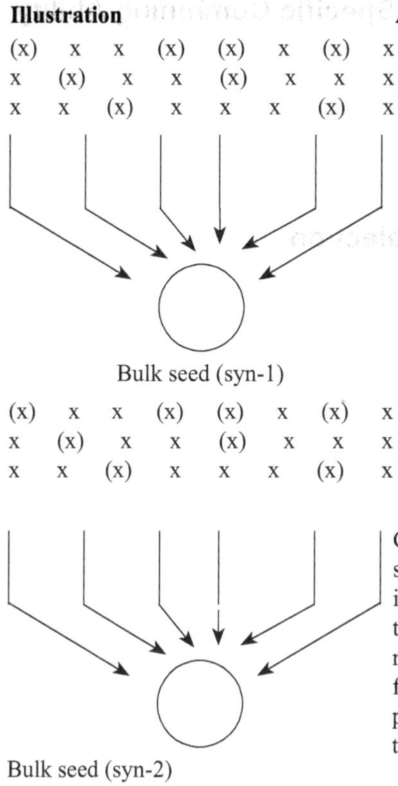

Year	Activities
First year	Self-pollinate (x) and select superior type plants at maturity
Second year	Grow progeny rows of superior plants selected in the first year, cross them in all possible combinations or grow them in isolation ensuring random mating, bulk equal quantity of seed from each cross, start second cycle of phenotypic recurrent selection from this new base population
First year	Self-pollinate (x) and select superior type plants at maturity
Second year	Grow progeny rows of superior plants selected in the first year, cross them in all possible combinations or grow them in isolation ensuring random mating, bulk equal quantity of seed from each cross, start second cycle of phenotypic recurrent selection from this new base population

Fig. 16.1. Phenotypic or simple recurrent selection.

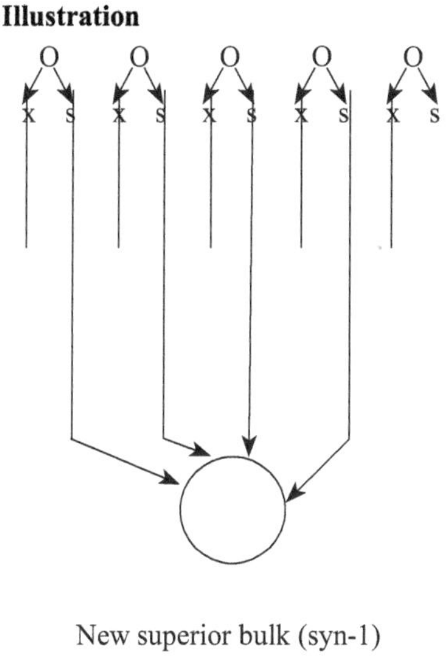

Year	Activities
First year	Grow base population, self-pollinate (s) and cross (x) the same plants to a heterozygous tester
Second year	Evaluate test-cross progenies, identify those that are superior
Third year	Plant those selfed seed kept in reserve whose test-cross progenies performed better in second year, make all possible crosses among them or grow them in isolation for random mating, mix equal quantity of seed from each cross into a bulk which is supposed to be superior to the original base population and start second cycle of this selection in the new population, if needed

Fig. 16.2. Recurrent selection for general combining ability.

Recurrent Selection for Specific Combining Ability

Recurrent selection for specific combining ability proposed by Hull in 1945 is similar to recurrent selection for general combining ability except that the tester used is a homozygous, stable inbred line to provide a measure of specific combining ability.

Reciprocal Recurrent Selection

Reciprocal recurrent selection was proposed by Comstock, Robinson and Harvey in 1949. Two populations are involved usually designated as A and B. Each population is handled in the same manner as in recurrent selection for general combining ability except that population A is used as tester for population B and population B as tester for population A. With each cycle, new versions of A and B become testers. The two starting base populations should be genetically divergent and should combine well together. The starting populations designated as A and B could be two varieties, two synthetics or two single crosses involved in the making of a successful double cross.

Comstock, Robinson and Harvey compared its efficiency on theoretical grounds with that of recurrent selection for general combining ability and recurrent selection for specific combining ability and concluded as follows (Allard, 1960):

(i) If dominance is incomplete, recurrent selection for general combining ability and reciprocal recurrent selection are about equal to each other and both are superior to recurrent selection for specific combining ability.

(ii) If dominance is complete, all three methods are essentially equal in efficiency.

(iii) If overdominance is important, recurrent selection for specific combining ability and reciprocal recurrent selection are equal to each other and both are superior to recurrent selection for general combining ability.

Operationally a number of plants from source population A are self-pollinated and crossed with a sample of plants from source population B. Likewise, a number of plants from source B are selfed and crossed with a sample of plants from source A. Test cross progenies are evaluated in a replicated yield trial in second year and superior progenies are selected and the selfed seed kept as reserve which gave rise to these superior progenies are planted third year and all possible inter-crosses are made within each group represented by population A and B. Equal seed from each cross are composited and two resulting populations designated as A' and B' serve as source population to initiate the next cycle of reciprocal recurrent selection.

The procedure is illustrated in Fig. 16.3 (Allard, 1960).

Illustration

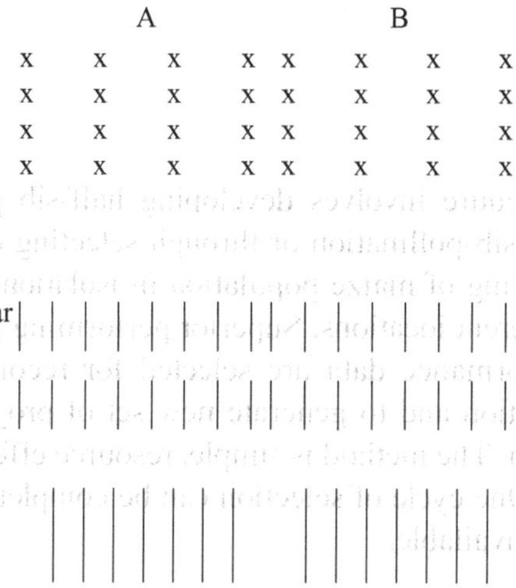

Year	**Activities**
First year	Outcross each of a group of plants selected for desirable characteristics from source population A with several randomly chosen plants from source population B and vice versa. Self-pollinate the selected plants from source populations A and B.
Second year	Conduct two field trials (one for plants from source A and another for plants from source b) of progeny of all crosses made in first year.
Third year	Plant seed produced by self-fertilization in first year using only seed from those plants in source populations A and B those were superior in evaluation trials of second year. Within each source population group, make all possible inter-crosses among progenies. Composite equal seed from intercrosses and the new bulks will be new source populations designated as A' and B' to initiate second cycle of selection.

Fig. 16.3. Diagrammatic representation of reciprocal recurrent selection.

Full-sib Recurrent Selection

This method as described by Vasal *et al.* (2004) involves developing full-sibs by making plant-to-plant crosses. Direct and reciprocal crosses can be attempted depending upon the seed requirement. Multi-location evaluation of progenies is done followed by the recombination of the superior families based on across-location performance data to complete a cycle of selection. One cycle of selection can be completed in one year provided two seasons are available. This method has been extensively used at CIMMYT to improve populations and to develop experimental varieties of maize. Selection intensity is kept around 4 per cent.

Half-sib Recurrent Selection

Half-sib breeding procedure and its various modifications have gained prominence beginning mid-1960s with the suggestion of modified ear-to-row (MER) method by Lonnquist in 1964. Ear-to-row method was introduced by Hopkins in 1896 to improve protein and oil content in maize. Various modifications of half-sib procedure are as follows:

Half-sib Selection: The procedure involves developing half-sib progenies either through controlled bulk sib-pollination or through selecting ears from randomly pollinated bulk planting of maize population in isolation. Half-sib progenies are evaluated at different locations. Superior performing progenies based on across-location performance data are selected for recombination to complete one cycle of selection and to generate new set of progenies for initiating next cycle of selection. The method is simple, resource effective and involves no hand pollination. One cycle of selection can be completed in one year provided two seasons are available.

Modified Ear-to Row Selection: This procedure has been suggested by Lonnquist in 1964 wherein half-sib progeny tests are conducted at 3-4 appropriate sites using single replicate. In the same season, a half-sib crossing block is planted at a different location in isolation. In the crossing block half-sib ears selected in the previous cycle are planted as separate families as female rows and male rows are planted with a balanced male composite comprising of all selected ears. The ratio of female to male is kept as 2: 1. Female rows are detaselled at flowering to permit pollination by bulk pollen from male rows. Between rows and within rows selection is practiced to select half-sib ears for the next cycle of selection. The data from the progeny test sites are used to select the best performing progenies in the crossing block. Within row selection is practiced in the selected rows to reach the number required for the next cycle of selection. MER is unique method of selection based on family performance permitting one cycle of selection in one year.

Modified Ear-to Row-Selection (Unreplicated Evaluation): Lonnquist's modified ear-to-row method having only half-sib crossing block without replication has been extensively used at CIMMYT in forming and subsequently improving gene pools. Using this procedure, one cycle of selection can be completed in one year.

Selfed Progeny Selection

In this method, following evaluation of selfed progenies, selected ones are recombined using the remnant seed. Thus a cycle of S_1 selection can be completed in three seasons, namely, season one for developing S_1 lines, season two for evaluation of S_1 progeny and season three for recombination of progenies grown from remnant seed where only those remnant seed are used which have given rise to better performing S_1 progenies. This selection is at times advanced to evaluation of S_2 and S_3 progenies. The generalized steps in breeding based on S_1/S_2 progeny performance are as follows:

(i) Select 50-100 plants in the source population.
(ii) Self-pollinate (S_0) and keep remnant seed.
(iii) Grow replicated S_1 progenies and identify superior progenies.
(iv) Composite equal amount of remnant seed of superior progenies and grow in isolation for random mating to create improved new population.

17

Synthetic Cultivars in Cross-Pollinated Crops

A variety produced by crossing *inter se* a number of genotypes selected for good general combining ability in all hybrid combinations with subsequent maintenance of the variety by open-pollination. According to Hayes and Garber (1919), the synthetic production of an improved variety by inbreeding and cross-breeding seems a reasonable plan. Synthetic cultivar is a commercial entity and a combination of number of inbred lines, sibbed lines, clones or other populations of cross-pollinated crops. The features of the components are as follows:

(i) They will have been tested for their combining ability.
(ii) They will be preserved for future syntheses of the synthetic variety.
(iii) The component lines will be combined in a manner so as to allow random mating.

Synthetic varieties are normally used in cross-pollinated crops in those situations where hybrid development and organized seed production are in the preliminary stages. Inbred lines are common components of synthetics because they are easy to maintain and can be used again and again as per requirement of reconstituting the synthetics. When selfing is not possible, sibbing may be used. In case of perennials, clones are used as components. Combining ability of the component lines is determined in different ways. General combining ability is desired more than the specific combining ability because there are several components to a synthetic variety and each is required to combine well with the rest other lines. Random crossing of components of a synthetic variety is assumed and each line should contribute the same amount of seed to the next generation. Because a synthetic variety is a mixture of many genotypes, it might change under selection pressure during multiplication.

The component lines are termed as syn-0 and the F_1 generation from random crosses of the components is termed as syn-1 and the F_2 generation as syn-2. A synthetic variety is used over a number of generations from syn-2 to syn-4.

Wright in 1922 developed a formula to predict the performance of a synthetic in the F_2 generation which is as follows:

$$F_2 = \frac{(\bar{F_1} - \bar{P})}{n}$$

where F_2 is the expected performance of the F_2, F_1 is the average performance of the F_1 hybrids from all combinations of the inbred lines, P is the average performance of the inbred lines and n is the number of inbred lines. This equation implies that the yield of the F_2 and the subsequent generations may be increased by:

(i) Increasing F_1 yields.
(ii) Increasing yields of parental inbred lines and
(iii) Increasing the number of lines.

On the basis of given formula, if there is no difference between the F_1 yield and the parental yield, the F_2 yield equals the F_1 yield. In other words, if there is no heterosis between the parents there is no inbreeding depression. Given random breeding and absence of differential selection against particular alleles, no further loss of vigour is theoretically to be expected in the generation mean after F_2 generation of a cross, since the gene frequency, and hence the genotypic frequency, will remain in equilibrium (Williams, 1964). Actual tests on F_2, F_3 and F_4 generations have confirmed this theoretical expectation. The question of the optimal number of selected lines that can be used as a basis for interbreeding populations, assuming the general validity of Wright's formula, has been investigated by Kinman and Sprague (1945). Theoretical yields were calculated for nine populations based on 2-10 lines drawn from among 10 inbred lines. The 10 lines as well as their 45 F_1 hybrids were assessed for yield, and actual yield figures used to calculate the F_2 yields of populations based on different number of lines. On the basis of these calculations, maximal yield was given by a population based on five or six of the best combining inbreds. Populations based on less than five of the best inbreds gave an excessive reduction in yield from F_1 to the F_2 generation, while those based on more than six lines lowered the F_1 mean to a level which nullified the advantages to be gained by increasing the number of lines.

Steps in Development of Synthetic Variety

Briggs and Knowles (1967) have outlined the following procedures to develop synthetic cultivars:

(i) Evaluate between 100 and 300 clones or inbreds over one-two years and select the best 25-50 for a polycross test.
(ii) Plant 25-50 selected clones or inbreds in a four to ten replicate polycross nursery ensuring completely random pollination. Seed from all replicates

of each clone or inbred is composited separately and it is called polycross seed.

(iii) The bulk seed from each entry of the polycross nursery is included with one or more check varieties in a replicated polycross progeny test at one or more locations. Superior polycrosses are identified and selected.

(iv) Clones or inbred lines from the superior polycrossses as identified in step (iii) are established in a replicated crossing block. This is called syn-0. The seed from the block is harvested in bulk.

(v) The crossed seed is increased in isolation under random mating and is termed as syn-1.

18

Hybrid Breeding in Cross-Pollinated Crops

Mangelsdorf (1951) has opined that hybrid corn has been the most far-reaching development in applied biology. Hybrid cultivars have contributed enormously to world's food and feed resources. Although dramatically successful at first in corn, hybrids have spread to several other crops both in cross-pollinated and self-pollinated species. Their use is likely to spread more in time to come. Hybrid varieties are those in which F1 populations are used to produce commercial crop. Parents of F1 may be inbred lines, clones, varieties, or other populations. Hybrid breeding is historically rooted in development of hybrids in corn and therefore most of the issues covered in this section relate to hybrid breeding in maize.

Maize

Single Crosses

F1 hybrid between two inbred lines is called as single cross. It was first proposed by G. H. Shull in 1909. While proposing the development of single crosses in maize, Shull recognized the necessity of carefully selecting inbred lines based on their performance in hybrid combinations and inbred lines being homozygous will reproduce true to the type and production of single cross will be repeated as and when required. While recommending the utilization of single crosses in maize, the scientists were conscious of the fact that the hybrid seed produced on low yielding inbred parents will have to be looked into carefully. The low yields of inbreds and that of hybrid seed were certainly major limitations in the initial stages. Additionally, the seed of the inbred lines was small and misshapen and not well suited to mechanical planting. Often seed germination was poor and the seedlings lacked vigour. Normally two rows of female were planted alternating with one row of male parent. Thus only two thirds of the field could grow single cross hybrid seed. However, it should be understood that now these limitations have been taken care of to a great extent and large numbers single cross hybrids of corn are under commercial production on a large scale including India where several private sector seed companies are involved in research and development on big scale.

Double Crosses

Double cross is derived by crossing two single crosses and the resulting F1 seed between the two single crosses is used to grow commercial crop. Jones in 1918 proposed the use of the double cross. Since the hybrid seed is produced on single cross which is high yielding and vigorous, the hybrid seed produced is relatively large in quantity on per plant basis and hence production of double cross hybrid seed is more economical. The normal planting ratio of female to male rows is 4 is to 1 (4 rows of female planted alternately with 1 row of male parent). With this arrangement, 80 percent of seed field is used in production of commercial hybrid seed. With improvement of inbred lines and selection of improved inbred lines as parents, now it has been possible to produce double cross hybrids that are yielding almost as the single cross hybrids. Double cross hybrids are more variable than the single cross hybrids but show wider adaptability than the single cross hybrids.

Three-Way Crosses

Three-way cross is derived from cross of a single cross as female with another inbred line as male. These are intermediate between single and double cross hybrids in their characteristics.

Inbred Line Development and Testing for GCA and SCA

Inbred lines are produced by selfing open-pollinated stocks which may be improved populations, varieties or the hybrids. Selfing with selection is continued for 5-7 generations to make the lines uniform and largely homozygous. Selfing may be required for a longer period to eliminate small variations. During selfing selection is practiced for desirable traits like maturity, plant height, ear characters, kernel characters, stem strength and resistance to major diseases. Some consideration is given to yielding ability also as use of productive inbred lines in hybrid seed production reduces the cost of growing hybrid seed. After developing the superior inbred lines through selfing accompanied with selection for desirable agronomic traits, the inbred lines are evaluated for general combining ability followed by for specific combining ability.

For estimation of general combining ability, all possible crosses need to be made among '*n*' number of inbred lines and this number is given by n(n-1)/2 excluding reciprocals. This number becomes unmanageable with large number of inbred lines. Therefore, a more efficient method as suggested by R. L. Davis in 1929 of inbred x variety or top cross is used. This gives a measure of general combining ability. It has been generally concluded that using top cross, half of the inbred lines can be discarded and remaining lines are tested for specific combining ability. In top crosses, normally inbred lines are used as female parent and open-pollinated variety is used in every fourth or fifth row as a pollen source. Inbred lines are detasseled. In case open-pollinated variety is used as female, at least ten plants should be used in crosses with the inbred line to sample adequately the variability of the open-pollinated variety. For determining GCA of inbred lines, mostly early testing is done to eliminate up to 80 percent of the material selected for selfing in open-pollinated materials as suggested by M. T. Jenkins as back as in 1935.

Specific combining ability, a term suggested by Sprague and Tatum in 1942 is the ability of inbred lines to combine in specific single, three-way, and double cross. The most important consideration here is the combining ability in double crosses which are largely marketed. A line may be of very high combining ability in single crosses but will have no commercial value if the single cross involving a particular line is not a good parent of a double cross. Number of double crosses to be made is given by n(n-1) (n-2) (n-3)/8.

With 30 inbred, the number of double crosses to be made becomes 82215 excluding reciprocals. Even if the number of inbred lines is reduced to 10, still 630 double cross combinations will have to be made which again is huge task. In order to avoid this virtually impossible task, M. T. Jenkins (1934) presented data on four methods of estimating the yield potential of four inbred lines in a double cross which were based on their average yields in:

(i) Six possible single crosses.
(ii) The four single crosses not used in making double crosses.
(iii) All possible combinations with ten other inbreds.
(iv) Top crosses.

Based on this and other research reports, it has become an accepted practice to to predict the yields of double crosses from the yields of the four non-parental single crosses. Experimental evidences have shown that in general, it is advisable to combine the similar or closely related inbreds in the single crosses and the different or distantly related inbreds in the double crosses (Briggs and Knowles, 1967). The inbred lines need continuous improvement also for which pedigree method, back-cross method, convergent improvement, gamete selection and DH technology are used depending upon the competence and resources.

Production of Hybrid Seed of Corn

Manual Detaselling and Hand Pollination Method: This is the common method of producing hybrid corn seed in India. In this system, female and male parental lines are grown alternately in ratio of 6 rows of female: 2 rows of male. The female rows are detaselled before the inflorescence sheds pollen grains and seed produced on these detasseled plants is the hybrid seed for commercial marketing and production of hybrid corn in the farmers' field.

Use of Cytoplasmic Male Sterility in Production of Hybrid Seed in Corn: Cytoplasmic male sterility has facilitated the production of hybrid corn seed because it has eliminated the need for detaselling which is reported to cause reduction in yield. Cytoplasmic male sterility is incorporated into the inbreds to be used as female parents of single crosses. Various combinations of sterile cytoplasm and fertility restorer genes have been used to produce double cross hybrid as given in Fig. 18.1 (Briggs and Knowles, 1967). It should be noted that the single cross and double cross hybrid yields drop between F1 and F2 and not between F2 and F3. Equilibrium is reached in most cases after one generation of random mating.

1. One inbred male-sterile and no fertility restorers

Inbreds rfrf + S cytopl x rfrf + N cytopl Detaselled rfrf + N cytopl x rfrf + N cytopl

↓ ↓

Single cross F1 rfrf + S cytopl x rfrf + N cytopl

↓

Double cross F1 rfrf + S cytopl (Mix 50% of this double cross with 50% of the same double cross with N cytopl)

2. One inbred male sterile and two with fertility restorer genes

Inbreds rfrf + S cytopl x rfrf + N cytopl Detaselled RfRf + N cytopl x RfRf + N cytopl

↓ ↓

Single cross F1 rfrf + S cytopl x RfRf + N cytopl

↓

Double cross F1 Rfrf + S cytopl
Male fertile

3. Two inbreds male sterile and one with fertility restorer gene

Inbreds rfrf + S cytopl x rfrf + N cytopl rfrf + S cytopl x RfRf + N cytopl

↓ ↓

Single cross F1 rfrf + S cytopl x Rfrf + S cytopl

↙ ↘

Double cross F1 Rfrf + S cytopl rfrf + S cytopl
50% male fertile 50% male sterile

Fig. 18.1. Alternative methods of using cytoplasmic male sterility to produce double cross hybrids of corn (S= sterile cytoplasm, N = normal fertile cytoplasm, Rf is a restorer gene that restores fertility in sterile cytoplasm).

Other Cross-Pollinated Crops

Development of hybrid cultivars has been done in several other crops like sorghum, pearl millet and sunflower using cytoplsmic genetic male sterility or three-line system termed as ABR or ABC line system. The details on these have been given in the section in breeding methods in self-pollinated crops, for example, rice. So the same is omitted here.

Castor

Release of Pioneer hybrids CGH 4 and CGH 5 has been a breakthrough in castor production and productivity in the irrigated areas of Gujarat and Rajasthan. Unlike in other field crops, wherein either cytoplasmic –genetic or genetic or manual pollination control system is the practice, hybrid technology in castor is based on pistillate sources available in the otherwise monoecious crop. Multiplication of pistillate line and hybrid seed production are done by adjusting flowering phase with periods of conducive temperature.

There are three different pistillate sources, namely, N, S and NES types. N type pistillate line governed by a recessive sex-switching gene (*ff*) is maintained by sib-mating in a progeny of pistillate plants (seed set by interspersed staminate flowers) which segregate in the ratio of 1 pistillate to 1 monoecious plant. The S-type pistillate line is characterized by the phenomenon of sex reversal i.e. female to monoecious at any time after the appearance of first raceme. Repeated inbreeding of sex reversal types results in a spectrum of types each reverting at different times of development and monoecists. The third type (NES) is homozygous for N-pistillate gene along with environment sensitive gene(s) for interspersed staminate flowers. Yet another system is an induced mutant having male flowers with fused perianth which prevents pollen shedding and hence usable as stable female line in commercial seed production. In India, S-type pistillate lines are widely used female parents in many of the commercial hybrids. However, the hybrid technology needs lot of refinement for effective use in hybrid seed production (Siddiq *et al.* 2004).

19

Breeding for Disease Resistance

This chapter starts quoting first para of the book entitled Disease Resistance in Plants by Van Der Plank (1968). "In 1958 the American Phytopathological Society met to celebrate its Golden Jubilee and review the problems and progress in plant pathology during the 50 years, 1908 to 1958. J. C. Walker was elected to review the problems and progress in controlling plant diseases by host resistance, a fitting tribute to one who perhaps more than any other had contributed during those 50 years both to understanding of nature of resistance and to the practical application of resistance to problems in horticulture and agriculture. In his summary Walker wrote "By and large the development of resistant varieties must be looked upon as a continuing program. The potential variability of most pathogens (including viruses) will not permit any currently successful variety to remain so for an indefinite period". This opening para tells the importance of breeding for resistance to diseases in crop plants on a continuous basis as the fight between host and the pathogen will go on and on.

Some Definitions

Monogenic Resistance: This resistance is governed by one gene. The gene in monogenic resistance is usually studied in detail. It is identified individually by letters or numbers or both. Thus, the gene R_1 gives the potato resistance to blight. The gene Sr_6 gives stem rust resistance in wheat. In monogenic resistance, it can be determined whether the gene is dominant or recessive. In this case, resistance is clearly defined and in the segregating populations, the plants can be divided by their resistance into clear and natural groups. Often, location of these genes is determined on a particular chromosome, for example, gene Sr_6 is found to be located on chromosome XX of wheat.

Polygenic Resistance: Polygenic resistance is conferred by too many genes to be counted and to be identified individually. The plants in segregating populations show a continuous variation for resistance without falling into clearly defined groups. No single gene has an effect large enough to be followed or for the gene to be identified and located on a particular chromosome. One knows the combined effect of the genes as a whole. However, now it has been possible to locate the polygenes through quantitative trait loci (QTL) mapping using the techniques of molecular genetics and breeding.

Oligogenic Resistance: In between monogenic ad polygenic resistance, there are intermediate sorts of resistance governed by digenic or trigenic resistance which are neither monogenic nor polygenic. This resistance is termed as oligogenic.

Major and Minor Gene Resistance: The terms "major" and "minor" are sometimes used to designate oligogenic (including monogenic) and polygenic resistance.

Vertical Resistance: When a variety is more resistant to some races of a pathogen than to others, the resistance is called as vertical or perpendicular resistance. This is also called as race-specific resistance, pathotype-specific resistance, major gene or oligogenic resistance, non-durable or qualitative resistance. Vertical resistance is a differential resistance or host x isolate interaction of cultivars against different races of the pathogen. The vertical resistance brings delay in start of epidemic. This type of resistance is easy to breed because the major genes are easy to identify and transfer through simple crosses. These genes control specific races or genotypes of pathogen and hence do not provide protection against new races of the pathogen. Cultivars equipped with this narrow based genetic protection are often vulnerable when a genetically variable population of a pathogen occurs in the environment. This resistance is also known as non-durable resistance.

Horizontal Resistance: When the resistance is evenly spread against all the races of the pathogen, it is called as horizontal or lateral resistance. This is also called as race-nonspecific resistance, slow rusting, partial resistance, minor gene reactions or polygenic resistance. Horizontal resistance slows down the epidemic. There is lack of differential reaction of isolates on different cultivars. The resistance is effective against all genotypes of pathogen without cultivar x isolate interaction. Following the initial establishment of the pathogen, plant may resist its spread and reproduction such that the disease develops a slower rate. Unlike vertical resistance, evolution of new races due to selection pressure against any specific race is absent under horizontal resistance and this makes the horizontal resistance as more stable and long lasting. Horizontal resistance is controlled by polygenes where each gene contributes toward the resistance and hence the resulting resistance is also called as minor gene resistance. Polygenic resistance is wide spread and provides protection against a wide range of races and is less susceptible to being overcome by a new race. This resistance protects plants by slowing down the spread of the disease and development of epidemics in the field. Breeding for polygenic resistance is more challenging. Many minor genes cannot be identified individually and consequently they cannot be transferred through crossing in a predictable fashion. However, heritability of polygenic resistance is high. Horizontal resistance is said to be durable.

Partial Resistance: J. E. Parlevliet (1975) used the term "partial resistance" to characterize slow rusting to leaf rust in barley. According to him, partial resistance is a form of incomplete resistance characterized by a reduced rate of epidemic development despite a high or susceptible infection type. He further showed that partial resistance to leaf rust involves interaction of minor genes and the components of resistance may be under pleiotropic genetic control.

Durable Resistance: Durable resistance as defined by Johnson (1978) is that which has remained effective in a cultivar during its widespread cultivation for a long sequence of generations or period of time in an environment favourable to disease or pest. This term has received wide acceptance and is popularly used. Improper use of the major, race specific-type of resistance to control rapidly evolving pathogens (e.g., the rust of wheat) has led to boom-and-bust cycles that make it necessary to replace cultivars at a short time after their release. Durable resistance on the other hand is based on interaction of minor, additive genes. In case of leaf and yellow rusts of wheat, accumulation of 4-5 slow rusting genes results in a high level of resistance that approaches immunity. Breeding to achieve durable slow rusting resistance in wheat was undertaken in CIMMYT during the early 1970s by S. Rajaram. He selected plants and lines in segregating populations that would show up to 20% of disease severity under high disease pressure, when the susceptible check cultivars would have shown complete susceptibility. This major thrust toward accumulating slow rusting genes in CIMMYT germplasm resulted in the release in Mexico of cultivars such as Torim 73, Cocoraque 75, Pavon 76 and Nacozari 76 in the mid-1970s. The moderate resistance level they carried has remained durable. Since then, the understanding of the genetic basis of durable resistance to rust diseases in wheat germplasm of CIMMYT origin has been enhanced and a simple breeding methodology formulated. This is the type of resistance needed in 21st century to ensure food security (Singh *et al.*, 2004).

Tolerance: Unlike avoidance and resistance mechanisms that operate to reduce the level of infection by the pathogen, tolerance (or endurance) operates to reduce the extent of damage inflicted and the infected host attempts to perform normally in spite of the biotic stress. In this situation, a host may be highly susceptible and yet exhibit little reduction in economic yield. Because tolerance is measured in terms of economic yield, it may not be applicable to those cases where the economic yield, for example, grain itself is infected. This mechanism may be applicable to a situation where a plant recovers quickly after infection. From the point of view of virologists, plants with few or no symptoms are described as tolerant. The exact mechanism of tolerance is not understood but it is believed that plant vigour, compensatory growth and changes in photosynthate partitioning may be the contributory factors to tolerance.

Pathogen and Host

A pathogen is a living organism capable of causing a distinct disease or disorder in another organism (host). The capacity of a pathogen to cause disease or disorder in a particular host is called as pathogenecity. The extent of disorder in disease development is called as virulence. The pathogencity and virulence of a pathogen vary among pathogen types (pathotypes or races). Races or pathotypes that fail to cause disease symptoms are said to be avirulent. It should be noted that mere presence of a pathogen on a host is not enough o cause disease symptoms to occur. A third factor-favourable environment- is needed to cause the disease by the pathogen on the host. This typical scenario of the trio (pathogen plus susceptible host plus favourable environment) is referred to as disease triangle. Races of pathogens may also be described in terms of aggressiveness or non-aggressiveness in relation to the rate at which they produce disease symptoms.

The host is the plant/organism in which a pathogen can produce disease symptoms. A susceptible host is one in which a pathotype /race can express the disease symptoms. A host may employ one of several mechanisms to resist pathogen. The defense mechanisms may be pre-existing (morphological features that pose barriers to the penetration of the pathogen into the plant, like presence of lignin, cork layer, callose layer) or secondary metabolites like phenols, alkaloids, hydrolases, glycosides, etc. Besides, pre-existing features, there could be infection-induced defense mechanisms which include production of chemicals like peroxidases, hydrolases, phytoalexins, etc. upon infection.

Defense Mechanisms in Plants Against Pathogens

Avoidance: Avoidance also described as escape is a mechanism that reduces the probability of contact between pathogen and the plant. This mechanism is rare in pathogens but common in insects.

Resistance: The mechanism of resistance operates once the host has been attacked by the pathogen. It curtails the invasion by the pathogen and reduces the growth and development of the pathogen after the infection. The nature of mechanism may be biochemical, anatomical, or morphological. The equivalent term to describe this mechanism in insects is called as antibiosis. Genetically resistance may mean the ability of the plant to reduce the adverse effect of the attack by the pathogen because of genotype of the plant. Resistance reactions are of the following types:

Hypersensitive Reaction: This mechanism acts to prevent establishment of the pathogen inside the host. An infection by the pathogen evokes a rapid localized reaction whereby the cells immediately surrounding the point of attack die so that infection is contained and not allowed to spread further. The pathogen eventually dies leaving a necrotic spot. This kind of infection inhibition is most common and is evoked in diseases like leaf spots, blights, cankers and decays.

Overdevelopment of Tissue: Meristematic activity may be stimulated resulting in excessive abnormal tissue growth as can be seen by galls and leaf curling. In some cases, a layer of suberized cells forms around the point of infection and this helps in further spread of the pathogen.

Underdevelopment of Tissue: Affected plants become stunted in growth or the organs are partially developed. This is common in case of virus infection.

Gene for Gene Relationship

H. H. Flor (1956) was the first scientist to initiate simultaneous inheritance studies involving the pathogenecity in flax rust (*Melompsora lini*) and resistance to its host flax (*Linum usitatissimum*). Based on these studies, Flor was able to formulate 'Gene-for-Gene Relationship' concept which states that "for each gene conditioning rust reaction in the host, there is a specific corresponding gene conditioning pathogenecity in the parasite". As given in Table 19.1, resistance is expressed only when the host cultivar carries resistance alleles and the pathogen race possesses avirulence allele. In another three situations, the interaction phenotypes resulting from action of gene in host with that of parasite is susceptibility (Singh *et al.*, 2004). In other words, a resistance gene can act only if a locus in the pathogen carries a matching allele for avirulence. By themselves, resistance genes are ineffective if pathogen carries the appropriate allele for virulence (usually a recessive allele). Most researchers agree that specific interaction occurs between dominant resistance alleles and dominant avirulence alleles.

Table 19.1. Explanation of the gene-for gene hypothesis presented by Flor (1956).

Host	Pathogen	
	Avirulence allele (P1P1)	Virulence allele (p1p1)
	Interaction phenotype	
Resistant allele (R_1R_1)	Resistant	Susceptible
Susceptibility allele (r_1r_1)	Susceptible	Susceptible

The gene-for-gene theory deals only with pathogen-host relations. Flor's work received more attention when Person in 1959 provided a theoretical analysis of the gene-for-gene relationship and concluded that such relationship should occur as a general rule in host-parasite system as a result of selection pressure during evolution. Soon after this publication, the gene-for-gene concept was demonstrated or hypothesized in a number of host-parasite system including the rusts.

General Considerations in Breeding for Resistance to Diseases

Acquaah (2007) has listed the following points while contemplating a project on disease resistance breeding:

1. Breeding is an expensive and long duration undertaking and therefore disease resistance breeding should be undertaken only when the pathogen is causing a major impact on a crop of economic significance.
2. Natural resistance is not always available in cultivated types, but available in un-adapted gene pool requiring additional time and cost for pre-breeding.
3. Breeding for resistance varies in complexities depending upon the nature of pathogen. Resistance to vascular pathogens, viruses, smuts, rusts, and mildews is relatively easier to carry out than breeding against pathogens that cause rots (root rot, crown rot, storage rot) and nematodes.
4. Instability of disease resistance is a key consideration in disease resistance breeding because new pathogenic races may arise or even the environment may modify the expression of the resistance.
5. Molecular breeding approaches particularly those related to marker-assisted selection my help disease resistance breeding in terms of precision and selection of desirable genotypes while advancing the segregating generations.
6. For horticultural crops where no blemishes on fruits are accepted by the consumers/markets, breeding for major gene resistance with complete expression is desirable.
7. When a crop is grown for food and feed, breeding for mechanisms that increase the levels of chemical toxins in plants is not desirable.

Screening Techniques in Disease Resistance Breeding

Expression of disease depends on the presence of pathogen, a host and a favouravle environment for expression of the disease. These conditions are met sometimes under natural conditions at a particular place where the disease occurs regularly in severe form and these locations are called as 'hot-spot' locations for disease screening. This kind of field screening has the advantage of representing the conditions under which the resistant cultivar will be produced. However, it has its limitations. The weather (or the environment component of the disease triangle) remains un-reproducible and unpredictable making it difficult to have uniformity and consistency of the pathogen. In some years, the weather may not be favourable and thus adequate build-up of pathogen may not happen resulting into a situation where even susceptible plant may appear to be resistant. Controlled environmental tests provide reliable, uniform and consistent evaluation of disease. Therefore, screening under controlled environmental conditions in greenhouses with suitable temperature, humidity and inoculums build up, for resistance to diseases is preferred. Of course this resistance has to be correlated with field resistance at advanced stages of the material under development. The disease screening under artificial epiphytotic conditions will vary from crop to crop and disease to disease.

Breeding Methods for Disease Resistance Breeding

The basic requirements for breeding disease resistant plants are sources of resistance and methodology for combining resistance with commercially or aesthetically acceptable plant types. These requirements have been met in many ways. Early practices of Bolley, Orton, Essary, Jones, and Norton are still pursued. Variations of these practices are introduced and new methods in breeding and pathology are invented, new sources of resistance are found, and changes occur in cultural practices, harvesting methods, and consumer preference. For these reasons, breeding disease-resistant cultivars remains a continuous process. To produce new cultivars, there must be a continual search for germplasm with disease resistance traits that meet the changing needs. The search for needed genes goes on in several ways; the emphasis varies with crop or need in question. Selection, plant introduction, native wild species, and induced mutations represent the primary sources; each has played a significant role (Roane, 1973).

Historically in the past, selection has been the most commonly used and effective method of breeding against diseases. Orton developed varieties of cotton, cowpeas and watermelon resistant to Fusarium wilt. Similarly, Bolley developed Fusarium resistant cultivars in flax using selection. Initially selection was practiced by isolating resistant survivors of natural epiphytotics. Later, artificial epiphytotics were created in disease nurseries to facilitate selection of resistant individuals rather than escaped suscepts. Later on, pedigree selection has been widely used method to breed against diseases in large number of self-pollinated crops. Next in popularity has been the bulk method of breeding. Back-cross method of breeding has been the next in order to be used in disease resistance breeding. But now back-cross method of breeding is quite commonly used. For cross-pollinated species, recurrent selection is effective for increasing the level of resistance in a population of genetically heterogeneous population.

Back-cross method of breeding for disease resistance has several desirable features. It requires less time as compared to other methods of breeding. Further, back-cross method of breeding permits addition of more disease resistance genes into a variety without affecting the previously present genes provided the loci are different for the sets of genes.

Mutation breeding seems quite appealing to be used in disease resistance breeding because it opens up the possibility of adding resistance without affecting the characteristics of the variety being grown and already adapted. Where no genes for resistance to a race or races of a pathogen are available, mutation breeding is the only option. However, now the development of transgenic crops has added a new dimension to disease resistance breeding where genes from altogether unrelated sources can be introduced into the crop species breaking the crossing barriers and this has been put to use in few cases like virus resistance in papaya. The first disease resistance gene, ***Pto*** (binds with products of pathogen to give resistance) was cloned in 1993 by Greg Martin. The *Pto* gene encodes a serine-threonine kinase that confers resistance in tomato to ***Pseudomonas syringae*** pv tomato strains expressing the avirulence gene *avrPto*.

Engineering transgenic plant with resistance to virus is accomplished by the method called coat protein-mediated resistance. First the viral gene is reverse transcribed (being RNA) into DNA from which a double-stranded DNA is then produced. The product is cloned into a plasmid and sequenced to identify the genes in the viral genome. A chimeric gene is constructed to consist of open reading frame for coat protein to which a strong promoter is attached for high level of expression in the host. This gene construct is transferred into plants to produce transgenic plants. Success with this strategy has been reported in summer squash (the first product developed by this approach) and for resistance to papaya ring spot virus (a lethal disease of papaya) among others. GM technology holds huge potential to breed crop cultivars having resistance to diseases.

Gene pyramiding has become now a common breeding strategy against diseases where several specific genes are transferred into one variety through back-crosses facilitated by marker –assisted selection. Three major genes conditioning resistance to blast in rice, *P*i-*1*, *P*i-*2* and *P*i-*3* were pyramided through pair-wise crosses of isogenic lines carrying the genes.

Breeding for horizontal resistance is becoming more acceptable as vertical resistance which is pathotype specific and governed by major one or few genes, is likely to be broken down over a period of time due to evolution of new races of the pathogen. Further, vertical resistance breeding is not suited to perennials. Tree breeding is a much longer process than breeding of annuals and errors are costly to correct. Should vertical resistance be overcome, it cannot be quickly replaced with new genotype with new resistance genes. Therefore, quite often breeders are suggesting first to breed for a high level of horizontal resistance in a genotype and then cross it with one that has high vertical resistance. Combining vertical and horizontal resistance seems to be better strategy to deal with disease resistance. However, here it will be worthwhile to mention that there is chance of erosion of horizontal resistance while breeding for race-specific vertical resistance which is called as the vertifolia effect (after the potato cultivar Vertifolia) in which the major gene is so strong that while the breeder focuses on vertical resistance, no evaluation and selection for horizontal resistance is possible eventually leading to the loss of horizontal resistance. However, vertifolia effect is not of universal occurrence. There are reports of combining race-specific resistance in addition to high-level polygenic resistance to leaf rust in barley. To reduce the occurrence of vertifolia effect, it has been suggested to discard susceptible plants in the segregating generations than selecting highly resistant genotypes. Also, there are suggestions to first breed for a high level of horizontal resistance in a genotype followed by crossing the same with one that has high vertical resistance (Acquaah, 2007).

Wild relatives of cultivated crops have been a source of resistance to diseases in several crops. Success of transferring such genes into the cultivated cultivars has been varying from crop to crop. Transfer of disease resistance genes from wild types to the cultivated types has been quite successful in tomato. In tomato, wild relatives have provided genes for resistance to tomato yellow leaf curl virus. *Lycopersicon esculentum* and *L. pimpinellifolium* are easily crossable. Resistance to devastating late blight of potato has been found in wild species. Similarly, resistance to root knot nematode in peanut has been obtained from three wild species. A wild relative of rice (*Oryza nivara*) growing in wild in state of Uttar Pradesh in India has been found to have one single gene for resistance to the grassy stunt virus, a disease that devastated the rice crop in South and South East Asia in the 1970s.

20

Breeding for Resistance to Abiotic Stresses

When conditions in the plant growing environment are less than the optimum, the plants experience stress which is broadly termed as abiotic stress. Major abiotic stresses impacting crop yield significantly are as follows (Grover *et al.*, 2004; Acquaah, 2007):

1. **Drought:** This is the environmental condition caused by lack of adequate rainfall. This is also termed as water stress or desiccation stress. Effects of drought on crop plants are complex and variable and are greatly accentuated by a number of interacting factors. The onset of drought in general has been observed to reduce germination, emergence, hypocotyl length, water uptake and the mobilization of dry matter reserves even at the early growth stages. Drought stress causes a marked reduction in leaf area. Soil moisture availability at the seed germination, anthesis and at post-anthesis growth stages determines as to what extent production potential of crops would be expected in drought-prone situations.

   The yield losses in field crops may be attributed to inadequate water availability at the critical growth stages. Crops therefore differ in their susceptibility of growth stages to water stress. Stress exposed plants immediately lower down relative water content (RWC) of their leaves, leaf water potential, osmotic potential. Turgor pressure may also be decreased unless plants have finally adapted to drought by opening of stomata again and improving the water potential. It has been reported that low leaf expansion rate at the early stage and accelerated leaf senescence accompanied by mortality of older leaves at later stage decreases total leaf green area.
2. **Water Logging:** Excessive soil moisture resulting from prolonged rainfall causes roots to suffer from lack of oxygen. This is also known as submergence stress or flooding stress.
3. **Heat:** Heat stress occurs when the temperatures are high enough to cause irreversible damage to plant function. A temperature higher than the optimum decreases the growth rate, maturity period, and metabolic processes resulting in reduced plant productivity. High temperature stress

is a major cause for unstable and low yields of nearly 1 t/ha, which is far less than the potential yields as reported in peanuts, soybean, chickpea and pigeon pea, etc. Wheat yield in northern India is reduced by almost 4% for every 1^0 C rise in temperature over the range of 12.2-27.5^0 C. *Brassica* species if planted at more than 30^0 C mean temperatures, particularly in arid zones may show a high degree of seedling mortality leading to sub-optimal and patchy plant stand.

4. **Cold Stress:** Cold stress manifests when plants are exposed to low temperatures that cause physiological disruptions that may be irreversible. Low temperature affects agricultural crops to a considerable extent in many parts of the world. Though generally mild in northern India, cold shocks cause extensive damage to fruits, vegetables, food crops, ornamental and industrial crops. In general frosts in northern plains are fewer in December and common in January and rare after mid-February. Among leguminous crops, field pea and chick pea are sometimes damaged by frost in India. The oilseed *Brassica* and potato are susceptible to frost and a single night of frost can cause extensive damage to these crops in northern plains and hilly regions of India.
5. **Salinity:** Stress from salinity occurs when the dissolved salts accumulate in the soil solution to an extent that plant growth is inhibited. Another term for this stress is salt stress.
6. **Mineral Toxicity:** Mineral toxicity results when an element in the soil solution is present at a concentration that the plants are physiologically impaired.
7. **Oxidation:** Oxygen free radicals (or activated oxygen) are known to cause degenerative conditions in plant cells.
8. **Mineral Deficiency:** Inadequate amounts of essential soil minerals for plant growth cause crop injury.
9. **Pollution Stress:** This is caused by increased levels of pollutants.

Screening for Stress Resistance

Because of complexity of environmental stress, simple, practical and effective tests that can readily be used by breeders as selection criteria are not available. Acevedo and Fereres (1993) summarized the criteria for development and use of screening tests as follows:

1. Genetic variation should occur in the germplasm pool for the traits under consideration.
2. The heritability for the trait should be greater than the heritability for yield *per se*.

3. The trait should be correlated with a yield-based stress-resistance index.
4. It is desirable that the designated trait should be causally related to yield.
5. The screening test should be easy, rapid and economical to use.
6. A few typical traits correlated with stresses are as:
   (i) **Drought Resistance:** Early flowering, small area leaf index, short anthesis-to-silk interval (ASI), more root depth, shoot developmental characteristics, osmotic adjustment, reduced stomatal activity, wax load on leaves, presence of pubescence on leaves, yellow leaf cultivars in wheat, delayed senescence or stay-green, high chlorophyll content and leaf reflectance, more carbohydrates in the stem.
   (ii) **Salt Tolerance:** Rate of Na^+ or Cl^- accumulation in leaves, vigorous plant habit in rice, an ability to limit sodium uptake from the roots, to distribute ions in older leaves rather than younger leaves and to tolerate the ions accumulated in the leaves.
   (iii) **Heat Stress:** Leaf electrolyte leakage, a surge of proline translocation towards the pollen grain, stable superoxide dismutase activity, reduction in chlorophyll content in leaves, lower malonidixldehyde (MDH) in leaves, longer and thicker leaves, thicker palisade tissues, more leaves/plant, taller and narrow plants, early bolters,

General Breeding Approaches to Breed Against Abiotic Stresses

There are large numbers of crop specific reviews and research papers to deal with this issue. However, they all are avoided as the scope of this chapter is to give a generalized concept to the readers so that they could understand the basic breeding approaches. Two basic genetical approaches being utilized to improve stress tolerance include (i) exploitation of natural genetic variations, either through direct selection in stressful environment or through mapping of QTLs and subsequent marker-assisted selection (MAS) and (ii) generation of transgenic plants to introduce novel genes or alter expression levels of the existing genes to affect the degree of stress tolerance (particularly salt tolerance) as reviewed by Blumwald *et al.* (2004) and Witcombe *et al.*, (2007).

Kumar *et al.* (2014) reported on the progress achieved in developing drought tolerant rice varieties through conventional breeding approaches based on direct selection for grain yield under drought, the identification of large effect QTLs for grain yield under drought and marker-assisted breeding using the identified QTLs to improve the grain yield of popular high yielding but drought susceptible varieties under drought. Whole genome genotyping (WGG), selective genotyping (SG), and bulk segregant analysis (BSA) were used to identify QTLs for grain yield under drought. Linkage maps for populations were developed using MapManager QTX. To detect the relationship between markers and trait value, QTL cartographer 2.5.009, Q gene 4.3.10 and QTL Network 2.1 were used. Single marker analysis (SMA) followed by composite interval mapping (CIM) were conducted to identify significant QTLs. All three software programmes allow permutation tests to determine the significance threshold to identify significant QTLs. Some 500-1000 permutation tests were conducted to determine threshold levels. QTL and cartographer and Q gene provided the LOD score to describe the significance of a QTL, while QTL Network used mixed model analysis for detecting QTLs and significance of QTLs was provided in terms of 'F' value. Efforts were made to identify at least three major QTLs in the background of popular high yielding varieties from Asia. The initial introgression approach was to pyramid 2-3 QTLs in the background of a popular high yielding variety for which they had been identified to obtain an economic yield advantage of 1.0-1.2 t/ha under drought in farmers' field.

According to Kumar *et al.* (2014), direct selection for grain yield under reproductive-stage drought has been found effective. The suitability of grain yield as a selection criterion allowed the initiation of large-scale breeding programmes aimed at developing high yielding and drought tolerant varieties with good grain quality and tolerance to major diseases in rice. The conventional breeding programme at IRRI included screening of about 5000 F2 plants for rice blast. The selected tolerant plants were transplanted in the field and evaluated for reaction to bacterial blight at 30 days after transplanting (DAT) by clipping the leaves with scissors infested with PX061 and PX086, two strains of *X. oryzae* pv *oryzae*. Plants showing scores of >3 based on Standard Evaluation System (SES) were rejected. Single plant selections were screened under reproductive-stage drought stress conditions in F3 generation and tolerant plants were selected based on grain yield. In F4 and F5 generations, selection was carried out in non-stress conditions with grain quality traits evaluated in F5 generation. In the F6 generation, lines were screened under both reproductive-stage drought stress and non-stress conditions. The lines with high yields under both conditions were advanced to an observation yield trial (OYT) and selected lines were advanced to advanced yield trials (AYT). The AVTs were conducted in target environments and the lines performing well in the target environments were recommended for release by the national systems. This approach has led to the successful development and release of 17 high yielding and drought tolerant varieties in South Asia, Southeast Asia and Africa.

Blum (PlantStress.com, http://plantstress.com/articles/drought._m/drought_m htm, accessed 17 Aug 2015) has summarized his views on utility of yield as a selection criterion for drought resistance. According to him, the issue of selection for yield, the impact of environment on genetic gains from selection and selection efficiency are under continuous debate as a central issue in conventional plant breeding. The comparative performance of two genotypes with respect to one another can vary from one environment to the other and this is fundamentally termed as genotype by environment interaction (G x E) for yield. Generally, the genotype preferred by most breeders and breeding text books is the one that expresses minimal G x E and yield is then considered stable across all the test environments. However, while yield under stress is the target of breeding programmes, selection for yield under stress is generally inefficient. Yield is a complex trait that has low heritability. There are the various developmental and physiological processes which make up yields that are inherited. The heritability of yield becomes still low especially under stress. It has been general and repeatable observations of the plant breeders that using yield as a selection index to improve drought resistance is generally not efficient. To compensate for low efficiency, breeders screen very large populations with the expectation that the "number game" will allow to identify the desirable genotypes. This expectation is not unfounded. While this approach has been generally successful, it is expensive. The use of molecular markers to tag and select for certain yield related quantitative trait loci (QTLs) can increase the efficiency of selection for yield but less effective under stress. Again, QTL by environment interaction seems to be the rule rather than exception. However, when breeding populations contain effective genes for drought resistance (say segregation for deep roots in upland rice populations), the efficiency of selection for yield under drought stress can be increased provided all precautions are taken to minimize spatial variability at the selection site. If selection for yield under stress is practiced, a positive G x E for specific drought stress condition is a strong indicator of resistance. There are many statistical models and methods that estimate G x E interaction, in most cases of a planned breeding programme for drought resistance, evaluation of G x E simply requires a comparison of yield performance under stress and non-stress (fully irrigated) conditions.

Alternatively, the "Fischer and Maurer" stress resistance index (RI) can be computed as follows:

$$RI = (Gs/Gn)/(Ms/Mn)$$

where genotypic yield under stress (Gs) and non-stress (Gn) is normalized for the mean yield of all the genotypes under stress (Ms) and non-stress (Mn). Values above 1 indicate a relative resistance as compared with the mean of the population. It has been argued that this index is flawed because it is affected by yield potential. This is true and as it has been seen indeed genotypes of high yield potential tend to be more susceptible to drought stress. This is a reality and not mathematical flaw. A major impediment in comparing genotypic response to a managed water stress environment is the variation among genotypes in their phenology. With such variation, different genotypes may be stressed at different growth stages. This has been a major pitfall in many drought resistance mapping exercises using populations such as recombinant inbred lines. The solution is to divide the population into several phenology sub-populations and compare the effect of stress only within the sub-populations of similar phenology. Alternatively, staggered planting dates can be attempted where the entire materials are planted later as compared with late flowering materials. However, in most breeding programmes, when tests for drought resistance are performed in the field at more advanced generations (for example, in F5 and beyond), the populations of lines do not express large variability for phenology.

Witcombe *et al.*, (2008), have written a detailed review on this subject and have concluded that the simplest approach in breeding for stress resistance is also the most effective. It is to select for yield directly which is the integrating trait and to carry out the selection in a target stress environment. It can be enhanced by the use of screens for stresses such as low N, drought or salinity where the stresses are carefully managed and by carefully choosing parents of crosses so that various physiological traits can be pyramided. This implies a reduction in number of crosses that are made so that larger population can be employed, an approach that has been effective in the breeding of stress tolerant, and widely adapted rice. Marker-assisted selection can be effective in enhancing efficiency but, at least so far, selection for markers linked to component traits of low heritability has not produced predicted outcomes. The introgression of QTLs for yield under stress is proving to be a more successful strategy in pearl millet. Fine mapping and map based cloning of QTLs will allow them to be used more effectively in breeding by eliminating the effects of unwanted linked alleles. MAS need to be fine-tuned so that precise combination of alleles can be combined for maximum effects. Transgenic and functional genomics approaches hold tremendous promise assuming that realistic evidence-based approaches to their employment emerge for informative assessment of stress tolerance.

Plant genetic engineering owes its growth to (i) plant tissue culture and (ii) plant molecular biology disciplines. The methods involved in tissue culture and regeneration response have been standardized for a large number of crops. The protocols to introduce the desired gene in crops have been perfected to a great deal. A range of different vector systems has been constructed for introduction of the transgenes in both dicots and monocots. The methods for stable genetic transformation (using a variety of approaches such as ***Agrobacterium tumefaciens***-based method, direct DNA uptake by electroporation and particle bombardment) as well as for regulation of the introduced transgene (using a variety of different approaches such as constitutive or induced promoters) have been optimized. Transgenic plants over expressing a number of different genes have been produced leading to commercialization of GM crops, namely, corn, canola, soybean, cotton, alfalfa, sugar-beet, papaya, summer squash, tomato, etc. A total of 24 countries planted GM crops in 189.8 ha in 2017. The GM traits included insect resistance (IR), herbicide tolerance (HT), stacked traits (IR + HT) and virus resistance.

Drought tolerant water efficient maize for Africa (WEMA) developed as transgenic crop is in advanced stages of testing in several countries in Africa. The Water Efficient Maize for Africa (WEMA) project has concluded its first decade of research, achieving significant advances that are benefitting small-holder maize (corn) farmers in sub-Saharan nations. Hybrids developed under the WEMA project can help farmers better respond to droughts, and the associated loss in productivity, caused by climate change.

The public-private partnership (USDA and Bill Gate Foundation) is led by the nonprofit African Agricultural Technology Foundation (AATF). Other partners are the International Maize and Wheat Improvement Center (CIMMYT), Monsanto (now taken over by Bayer) and the National Agricultural Research Systems (NARS) from the participating nations: Kenya, Uganda, Tanzania, South Africa, Ethiopia and Mozambique.

The WEMA project has developed conventional drought-resistant maize hybrids, which are sold under the brand name Drought TEGO, as well as genetically modified (GM) varieties-to be marketed under the brand name TELA — that offer both drought tolerance and insect resistance.

21

Molecular Breeding

Spectacular productivity gains have been achieved in wheat and rice in south and south-east Asia and also several other developing countries in Latin America and central Asia due to introduction of dwarfing genes in these two crops purely through conventional plant breeding approaches and these developments were christened as "Green Revolution" (Reynolds and Borlaug, 2006). The wheat breeding progrmme was being spearheaded by Norman Borlaug at CIMMYT (International Maize and Wheat Improvement Centre, Mexico) supported by Sanjay Rajaram and the rice breeding programme was under leadership of Gurdev Singh Khush at IRRI (International Rice Research Institute, Manila, Philippines) coinciding with these exciting developments during mid-sixties to mid-seventies. In recognition of these quantum jumps in yield of wheat and rice, Borlaug was awarded Nobel Peace Prize for Peace in 1970. Later on World Food Prize was instituted by joint efforts of Borlaug and his colleagues to honour Agricultural Scientists and others on pattern of Nobel Prize who had made significant contributions to sustainable food production and food and nutritional security globally. M. S. Swaminathan happened to be Director General, Indian Council of Agricultural Research when Green Revolution was taking roots in India and because of his enormous contributions to provide effective leadership to ensure that Green Revolution happened in India, M. S. Swaminathan is credited as "Father of Green Revolution in India" and he was the first recipient of World Food Prize in 1987 for spearheading the introduction of high-yielding wheat and rice varieties in India. G. S. Khush along with Henry Beachell received 1996 World Food Prize for achievements in enlarging and improving the global supply of rice during a time of exponential population growth and Sanjay Rajram was the 1914 World Food Prize recipient for his scientific research that led to huge increase in world wheat production.

Now, the productivity increase in most crops including rice and wheat has slowed down and even seems to have reached a plateau in some cases. The scientists are looking for innovative approaches and precision breeding to add genes of interest from various untapped sources to realize further yield gains in relatively shorter span of time. This is where it is being recognized that molecular aided conventional plant breeding methods may be the main short-term options for increasing productivity. The current dogma is that efficiency and effectiveness of conventional breeding can be significantly improved through indirect marker-assisted selection (MAS). The conceptual details on molecular breeding are available in several standard publications (Sundaram and Rajendrakumar, 2008; Gupta *et al.*, 2010; Xu, 2010). By now thousands of marker-trait association studies including hundreds of quantitative trait loci (QTLs) mapping studies have been conducted in major crops and use of molecular markers in conventional breeding primarily as marker-assisted selection is on increase.

Genetic Markers

Genetic markers represent genetic differences between individual organisms or species. Generally, they do not represent the target genes themselves but act as 'signs' or 'flags'. Genetic markers that are located in close proximity of target genes (are tightly linked) are referred to as 'tags'. Such markers themselves do not affect the target phenotype as they are located only near or 'linked' to genes governing the target trait. All genetic markers occupy specific genomic positions within the chromosomes (like genes) called 'loci' (singular 'locus'). There are three types of genetic markers, namely, (1) morphological/classical/ visible markers which themselves are phenotypic traits or characters of interest in plant breeding, (2) biochemical markers which include allelic variants of enzymes called isozymes, and (3) DNA/molecular markers which reveal sites of variation in DNA. Morphological markers are obvious phenotypic traits which can be visually characterized/identified like flower colour, seed shape, growth habit and pigmentation. Isozyme markers are differences in enzymes that are detected by electrophoresis and specific staining. The disadvantages of morphological and isozyme/biochemical markers are that they may be limited in number and are influenced by environment or the developmental stages of the plant.

DNA markers are the most widely used markers predominantly due to their abundance. They arise from DNA mutations such as substitution mutations (point mutations), rearrangements (insertions or deletions) or error in replication of tandemly repeated DNA. These markers are selectively neutral because they are usually located in non-coding regions of the DNA. They are practically unlimited in number and are not influenced by the environmental factors and/or developmental stages of the plant. Apart from the use of DNA markers in the construction of linkage maps and tagging traits of interest, they have several other applications in plant breeding such as indicators of genetic diversity and establishing cultivar identity.

DNA Markers

DNA markers are commonly divided into three types based on method of their detection. These types are (1) hybridization based, (2) polymerase chain reaction (PCR) based and (3) DNA sequence based. Most molecular markers belong to the so-called anonymous DNA marker type and generally measure apparently neutral DNA variation. Suitable DNA markers should represent genetic polymorphism at the DNA level and should be expressed consistently across tissues, organs, developmental stages, and environmental variations. Their number should be almost unlimited. There should be a high level of natural polymorphism. The markers should be neutral with no effect on the expression of the target trait. Most DNA markers are co-dominant or can be converted into co-dominant markers. Desirable DNA markers should meet the following criteria (Xu, 2010).

(i) High level of genetic polymorphism.
(ii) Co-dominance so that heterozygotes can be distinguished from homozygotes.
(iii) Clear distinct allele features so that different alleles can be identified easily.
(iv) Even distribution on the entire genome.
(v) Neutral selection, without pleiotropic effect.
(vi) Easy detection so that the whole procedure can be automated.
(vii) Low cost of marker development and genotyping.
(viii) High level of reproducibility so that the data could be accumulated and shared between labs.

Table 21.1 lists the major molecular marker technologies that are currently available (Xu, 2010).

Table 21.1. DNA markers and related major molecular techniques.

Southern Blot Based Markers: -Restriction fragment length polymorphism (RFLP) -Single strand conformation polymorphic RFLP (SSCP-RFLP) -Denaturating gradient gel electrophoresis RFLP (DGGE-RFLP)
PCR Based Markers: -Randomly amplified polymorphic DNA (RAPD) -Sequence tagged sites (STS) -Sequence characterized amplified region (SCAR) -Random primer-PCR (RP-PCR) -Arbitrary primer-PCR (AP-PCR) -Oligo primer-PCR (OP-PCR)

-Single strand conformational polymorphism-PCR (SSCP-PCR)
-Small oligo DNA analysis (SODA)
-DNA amplification fingerprinting (DAF)
-Amplified fragment length polymorphism (AFLP)
-Sequence related amplified polymorphism (SRAP)
-Target region amplified polymorphism (TRAP)
-Insertion/deletion polymorphism (Indel)

Repeat Sequence-Based Markers:
-Satellite DNA (repeat unit containing several hundred to thousand base pairs (bp)
-Microsatellite DNA (repeat unit containing 2-5 bp)
-Minisatellite DNA (repeat unit containing more than 5 bp)
-Simple sequence repeats (SSR) or simple sequence length polymorphism (SSLP)
-Short sequence repeats (SSR)
-Tandem repeat sequence (TRS)

mRNA-Based Markers:
-Differential display (DD)
-Reverse transcription PCR (RT-PCR)
-Differential display reverse transcription-PCR (DDRT-PCR)
-Representational difference analysis (RDA)
-Expression sequence tags (EST)
-Sequence target sites (STS)
-Serial analysis of gene expression (SAGE)

Single Nucleotide Polymorphism-Based Markers:
-Single nucleotide polymorphism (SNP)

Essentially DNA markers may represent genetic differences that can be visualized by using electrophoresis and staining with ethidium bromide or silver or detection with radioactive or colourimetric probes. DNA markers are particularly useful when they reveal differences between individuals of the same or different species. Such markers are called as polymorphic while those which do not discriminate between genotypes are called as mono-morphic markers. Markers may also be described as co-dominant or dominant. Co-dominant markers can discriminate between homozygotes and heterozygotes. Co-dominant markers indicate differences in size while dominant markers indicate presence or absence. Different forms of a DNA marker (different size bands on gels) are called marker 'allele'. Co-dominant markers may have different alleles whereas a dominant marker has only two alleles.

M. Mohan and colleagues in 1997, listed following DNA markers:

AFLP	Amplified fragment length polymorphism
ALP	Amplified length polymorphism
AP-PCR	Arbitrarily primed PCR
AS-PCR	Allele specific PCR

CAPS	Cleaved amplified polymorphic sequence
DAF	DNA amplification fingerprinting
ISA	Inter-SSR amplification
RAPD	Random-amplified polymorphic DNA
RFLP	Random fragment length polymorphism
SAP	Specific amplicon polymorphism
SCAR	Sequence characterized amplified region
SSCP	Single-strand conformational polymorphism
SSLP	Microsatellite simple sequence length polymorphism
SSLP	Mini-satellite simple sequence length polymorphism
SSR	Simple sequence repeats
STS	Sequence tagged sites
SNP	Single nucleotide polymorphism (added later on)

Construction of Linakge Map

A linkage map may be considered as a 'road map' of the chromosomes derived from two different parents. Linkage map indicates the position and relative genetic distance between markers along chromosomes which in turn amounts to signs or landmarks along a highway. The most important use of linkage map is to identify chromosomal locations containing genes and QTLs associated with traits of interest. Such maps are called as 'QTL' or genetic maps. QTL mapping is based on the fundamental principle that the genes and markers segregate via chromosome recombination (called crossing over) during meiosis wherein progeny analysis is required. Genes or markers that are close together or tightly linked will be transmitted together from parent to progeny more frequently than genes or markers that are located further apart. In a segregating population, both the parental types and recombinant types are available. The frequency of recombinant types is used to calculate recombination fraction which in turn is used to infer genetic distance between markers. By analyzing the segregation of markers, the relative order and distance between markers can be determined. Lower the frequency of recombination between two markers, closer they are located on a chromosome and vice-versa. Markers that have a recombination frequency of 50% are described as 'unlinked' and are assumed to be located far apart on the same chromosome or on different chromosomes. Mapping functions are used to convert recombination fractions into map units called centi-Morgans (cM). Linkage maps are constructed from the analysis of many segregating markers. For marker-assisted selection to succeed, it is absolutely essential to identify closely linked markers which are flanking the gene of interest and validate them for future use (Sundaram and Rajendrakumar, 2008). Following steps are needed in gene tagging and mapping (Collard, 2005; Sundaram and Rajendrakumar, 2008).

Selection of Target Trait

This is the first step in gene tagging. Genes of interest could be qualitatively or quantitatively inherited. Qualitatively inherited traits are conditioned by one or few genes which have major effect on the trait of interest. These genes typically follow a Mendelian segregation and are easy to tag and map since they are not influenced by environment and the genetic background. Such traits in rice, for example, include gall midge and bacterial leaf blight resistance. In contrast, quantitatively inherited traits are controlled by many genes/loci. Each gene has a typically small effect over the trait of interest and the cumulative effect of alleles at all the loci controlling the trait determines trait expression. These traits show a continuous variation or distribution in terms of expression in segregating populations and are usually difficult to tag and map as they are highly influenced by the environment and genetic background. For example, yield and tolerance to drought, water-logging, salinity, etc. could be considered in this category in rice and other crops.

Identification of Parents Differing in the Trait of Interest

For success of gene tagging, it is essential to have two parental lines which show marked difference for the trait of interest. For example, in rice it is convenient to use donors like Phalguna, W 1263 and RP 2068 having single gene for resistance to gall midge (*Gm1*) and the recipient lines like TN 1, BPT 5204 and Swarna which do not have any gene for resistance to gall midge. For quantitatively inherited traits like drought tolerance, the selection of parental lines is somewhat complicated. In such situations, it is necessary to carry out hybridization of the donor line with two or more recipient lines devoid of drought tolerance genes.

Development of Appropriate Mapping Populations Segregating for the Trait of Interest

The commonly used mapping populations are recombinant inbred lines (RILs) at F6/F8 generations, near isogenic lines (NILs) or double haploid lines (DHLs). Generally in self-pollinated species, mapping populations originate from parents that are both highly homozygous (inbred). Mapping populations for cross-pollinated species may be derived from a cross between a heterozygous parent and a haploid or homozygous parent. Several different populations may be derived in a self-pollinated species like F2 populations derived from F1 hybrids and back-cross populations resulting from crossing of F1 hybrid to one of the parents. The main advantage of these populations is that they are easy to construct and require only a short time to produce. Inbreeding from individual F2 plants allows the development of recombinant inbred lines (RILs) which include a series of inbred lines each having a unique combination of chromosomal segments from the original parents. The time required to produce RILs is long as the selfing has to go up to six to eight generations and this is the main disadvantage of this method. Double haploids may be produced by regenerating the plants by induction of chromosome doubling from pollen grains. However, this method is possible only in those species which are amenable to tissue culture like rice, barley and wheat, etc. The major advantage of RILs and DHLs is that they are homozygous and true breeding lines and they can be multiplied without any genetic change over the years/generations. Thus, both these populations represent 'eternal' resources for QTL mapping. Additionally, seed from individual recombinant inbred lines or double haploid lines can be transferred between different labs for further linkage analysis and addition of markers to the existing maps ensuring that the experimental material is same for all the collaborating labs. Population sizes used in preliminary genetic mapping studies generally range from 50 to 250 individuals. However, large populations are required for high resolution mappings. In case the map is to be used for QTL studies, which usually is the case, then mapping population must be phenotypically evaluated before subsequent mapping.

Screening the Population for Target Trait (Phenotyping)

The phenotyping differs between qualitatively and quantitatively inherited traits. For most of the qualitatively inherited traits like resistance to diseases and insects, phenotyping involves exposing the individuals of the population to a particular biotype/pathotype of the pest/disease and scoring the plants for resistance/susceptibility after appropriate time interval when the infestation can be easily detected.

Parental Polymorphism Survey with Markers and Identification of Markers That Co-segregate With Target Gene(s) in the Individuals Constituting the Population

After phenotyping in the designated population, the next step is to identify markers that co-segregate with the target trait. First, parental lines are analyzed for polymorphic markers. Usually, if mapped and co-dominant markers like SSRs are used then it is necessary to scan the parental lines with a set of uniformly spaced SSR markers (12-16 per chromosome) and identify at least 6-8 polymorphic markers per chromosome. It is always better if the polymorphic DNA markers are uniformly distributed on a chromosome. Once a set of polymorphic markers between the parental lines has been identified, next step is to carry out co-segregation analysis for these markers. For this 'bulk segregant analysis (BSA)' is done. This can be explained taking the example of gall midge resistance gene in rice. According to Sundaram and Rajendrakumar (2008), a set of resistant and susceptible F2 plants (usually 10-15 plants in each case) are bulked separately. These bulks are analyzed for parental polymorphic markers. If a fragment (hereafter called as marker) is present in the resistant donor, absent in the susceptible recipient, and simultaneously present in resistant bulk and absent in susceptible bulk, then in all likelihood, the marker is associated with resistance. The marker is then analyzed individually in all the plants forming resistant and susceptible bulks. If the marker is present in a majority (>70%) of individuals constituting resistant bulk and absent in majority of individuals constituting susceptible bulk, then it can be assumed that the marker is linked with gall midge resistance.

The next step is to carry out co-segregation analysis with all the individuals constituting the population and then determine linkage distances based on the extent of resistant individuals showing amplification of the resistance linked marker. Similarly, markers co-segregating with susceptibility can be identified. Once a marker is confirmed to be linked with the target trait/gene, the next step is to identify the chromosomal location. A majority of rice SSR markers has been genetically and physically mapped on the rice chromosome. Based on linkage between a marker and the target gene (gall midge resistance in rice here) and with prior knowledge of location of that marker on a particular chromosome, the tentative location of target gene on a particular chromosome can be determined. For example, if the linked marker is on chromosome 9, then the target gene is also on chromosome 9. More SSR markers in the vicinity of co-segregating markers can be used to locate the exact position of the target gene. For example, in the experiment on tagging the gall midge resistance gene *Gm1* and SSR marker RM 219 were found to co-segregate with the trait phenotype in the designated population. Since, marker RM 219 is known to be near the centromeric region of chromosome 9, it was assumed that the gene *Gm1* was located in its vicinity and the researchers looked for more SSR markers in the region. Out of 7 SSR markers near the centromeric region tested, two markers (RM 444 and RM 316) exhibited parental polymorphism and also co-segregated with the gall midge resistance. With the co-segregation data, Sundaram and Rajendrakumar (2008) constructed a linkage map of *Gm1* using the co-segregating markers.

Construction of Linkage Map

Construction of a linkage map requires a segregating population where the parents selected will differ for one or more traits. The population size will be generally in the range of 50 to 250 plants. F_2 plant populations derived from selfing the F_1 plants are most commonly used to have recombinant inbred lines (RILs). Near-isogenic lines) NILs) and double haploid lines (DHLs) are also used. Although linkage analysis can be performed manually for few markers, it is not possible to manually analyze and determine linkage between large numbers of markers that are used to construct maps. Under such situations, computer programmes are required. Linkage between markers is usually calculated using odds ratios i.e. the ratio of linkage vs no linkage. This ratio is more conveniently expressed as the logarithm of the ratio and called a logarithm of odds (LOD) values or LOD scores. LOD values more than 3 are typically used to construct linkage maps. A LOD value of 3 between two markers indicates that linkage is 1000 times more likely (i.e. 1000:1) than no linkage (null hypothesis). LOD values may be lowered in order to detect a greater level of linkage or to place additional markers within maps constructed at high LOD values. Commonly used software programmes for linkage analysis of markers and construction of genetic maps include Map-Maker/EXP, MapManager QTX, Cartographer, Linkage and JoinMap which are freely available on internet. The co-segregation data has to be fed to computer in excel format and the software automatically constructs linkage map after calculating LOD score for each pair of DNA markers. Final step in construction of linkage map involves coding data for each DNA marker on each individual of a population and conducting linkage analysis using computer programmes. The flow diagramme of this process is as follows:

Mapping population

↓

DNA extraction and PCR

↓

Marker scoring and coding

↓

Mapping programme

↓

Linkage map

Linkage markers are grouped together into 'linkage groups' which represent chromosomal segments or entire chromosomes. In reference to road map analogy, linkage groups represent roads and markers represent signs or landmarks. A difficulty associated with obtaining an equal number of linkage groups and chromosomes is that the polymorphic markers detected are not necessarily evenly distributed over the chromosomes but clustered in some regions and absent in others. Minimum plant population size for linkage map construction should be 50.

Sundaram and Rajendrakumar (2008) reported gene tagging and mapping exercise on gall midge resistance gene (*Gm1*) in rice. Three SSR markers in the centromeric region of chromosome 9, RM 219, RM 444 and RM 316 co-segregated with resistance. The markers RM 219, RM 444 and RM 316 were able to predict the presence of trait phenotype in 137, 138 and 133 F2 plants out of a population of 145 F2 plants, respectively. When the co-segregation data was analyzed using Mapmaker software, it was observed that the three markers were located approximately at a genetic distance of 5.9, 4.9 and 8.0 cM, respectively from *Gm1* and that markers RM 219 and RM 444 were flanking the gene.

Marker Validation

Once a marker or a set of markers is identified to be tightly linked with a particular gene of interest, next step is to validate the markers and their linkage distances in alternate populations. Alternate population can be developed by taking another donor line having the same resistance gene/gene of interest and crossing the same with the susceptible parent.

Utilization of Markers in Marker-Assisted Selection

Once closely liked markers (<2 cM from gene of interest) and/or flanking markers (< 5 cM on either side of the gene) are identified and validated in alternate populations, they are ready for use in marker-assisted selection in any breeding programme. Flanking markers have distinct advantage as compared to single marker since selection based on flanking marker will eliminate all false positives. For example, if a single marker is approximately 2 cM away from the gene of interest, then 2 out of 100 segregating plants can be false positives. However, if flanking markers are approximately 5 cM apart from the gene of interest, their combined recombination values will be 5/100 x 5/100 = 25/10000, i.e. false positive could be 2.5 out of 1000 segregating plants.

Principle of QTL Analysis

Collard *et al.*, (2005) have explained it in easy to understand manner. In simple terms, QTL analysis is based on the principle of detecting an association between phenotype and the genotype of markers. Markers are used to partition the mapping population into different genotypic groups based on the presence or absence of a particular marker locus and to determine whether significant differences exist between the groups with respect to the traits being measured. A significant difference between genotypic means of the groups (either 2 or 3) depending upon the marker system and type of population, indicates that the marker locus being used to partition the mapping population is linked to a QTL, controlling the trait of interest. The closer a marker is from a QTL, lower the chance of recombination occurring between marker and QTL and hence the QTL and the marker will usually be inherited together in the progeny, and the mean of the group with the tightly linked marker will be significantly different ($p < 0.05$) to the mean of the group of the plants without the marker. When a marker is loosely linked or unlinked to a QTL, there is independent segregation instead of co-segregation of marker and QTL and in this situation there will be no significant difference between means of genotype groups based on the presence or absence of the loosely liked marker. Unlinked markers located far apart or on different chromosomes to the QTL are randomly inherited with the QTL and no significant difference between means of the genotype groups will be detected.

Marker Assisted-Selection (MAS)

Selecting plants having desirable traits in a segregating generation is an important and critical task in any crop breeding programme. In this exercise, the plant breeder has to deal with hundreds and thousands of individual plants. Marker-assisted selection/marker-assisted breeding/marker-aided selection may greatly increase the efficiency and effectiveness of the selection compared to normal phenotypic selection in any conventional breeding programme. Once markers that are tightly linked to genes or QTLs of interest have been identified, breeders may use the markers as diagnostic tool to identify the plants carrying the genes/QTLs prior to field evaluation.

MAS for early generation selection is commonly practiced for disease resistance where a susceptible parent is crossed with the resistant parent and F1 plants are self-pollinated to produce a F_2 population. By using a marker to assist selection, plant breeder may substitute large field trials and eliminate many unwanted genotypes and retain only those plants that possess the desirable genotype based on the presence of tightly liked marker. 75% of the plants may be eliminated after one cycle of MAS for a recessive disease resistance gene from the F_2 population. This is important because breeders typically use very large population consisting of about 2000 F_2 plants derived from a single cross and may use populations derived from hundreds of crosses per year.

Advantages of MAS

The major advantages of MAS include:

- Time saving from substitution of complex field trials with molecular tests.
- Elimination of unreliable phenotypic evaluation associated with field trials due to environmental effects.
- Selection of genotypes at seedling stage.
- Gene 'pyramiding' combining multiple genes simultaneously.
- Avoiding the transfer of undesirable or deleterious genes (linkage drag) during introgression of genes from wild species.
- Effective selection of traits with low heritability.

Limitations of MAS

- MAS results may not be easily published.
- Reliability and accuracy of QTL mapping.
- A loose association may exist between marker and gene/QTL.
- There are limited markers and limited polymorphism for markers in breeding material.
- Effect of the genetic background.
- QTL x environment interactions.
- High cost of MAS.
- Application gap between research laboratories and plant breeding institutes.
- Knowledge gap between plant breeders and molecular biologists or scientists in other disciplines.

Marker-Assisted Back-Crossing (MAB)

Backcross breeding is a traditional breeding method routinely used for introgression of a desirable trait from a donor to a recipient variety which is otherwise suitable but for the trait under introgression. The target traits are introgressed from a donor genotype into the genomic background of an elite variety (recipient genotype) through backcrossing. Traditional backcrossing programs are planned on the assumption that the proportion of the recurrent parent genome is recovered at a rate of $1-(1/2)^{n+1}$ for each of n generations of backcrossing. This means that usually six to seven generations are required to achieve with > 99% genetic similarity to the recurrent parent. However, the use of markers can accelerate backcross breeding through the precise transfer of genomic regions involved in the expression of target traits (foreground selection) and by speeding up the recovery of the recurrent parent genome (background selection). This method, described as marker-assisted backcrossing (MAB), is particularly useful for pyramiding genes or QTL for resistance against a pathogen or pest and for traits that are highly influenced by the environment. In India, using this approach in rice, bacterial blight resistance from a non-Basmati donor has been combined with Basmati quality traits.

When the donor is an un-adapted or wild genotype having several undesirable genes also, the desirable genes can be introgressed through simple backcross breeding using MAS. This method helps select rare progenies having only targeted regions from a donor parent using linked markers (foreground selection) and rest of the genome from a recipient parent using whole genome markers (background selection). The use of markers that flank a target gene (<5cM on either side) can minimize the linkage drag which is common in conventional breeding methods.

The indirect selection of desirable plants using MAS, as discussed, helps to save time by reducing the number of backcross generations (3-4 generations) and also increases genetic grain. MAS may also improve mass selection and increase efficiency through progeny testing and decreasing the number of replications and increasing selection intensity. Marker-assisted backcross breeding has been used successfully to incorporate genes or QTL for both qualitative and quantitative traits in a number of crop species and in some cases leading to the development of improved cultivars.

Single Large-Scale MAS

Marker-assisted backcross breeding (MABB) perhaps invariably leads to successful introgression of desirable alleles at the targeted loci, but often it also leads to loss of genetic variability in other regions of the genome. Scientists have suggested a novel approach, described as single large-scale MAS (SLS-MAS), to overcome this limitation. In this approach, MAS for up to 3 QTLs is performed as a single step in an early generation (F_2 or F_3 population derived from elite parents). Flanking markers at a distance of <5cM on both sides of target locus are recommended for SLS-MAS where three steps are involved (Gupta *et al*., 2010):

1. Selection of parental lines for the trait of interest with the best allelic complementarities from among outstanding elite material, achieved through diallel/factorial cross or by fingerprinting.
2. Selected parental lines are then crossed with a tester (elite line lacking the target trait) for QTL identification and are also intercrossed among

themselves to generate segregating populations (each consisting of thousands of plants).

3. Segregating populations are subjected to SLS-MAS for the desirable alleles at the QTL identified through test crosses.

Examples of hybrids/cultivars released for cultivation developed through MAS in major crop plants are listed by Ram (2019).

Gene Pyramiding

Gene pyramiding is defined as an assembly of multiple desirable genes from multiple parents into a single genotype which is often cited as one of the major applications of MAS, since gene pyramiding through conventional plant breeding is difficult. Several possible breeding schemes can be used for gene pyramiding involving MAS. A summary of successful examples of gene pyramiding is presented in a review by Gupta *et al.* (2010). Gene pyramiding may involve assembly of multiple genes either for the same trait or for a number of traits. Most important examples for individual traits are, however, available in rice, where pyramiding of genes was successfully achieved for such traits as root length and resistance against bacterial leaf blight and blast. The possibility of introducing the genes for several traits, including simple and complex ones, has been examined in several reports (Gupta, 2007).

Looking Ahead

In recent years both the private and public sectors appear to be optimistic regarding the future use of MAS as an integral component of crop breeding programs in both the developed and the developing world. Collard and Mackill (2008) have listed six factors that may lead to a greater adoption of MAS in the coming decades:

1. Spread of marker technology, coupled with availability of enormous data on marker-trait associations.
2. Availability of AB-QTL, technology for development of introgression lines and other similar approaches, which combine QTL mapping with MAS.
3. Use of MAS in tracking transgenes during development and further use of genetically modified crops.
4. Rapid growth of genomics research and the generation of enormous data on genome sequences.
5. Development of high-throughput DNA extraction and marker genotyping platforms.

6. Availability of data and development of user-friendly databases on marker-trait associations.

According to Gupta *et al.* (2010), among these factors, the availability of information on marker-trait associations, reduction in cost, and the awareness among plant breeders about the utility of MAS will play a major role in facilitating adoption of MAS as an integral component of conventional plant breeding. The reduction in cost will also lead to the establishment of marker genotyping companies in the private sector and regional genotyping centers in the public sector in several countries. Public-private partnership is also being encouraged in several countries, so that in the future, plant breeders could outsource their genotyping work for MAS at an affordable cost.

The advances in genomic analysis would increase the possibilities of cracking down the genetic network underlying quantitative traits, which would help for precise introgression of a combination of desirable genes through MAS. The integrated marker-assisted selection (iMAS) system, with an integrated molecular breeding analysis platform (developed at ICRISAT, India), is an example that will facilitate integration of MAS with conventional plant breeding. The iMAS provides simple-to-follow guidelines embedded into the common front end of the system to help users choose the most appropriate experimental design and data analysis methods and it also provides users with a regularly updated selection of the currently most appropriate options. The iMAS performs the required data formatting for all included softwares. These developments will further encourage the plant breeders to utilize available genomic resources routinely in better way in crop improvement. It should also be recognized that in the past, only major crops have been used for marker-trait association studies, genomics research, and deploying MAS in breeding programs and now it is expected that orphan crops, like finger millet, foxtail millet, tef, lupins, lentils, clovers, beans, cassava, yams, sweet potato, melons, sunflower, rapeseed, safflower, sesame, mustard, and bananas/plantain crops will get the attention of biotechnologists and molecular biologists in studies involving marker development and whole genome sequencing. Studies in these crops will benefit from the knowledge generated in related major crops, and further studies will be undertaken in these orphan crops to develop crop-specific markers to be used for MAS in conventional breeding programs. Conventional plant breeding is on the threshold of a major change, and molecular marker technology is becoming one of its integral components. A number of examples are now available where MAS has been incorporated in conventional plant breeding leading to release of cultivars that have reached farmers' fields. There is no longer any doubt about the utility of this technology in plant breeding programmes. It leads to reduction in time and effort for the improvement of existing cultivars and for the development of entirely new cultivars. With the development of high-throughput marker genotyping technology, cost is no longer an issue for utilization of this technology in plant breeding. Therefore, one would expect that the plant breeders will gradually become receptive to the use of MAS and will also become proficient in using this technology.

22

Plant Genomics

There is need to find better ways and solutions to mitigate future agricultural challenges toward increasing productivity across the crops and equip them with the desirable genes for resistance to biotic and abiotic stresses and to mitigate the impact of climate change and also to make them nutritionally superior. Here comes "Plant Genomics"—a newly evolved discipline of plant sciences—targeting to decode, characterize, and study the genetic composition, structures, organizations, functions, and interactions/networks of all plant genes in a genome-wide scale. Being evolved from plant molecular genetics, biology, and biotechnology, plant genomics represents the key sub-divisions of structural, functional, comparative, evolutionary, physiological, and genetical genomics. Its development and advances, however, are tightly interconnected with plant science sub-disciplines, such as proteomics, metabolomics, epigenomics, phenomics, metagenomics, transgenomics, breeding-assisted genomics, bioinformatics and system biology as well as modern instrumentation and robotics sciences.

Plant genomics is a recent convergence of many sciences. Simply stated "Genomics" is the new science that deals with the discovery and noting of all the sequences in the entire genome of a particular organism. The genome can be defined as the complete set of genes inside a cell. Genomics, is, therefore, the study of the genetic make-up of organisms. Determining the genomic sequence, however, is only the beginning of genomics. Once this is done, the genomic sequence is used to study the function of the numerous genes (functional genomics), to compare the genes in one organism with those of another (comparative genomics), or to generate the 3-D structure of one or more proteins from each protein family, thus offering clues to their function (structural genomics). In crop agriculture, the main purpose of the application of genomics is to gain a better understanding of the whole genome of plants. Agronomically important genes may be identified and targeted to produce more nutritious and safe food while at the same time preserving the environment. Genomics is an entry point for looking at the other 'omics' sciences. The information in the genes of an organism, its genotype, is largely responsible for the final physical makeup of the organism, referred to as the "phenotype". However, the environment also has some influence on the phenotype. DNA in the genome is only one aspect of the complex mechanism that keeps an organism running – so decoding the DNA is one step towards understanding the process. However, by itself, it does not specify everything that happens within the organism. The basic flow of genetic information in a cell is as follows. The DNA is transcribed or copied into a form known as "RNA". The complete set of RNA (also known as its transcriptome) is subject to some editing (cutting and pasting) to become messenger-RNA, which carries information to the ribosome, the protein factory of the cell, which then translates the message into protein.

Aiming to develop genome-wide scaled high-throughput technologies and methodologies for plant science research and application, plant genomics has made significant advances over the past three decades in the area of inexpensive, high-throughput, and automated new generation DNA sequencing technologies and platforms. This has resulted in decoding of more than 100 plant genomes of 64 megabases to 23 gigabases in length within past few years. These advances have many broad implications for every aspect of plant biology while generating many unexpected challenges and grand tasks ahead. The challenges lying ahead require more extended collaborations and integrated approaches and better computing and data storage and systematization capacity, ways to handle data with improved and novel analytical/bioinformatics tools, along with large investment and funding.

Plant genomics aims to develop high-throughput genome-wide-scale technologies, tools, and methodologies to elucidate the basics of genetic traits/characteristics, genetic diversities, and by-product production to understand the phenotypic development throughout plant ontogenesis with genetic x environment interactions, to map important loci in the genome, and to accelerate crop improvement with precision and speed. Plant genomics research efforts have continuously increased in the past more than 30 years due to availability of cost-effective, high-throughput DNA sequencing platforms that produced fully sequenced more than 100 plant genomes with broad implications for every aspect of plant biology research and application.

Plant species are very diverse and there are about 300,000 plant species in the world of which more than 2000 species are grown on agriculturally suitable land covering an area of about 15.5 million square kilometer of land. Out of 2000 agriculturally plant species, 15 species provide more than 90 percent of food products. Besides feeding properties, plants have many other usages as fuel, fibre, medicines, building material etc. and balance agrobiosenosis and earth ecology to sustain life on our planet. Plant domestication coupled with need for more food and feed products has led to accelerated breeding and genetic efforts toward crop improvement. Early primitive selection efforts, recombination breeding through controlled crosses, understanding of Mendelian and quantitative genetics of phenotypic variation in plant genotypes equipped with optimized, targeted and efficient selection, phenotyping and application of statistical genetics, etc. have revolutionized the crop breeding efforts and these have resulted into steady improvement in global food production to meet the ever increasing global population, close to 7.3 billion people in 2019 which is projected to be 9 billion by 2050. Naturally this calls for massive efforts to increase total food production proportionately. These developments prompted the plant research community to enrich and power the conventional plant breeding and genetics methods with precision tools beyond the conventional hybridization and selection and cultivar development. Further, this also necessitated due to several other factors, like the long duration of conventional crop breeding impacted by the limitations in phenotypic evaluation, masking the effect of environment, polygenic nature of many key traits with many unnoticed minor genetic components, negative genetic correlations between many important agronomic traits, linkage drags and distorted segregation issues in hybridization between diverse types. To take care of all these issues, molecular biologists have attempted to decipher the molecular basis of genetic diversities by cloning and sequencing the genes encoding the traits of interest, and utilize them in plant breeding as tools in vertical or even via revolutionizing horizontal gene transfers. Progress made toward this goal has elucidated plant genome composition and led to decoding the entire DNA sequences of plant genomes conditioning entire plant ontogenesis. Here comes "genomics" that was derived from the term "genome" which means a haploid set of chromosomes—coined by Winkeler in 1920. First used in 1986, genomics defined "the enterprise that aimed to map and sequence the entire human genome (Abdurakhmonov, 2016)". Similarly, "Plant Genomics" is a discipline of plant science targeting to decode, characterize and study the genetic (DNA/RNA) compositions, structures, organizations and functions as well as molecular genetic interactions/network of plant genome. Plant genomics aims:

- To develop high throughput technologies and efficient tools and methodologies to elucidate the basics of genetic traits, genetic diversities, and by-product production.
- To understand the phenotypic development throughout plant ontogenesis with genetic by environment interactions.
- To map entire important loci throughout the genome.
- To accelerate crop breeding and selection in a genome-wide scale.

Plant and Crop Species Genomes

Challenges and Opportunities

The first fully sequenced plant genome has been the model plant *Arabidopsis* which was published in 2000 where about 25,000 genes expressed (Abdurakhmonov, 2016). Since then about 50 plant species have been fully decoded by 2013 (Michael, 2013). Plant science community has finished more than 100 plant genomes by 2015 (Michael and VanBuren, 2015). Further, plant science community has projected a sequencing vision of 1001 *Arabidopsis* accessions and sequencing of 1000 plant species that "will have broad implications for areas as diverse as evolutionary sciences, plant breeding and human genetics" while creating many unexpected challenges and grand tasks ahead. The numbers of publications on plant genomics research have drastically increased and reached 17210 by 2015 as indexed in PubMed database.

Compared to eukaryotes, plant genomes are more complex, coupled with difficulty in extraction of high quality DNA from plant tissues enriched in phenolics and other metabolic compounds having high affinity to DNA, widely different chromosome numbers, transposon/retrotransposon transcript retention property, highly varied ploidy levels, with many varied supergenes, pseudo-genes and repetitive elements including low-, medium-and high-copy number DNA sequences such as transcribed genes, rRNA genes, and retro-elements or short repetitive sequences, respectively. As a result, plant genome can be 100 times larger than genomes of animals or other model eukaryotic genomes. Sugarcane sequenced genome (12 Gbs) and hexaploid wheat sequenced genome (17 Gbs in size) represent 80% repetitive elements. Massive repetitive "junk" DNA sequences, organized as simple random tandem repeats, repeat single-copy interspersion, inverted repeats, and compound tandem array arrangements do masking of functionally vital single-copy genes and all these collectively create a formidable challenge to characterize and clone important individual genes. Polymorphic sites create additional challenge in sequence assembly due to high rates of nucleotides mismatches to the reference genome. Lastly, plants tend to have abundant copies of chloroplast genome with two inverted repeat organizations as well as large inversions with some exchanged regions between nuclear genomes. Lately specific methodologies and bioinformatics tools have been developed to minimize the impact of these challenges. These innovations include optimized DNA isolation from difficult plant materials, use of high density linkage maps, identification and sorting out of paralogous alleles using local patterns of linkage disequilibrium and sequencing diploid relatives or ancestor like genomes of polyploidy plants. The use of laser capture micro-dissection technique can help isolating individual cell types or chromosome or its arms to minimize the ploidy complexities. Further use of third-generation single molecule sequencing approaches and novel assembly methods such as optical mapping and long-range Hi-C interactions can also minimize the difficulties of some of the challenging cases with plant genomes.

However, plants offer some advantages also over the eukaryotes in gene sequencing studies. These include clonal propagation and indefinite seed storage properties through long-term base collections offering opportunities for repeated collection of same DNA samples for sequencing and studying same phenotype multiple times in many generations across replicated environments, freedom from ethical issues associated with multiple uses of plant materials, opportunities to create highly homozygous samples and reducing heterozygosity, opportunities to obtain double haploids, and existence of collinearity and synteny of plant genomes being useful to use reference genomes of model species to study homologous and orthologous genes from yet un-sequenced genomes.

Sequenced Plant Genomes

The ability to sequence DNA molecules, which was made possible in the 1970s with the introduction of the "plus and minus" sequencing technique of Sanger and Coulson; and Maxam and Gilbert, is generally considered to be the starting point of genomics sciences. Later, the simple, long-read chain-terminating dideoxynucleotide DNA sequencing method became a method of choice to decode genetic sequences. DNA sequencing eventual automation extended the capacity and power of this chain-termination sequencing methods to decode the entire genome sequences of living organisms. Because of technological advances and automated sequencing instrumentations, a large-scale sequencing of cDNA libraries made it possible to perform serial analysis of gene expression (SAGE) and expressed sequences tags (ESTs). These were the first genomics technologies for all organisms, including plant genomes. Furthermore, these advances powered by microarray tools routinely used by many individual laboratories worldwide have helped to identify the genome structures and functional and regulatory elements across genomes and have facilitated to develop high throughput reliable molecular markers for genome/ trait mapping studies. The development and generation of massively parallel sequencing technologies provided cost-effective, new-generation sequencing (NGS) platforms that have helped to completely decode the entire genome of many different organisms within a short period. For example, in plants, the first sequenced genome was a model plant *Arabidopsis thaliana* with 125 Mbs in size, 25,489 individual genes, and 14% repetitive elements, which was published in 2000. Further, more than 109 plant genomes have been fully sequenced by 2015 including 21 monocots and 83 eudicots, 10 model and 15 non-model plant genomes, five non-flowering plants, and 69 crop species with 6 crop model genomes and 15 wild crop relatives. Following the Arabidopsis model, several rice (*Oryza sativa*) genomes in 2002 to 2005, black cottonwood (*Populus trichocarpa*) genome in 2006, and grape (*Vitis vinifera*) genome in 2008 were fully sequenced. Sequencing whole plant genomes has increased in subsequent years, and 10 plant genomes had been sequenced in 2011. About 80% of sequenced genomes were accomplished in the year, 2012-2014 and the number is continuously increasing and has gone beyond 100.

The smallest plant genomes sequenced so far are the two eudicot plants: corkscrew (*Genlisea aurea*) with 64 Mbs genome size and 17,755 genes and bladderwort (*Utricularia gibba*) with 77 Mbs genome size and 28,500 genes. In contrast, the largest genomes sequenced are from gymnosperm plants, including Norway spruce (19,600 Mbs), white spruce (20,000 Mbs), and loblolly pine (23,200 Mbs). The largest genome sequenced from crop species is the hexaploid wheat (*Triticum aestivum*) with a genome size of 17,000 Mbs. An average size for all published plant genomes is 1850 Mbs. The gene numbers of the smallest to largest genomes are within the range of 17,755 (corkscrew) to 124,201 (hexaploid wheat) with an average of 40,738 genes for all sequenced genomes. Repetitive elements are highly variable among published genomes that varied from 3% (bladderwort) to 85% (maize, *Zea mays*) with an average estimate of 46% per genome. These sequenced plant genomes not only provided an updated knowledge on structural compositions and complexities of plant genomes but also elucidated the evolution of gymnosperm and angiosperm plants and specific gene families contributing to the radiation of flowering plants. There is some direct correlation between genome sizes and gene numbers/repetitive elements, although it does not strictly follow the rules, which was evidenced by several exceptions. For example, one of the largest genomes, Norway spruce, has ~28,000 genes, which is similar to the smallest genome bladderwort. Moreover, medium-sized maize genome (2300 Mbs) or wild tomato (1200 Mbs) contains more or approximately the same (<80%) contents of repetitive elements compared to the largest sequenced genome of loblolly pine (23,200 Mbs). The results of the first draft of the full genome sequence and assembly of a fresh salad plant leaf chicory (*Cichorium intybus* subsp. *intybus* var. *foliosum* L., $2n=2x=18$, and 1.3 Gbs genome size), named as Radicchio in Italian is available. The results of decoding the full genome of leaf chicory will "extend the current knowledge of the genome organization and gene composition of leaf chicory, which is crucial for developing new tools and diagnostic markers useful for our breeding strategies in Radicchio" and will be an important addendum to the list of sequenced plant genomes (Abdurakhmonov, 2016).

Numerous crops providing hundreds of thousands of sequence files have been condensed into contigs of related sequences. For examples, ***Hordeum vulgare*** (barley) EST library contains more than 440,000 EST sequences that resolve into 28,000 contigs (http://harvest.ucr.edu/), each representing a gene or closely related family of genes (Thiel *et al.*, 2003). Further, cloning is also an essential technique for the discovery of gene structure and function. It is an important step for all of the genome sequencing projects.

Sequencing "1001 Genotypes" and "1000 Plant Species"

The availability of a few whole reference genomes limits our full understanding of ecotypic variations that affect the function and adaptive evolution of plant species in various climatic conditions. It reduces the power of genome-wide tagging of biologically meaningful natural variations. In other words, the general perceptions are that "a single reference genome is not enough" for plant biology to explain and understand the existing natural variations in particular plant species and its populations. To address this, the Arabidopsis plant research community has developed a vision of sequencing a larger number of Arabidopsis genotype accessions, including various ecotypic and experimental population samples. As on 2016, the "1001 genome sequencing project of Arabidopsis accessions" has completed the full genome sequencing of 1100 Arabidopsis accessions "to record the genetic variation in the entire genome of many strains of the reference plant ***Arabidopsis thaliana***" and with the future objective to develop efficient genome analysis tools and software.

To understand the tree of life of the Plant Kingdom and study its evolutionary aspects in comparison to other life forms, the international multi-disciplinary consortium of "The 1000 Plants (one KP or 1KP) Initiative" has generated a large-scale gene sequencing data for more than 1000 various plant species. Rather than concentrating on single species accessions as in the 1001 Arabidopsis whole-genome sequencing project, the "1KP" project targeted 1000 distinct plant species with the objective of generating only functionally expressed (i.e., transcriptome) gene sequences. The plant species selected for the project had no restriction, and the samples were "chosen to represent every species known to science, across the Plant Kingdom, at some phylogenetically or taxonomically defensible levels". The 1KP sample list consists of 1328 entries, broadly grouped by phylogenetically (angiosperm, non-flowering, and green algae species) and by application (agriculture, medicine, biochemistry, and extremophytes). Most of these species have been sequenced for the first time (Table 22.1, Abdurakhmonov, 2016)). To date, an average of 2000 Mbs transcriptome sequence data have been generated for these 1KP plant species using 28 Illumina Genome Analyzer next-generation

DNA sequencing machines at the Beijing Genomics Institute, BGI-Shenzhen, China (Leebens-Mack, 2015). Ultimately, the obtained genomic sequence data will be utilized to analyze the phylogenetic, taxonomic, and evolutionary relationships of plant species, to study plant speciation, and to determine the timing of gene duplications during speciation events. However, the biggest limitation is associated with sequencing only transcriptomes rather targeting the whole genome, which limits obtaining many non-coding and repetitive portions of genomes. The results of "1001" and "1KP" sequencing efforts will undoubtedly open a new paradigm for plant genomics and related sub-disciplines. The results should not only accelerate crop improvement and boost the agricultural and medicine production worldwide but also help to understand the basics of plant life, evolution, speciation, and plant adaptations to the extreme environments in the era of global climate change and technological advancements.

Table 22.1. Plant species samples chosen for the "1KP" plant genome sequencing project.

Plant species *	Number of samples
Phylogenetic groups	
Angiosperms	830
Angiosperms: Onagraceae samples	50
Non-flowering plants	257
Green algae	241
Application groups	
Medicinal samples	142
Medicine - Alkaloid samples	30
Medicine - Chemotherapeutic samples	12
Biochemistry - Lipid biosynthesis samples	15
Agriculture - C3/C4 samples	93
Agriculture - Weeds	25
Extremophyte samples	31
Halophytes samples	18

*The number of samples overlaps among groups.

Several approaches allowing exploration of gene expression have advanced on pace with DNA sequence analysis methods. Although Northern blot analysis delivers semi-quantitative analysis of gene expression, microarrays, serial analysis of gene expression (SAGE), Illumina and quantitative RT-PCR technologies simultaneously permit analysis of most of the genes expressed in a plant tissue. Among the most unforeseen findings in biology was the elucidation of small RNA mediated gene suppression, first identified in plants by Baulcombe in 2004. Further, the development of RNA interference pathway has led to exploration of new methods for plant disease management in addition to providing an entirely novel way by which one analyzes gene function. Microarray analyses also provided the first glimpse of global gene expression patterns. Proteomics and metabolomics begin to define the various components of the genome. Furthermore, massively parallel sequencing technologies are established to increase the amount of DNA sequence that can be collected in a short period that will intensely decrease the genome sequencing cost. Plant genetic resources also deliver untapped reservoirs of genes that can contribute to better local adaptation and productivity under various climate changes. Plant genomics research offers insight into the genetic diversity that will be desired to prosper in the 21st century.

Crop Improvement in the Genomics and Post-Genomics Era

Genomics-Assisted Selection or Genomic Selection

At present, the reference genomes for many agricultural plants including specialty crops have been sequenced, as reviewed by Michael and VanBuren (2015), which created a new paradigm for modern crop breeding. Crop breeding, which is powered and enriched by molecular markers, genetic linkage maps, QTL mapping, association mapping, and marker-assisted selection methods in the past century, has now greatly accelerated and become potentially productive and efficient in the plant genomics era. This is due to the (i) availability of large-scale transcriptome and whole-genome reference sequences; (ii) high-throughput SNP marker collection and cost-effective, automated, and high-throughput genotyping platforms (HTP) and technologies (e.g., genotyping by sequencing or GBS), allowing breeders to screen multiple genotypes within a short time; (3) identification and use of expression QTLs (genetical genomics) in breeding; and (iv) opportunity to perform genome-wide selection (i.e., genomic selection). The biggest driving force for genomics-assisted crop breeding in the plant genomics era has been the inexpensive sequencing and re-sequencing opportunity for population of individuals of genetic crosses and breeding lines. This helps to precisely identify and link genetic variations to the phenotypic expressions, taking into account the rare and private allelic variations that are abundant in crop line population or germplasm resources. Furthermore, the availability of SNP marker collections and automated genotyping platforms provided a better genome converge to perform genome-wide genotype-to-phenotype associations (GWAS). Also, when whole-genome sequences are not available and SNP markers are present in a limited number, the breeders using GBS and HTS platforms can readily genotype their mapping population and can provide genomic selections for the targeted crops of interest. Although it was first applied for animal breeding, recently genomic selection has been successfully applied to a number of plant species, including studies using GBS in the context of genomic selection. Most importantly, the application of available genomics tools and a large number of high-throughput DNA markers and new-generation genotyping platforms have made the "breeding by design" possible and have developed "virtual breeding" approaches for efficient crop improvement as reviewed by Abdurakhmonov (2016). The crop improvement applications also include plant resistance genomics and molecular breeding against biotic/ abiotic stress tolerance in agriculture crops. The availability of genome sequences and a large number of SNP marker collections also provided the analysis of copy number variations (CNVs) in crop genomes, and their links to the key traits have greatly enhanced the efficacy of crop improvement programmes. Furthermore, although challenges are evident, the opportunity provided by post-genome sequencing advances has helped to integrate and enrich genomic selection with key proteome and metabolome markers. This has significantly powered up the breeding of complex traits of crops. Consequently, the knowledge gained through plant genomics coupled with proteomic and metabolomic advances has facilitated the emergence of an innovative approach of "personalized" agriculture through the utilization of chemical genomics. This requires the translation of knowledge and expertise of the pharmaceutical industry on the development of "personalized medicine" to treat each person based on its reaction to the medical drugs into the agriculture. Because of high-throughput genome analysis, it is possible to date that many plant compounds, including herbicides, growth regulators and phytohormones, elicitors, low molecular metabolites (e.g., salicylic acids), and/or synthetic hybrid chemicals, can be screened for genetic response of individual crop genotypes and to study their mechanism of actions contributing to agricultural productivity. Once identified, highly genotype-specific chemical compounds can be developed that impact better than traditionally applied "fit for all" chemicals/growth stimulators and fertilizers. A combination of such chemical genomics approach, proteomics and metabolomics with genetic engineering, and genomic selection will further provide a way for "personalized" agriculture that sustains crop production.

The Center of Excellence in Genomics' (CEG) breeding and research facility established at International Crops Research Institute for Semi-arid Tropics, Hyderabad, India caters to the molecular breeding needs particularly of developing countries through its three main activities: (1) applied genomics research and sequencing, (2) high quality marker genotyping services and capacity building in modern genomics, and (3) molecular breeding tools. It provides genotyping services on cost-to-cost basis to all its partners, such as SSR, DArT and SNP genotyping platforms; and MiSeq-based next generation sequencing. These services are offered for all crops and from clients coming from the NARS, private sector and research community across the globe. Recently, CEG acquired the Illumina HiSeq 2500 – the world's most powerful sequencer – to analyze, store and share huge sequence data and make full utilization of modern genomics tools in breeding and research programs at ICRISAT and in other partner institutes in developing countries of Asia and sub-Saharan Africa. It is a high performance computational genome analysis (HP CGA) facility with two servers, each having 80 cores (total 160 cores), running on 1,000 Gb and 500 Gb RAM, respectively, with a total storage capacity of 100 Tb.

Novel Transgenomics Tools and Biotech Crops

Crop improvement is also greatly impacted by novel transgenomics and genome editing technologies developed as a result of plant genome characterization and understanding in the era of plant genomics. In the past two decades, a variety of novel transgenomics technologies have been developed to replace or enrich the traditional transgenesis-based genetic engineering and plant molecular biotechnology as reviewed by Ricroch and Henard-Damave (2015). These novel technologies include antisense, RNA interference (RNAi), artificial microRNA expression (amiR), virus-induced gene silencing (VGS), zinc-finger nuclease (ZFN), transcription activator-like effects nucleases (TALENs), oligonucleotide-directed mutagenesis (ODM) of Cibus Rapid Trait Development System (RTDS), and clustered regularly interspaced short palindromic repeats/Cas9 (CRISPR/Cas9) technologies. These novel transgenomics technologies including genome-editing tools, the latter also referred to as genome editing with engineered nucleases (GEEN), are widely developed and utilized to investigate the gene function and apply to solve problems in medicine and agriculture. These are being preferred for major functional genomics and biotechnological studies. RNA-mediated genome manipulation (RNAi) tools down-regulate the target genes due to gene silencing effects at transcriptional (TGS) or posttranscriptional (PTGS) levels, whereas GEEN systems help to insert, replace, or remove specific regions of DNA from a genome using artificially engineered nucleases that are referred to as "molecular scissors". The potential application of RNA-mediated gene silencing methods for crop improvement, including RNAi in plant biotechnology, is huge and the technology has already generated many successful examples in a wide range of food, and horticulture crops. For example, RNAi was used to improve crop yield, food/fiber quality, resistance to pests, and biotic/abiotic stresses, to develop new and innovative products. Employing ODM-mediated single nucleotide editing in Arabidopsis, targeting the BFP gene, has demonstrated a precise edition of CAC to TAC, converting histidine (H66) to tyrosine (Y66) in GFP protein that offered a non-transgenic breeding tool for crops. Similarly, GEEN tools have also provided a new strategy for "trait stacking," whereby several desired traits are physically linked to ensure their co-segregation during the breeding processes. The examples include *A. thaliana* and *Z. mays*, where ZFN-assisted gene targeting has helped to heritably insert herbicide-resistant genes (SuRA/SuRB and PAT) into the targeted sites in the genome. Although other GEEN technologies such as TALEN and CRISPR/Cas9 are just picking their application in plants, their utilization in Arabidopsis, maize, rice, potato, wheat, barley, and plum holds a great promise and potential for non-transgenic crop genome modification and improvement (Abdurakhmonov, 2016).

Different Tools and Approaches for Plant Functional Genomics

The Genome

Plant genomics originated with the establishment of model plant, *Arabidopsis* that allowed numerous scientists to dissect this plant using the various biological tools like genetics, physiology, molecular biology and biochemistry. The genome of the model flowering plant, *Arabidopsis thaliana* has one of the smallest plant genomes (120 Mbp) and it has several advantages like generation time is short, easy to grow, transform and use for genetic experiments (Garcia *et al.,* 2000). The *Arabidopsis* genome sequence provides us details of the basic set of genes that is needed for the growth and maintenance of other simple plants, but the systematic correlation of gene sequence and their specific function in the plant still requires to be done. Few of the remarkable insights derived from whole genome sequences are the number of gene clusters, large-scale duplications of chromosome segments, insertions of huge chunks of mitochondrial DNA, as well as 40% of newly discovered genes of unknown function. In *Arabidopsis* and its closest plant species relatives, similarity of genes and their order will facilitate the direct use of *Arabidopsis* genomics and functional information in the plants.

In contrast to *Arabidopsis*, rice with its relatively small genome (~430Mbp) has been selected to be the model monocot plant whose genome was sequenced by a multinational effort shows a high degree of synteny, i.e. conserved gene order, with other cereals like maize, barley and wheat. The completely annotated reference genomes of *Arabidopsis* and rice certainly serve as a preliminary point for the large-scale functional analysis of comparative plant genomics. Other plant such as *Medicago truncatula*, a nitrogen-fixing legume has a close phylogenetic relationship with many agronomically important legume crop plants such as pea, soybean and Lucerne (alfalfa). *M. truncatula* has a small diploid genome (~450 Mbp) and will be developed into a model for legume genetics and genomics. Once a plant whole genome has been sequenced to completion, all genes that can be recognized with their function may be presumed from algorithms that permit the comparison of all the sequence and apparent protein structure of an unidentified gene with genes of known function. These methods of *in silico* gene function prediction permit the prompt identification of candidate genes that can be investigated further. Gene function prediction by comparison of sequence, however, will not always lead to decisive results. For example, identifying the class of protein that belongs to a gene does not immediately tell scientists the exact function of a target gene. Gene functions may also vary even between close members of a gene family. To date approximately, 4000 genes currently have similarity only to uncharacterized proteins (i.e. hypothetical, predicted, unknown, etc) while 758 have no significant protein similarity to any proteins deposited in any database. Among these, 286 in addition have no supporting EST (expressed sequence tags)/cDNA evidence and may signify erroneous gene predictions integrated in earlier releases (Lamesch *et al.*, 2011). In these cases gene function must be inferred on the basis of gene product activity, and by investigating the gene expression in response to certain specific environments. This will permit us to assess a functional role for the gene in question.

EST Sequencing

EST sequencing reveals the transcribed regions of the genome in a specific tissue. It involves preparing cDNA of the transcribed messages and then randomly sequencing them from either the 3'-5' or 5'-3'. EST collections of about 160,000 are available in public databases from 19 plant species like Arabidopsis, rice, tomato, maize, soybean, cotton, loblolly pine etc., and provide an effective way of gene discovery in these plants. There are 37,746 Arabidopsis ESTs, 46,268 rice, 29,209 maize, 26,255 tomato and 17,738 soybean submitted in dbESTs and after comparisons it has been suggested that ESTs exist for around 55–60% of the Arabidopsis genes. Large stretches of repetitive DNA and transposable elements tend to be hyper methylated compared to gene coding regions which are unmethylated. The drawback to this approach is the redundancy of random EST sequencing. To avoid this, normalized cDNA libraries have been used to create a 'unigene' set of clones' representative of the plant genome. A comparison of EST databases from several plants represents the diversity in coding sequences between plants, yet offers a global perspective of the similarities in genes for specific tissues / conditions. Although similarity screens to databases can assign some degree of common function to genes for signal transduction or transcription factors, their exact role requires experimental approaches.

Despite the limitations, the different genome sampling approaches are still the best substitute to sequencing the entire genome for very large genome sequences. Using the highly parallel sequencing technologies, such as 454 or Solexa Illumina, these confines may be less relevant in the near future (Mardis, 2017). Several strategies are available for gene discovery: information obtained from ESTs, full-length cDNA, tilling and expression arrays, in silico gene prediction, MPSS (massive parallel signature sequencing), SAGE (serial analysis of gene expression) as reported by Coruzzi *et al.* (2018). The next step after gene discovery is gene functional annotation that can be done in several ways such as computational predictions, gene expression, mutant analysis and protein–protein interaction.

Epigenome

The study of heritable traits that do not involve alterations in the DNA sequence is called Epigenetics. These modifications alter the chromatin structure that affects regulation of gene expression. Henderson and Jacobsen (2007) showed two well-established forms of epigenetic regulation in plants are post-translational modifications of histones and methylation of the DNA molecule at the cytosine base. The heterochromatic regions of plants tend to be highly methylated and that's why sequencing methods such as methyl filtration provide sequences that are enriched in gene coding regions. To understand the importance of histone modifications for gene regulation in plants, it is necessary to understand the histone modification patterns across the plant genome and correlate them with gene expression data. Many scientists have carried out such studies in various plants involving isolating nuclear DNA and enriching for the methylated DNA fraction. To analyze global DNA methylation patterns, the methylated DNA is then hybridized to an Arabidopsis whole genome tiling microarray. These findings have uncovered numerous remarkable aspects of the epigenome in Arabidopsis. Many authors demonstrated that the heterochromatic regions are heavily methylated, for example approximately a third of Arabidopsis genes are methylated within the transcribed region. Arabidopsis genes that are methylated in the transcribed regions tend to be expressed at high levels and genes that are methylated in the promoter region tend to be tissue specific. Recently, several studies have been carried out to look at histone modifications (Yelagandula *et al.*,

2017). Epigenetics also includes another level to the regulatory complexity of gene expression (Zhou *et al.*, 2019). They revealed the first gene-expression correlated, organ specific, and whole genome survey of DNA methylation in plants that were sent to space as seeds, then germinated and grew in space. Authors provide fundamental insight into the strategies plants use in the physiological adaptation to the novel environment. Thus, explaining the importance and function of epigenetic modifications in the years to come will be essential to understanding how genomes respond to various perturbations (internal and external).

Transcriptome

The transcriptome study is the parts of the plant genome that are expressed as RNA transcripts in one or several populations of cells in a certain time and environmental condition. Many years ago, northern blots were the only choice for biochemists and molecular biologists. In early 1990s, transcriptional profiling using microarrays have developed into the most prominent tool for functional plant genomics and triggered a revolution in transcriptome studies. Synthesis of mRNA is the first step towards protein synthesis and, for several genes; variations in mRNA abundance are related to modifications in protein levels. Alterations in mRNA steady state levels are mostly accomplished by varying the transcriptional rate of genes. Such variations in relative mRNA amounts are reveals of modifications in developmental and environment programme or reflect reactions to all kinds of stimuli. To well-understand a gene's function it is not only critical to identify when, where, and to what extent a gene is expressed, but it is also important to discover other genes which are co-regulated with the gene of concern. Studying the transcriptome, by measuring mRNA concentrations of defined genes in a quantitative and multiparallel way allows assigning function to a multitude of unidentified genes. Endless development of the approaches will provide even more perspectives for the rapid and comprehensive analysis of plant gene function. The microarray expression analysis is based on large number of different genes, either anonymous or defined sequence, which can be expressed at the same time using a small amount of plant samples. In principle, DNA sequences representing all the genes in a plant can be placed on miniature solid supports and used as hybridization substrates to quantify the expression of all the genes represented in a complex mRNA sample. Microarray analysis of transcript levels permits the identification of low-abundance mRNAs at a copy number of only a few molecules per cell. Currently, two types of microarray formats are in uses that differ in the way the target molecules are produced and deposited on the chip surface, the nature and chemistry of the solid support, and the number of target molecules per area. The first format is based on DNA-fragment target molecules which can be either gene-specific PCR products (e.g. EST clones), plasmids or oligonucleotides and the second format is oligonucleotide-based microarray.

Most experimentations use reverse transcription for labeling the cDNAs with different fluorophores (e.g. fluorescein, Cy5, Cy3). After hybridization to the arrayed target molecules the relative measure of gene expression for each gene analyzed is determined with the help of a high-resolution laser scanning device. Microarray technology is also being continuously developed for various genomics applications other than expression profiling. Expression profiling experiments employing microarrays can be used to analyze diverse biological difficulties. By observing the varying expression patterns in response to a particular treatment or stimulus, one can get results of a defined biological process. According to new insights into the orchestration of certain plant responses, the complete set of genes (cluster) can be recognized that is characteristic of a certain biological process. Characteristic patterns of gene expression which are derived from such clustering analysis may serve as a transcriptional fingerprint of a particular transition phase of the cell. Outlining such sets of genes, which are co-regulated under defined conditions, also allows for the identification of "marker genes" that are diagnostic for certain developmental or environmental processes.

Some varieties of innovative techniques have been also developed such as differential display (DD), serial analysis of gene expression (SAGE), and cDNA-amplified fragment length polymorphism (cDNA AFLP), which alternatively offer great potential in analyzing the relative frequencies of mRNA transcripts in a given polyA(+) RNA preparation. Some other techniques use high-throughput sequencing approaches such as comparative expressed sequence tag (EST) sequencing of randomly chosen cDNA clones from different cDNA libraries or monitor gene expression by massively parallel signature sequencing (MPSS) of cDNAs cloned in vitro on micro-bead arrays to obtain signature sequences for each cDNA clone of a given library. The aforementioned methods have successfully been useful to decipher diverse problems of plant biology (Scheben *et al.*, 2017). Using microarrays for transcript profiling on a genome-wide scale will undoubtedly become the standard approach. Recent microarray analyses in plants have approached various biological questions in diverse plant species by integrating a rising number of genes. A crucial part of microarray analysis is the approach that data are collected and evaluated with image analysis software. Some of these tools for data mining, integration, and clustering analysis have already been developed and new agendas are still needed in order to refine data analysis. One of the biggest complications for comprehensive microarray data analysis is still the quality control and normalization of gene expression data between multiple experiments. Several ways are currently applied to decrease the noise in array experiments and develop the quantitative evaluation process (Drăghici, 2016). Transcriptional profiling experiments are not applied quickly and need a great compaction of data verification. Also, the equipment for microarray production, hybridization and read-out is still expensive. However, with the development of sequencing technologies, it is possible now to measure miRNAs and other small RNAs (sRNAs), quantify their expression and to understand their functional roles in plants. Transcriptomics, however, is already a frequently used approach for simultaneous analysis of a large number of genes. Innovative methods are being developed, some of which will eventually lead to considerable decrease of cost in the near future. This will make the transcriptome technology more consumer-friendly and accessible to many more plant biologists.

Proteome

The proteome has been defined as the set of all the proteins present in a cell, organism, or tissue that are expressed in a given time and under a defined environmental condition. Proteomics approaches have been employed according to the objective of the study and are based on certain criteria such as classification of proteins, comparative proteomics for comparison of protein profiles, post-translational modifications, protein-protein interactions, proteinomics for studying protein structure and functional groups, and translational proteomics that entails transfer of technology and information for crop development (Ramalingam *et al.*, 2015). Results obtained through above mentioned methods together with bioinformatics provide significant information on biological processes and stress tolerance mechanisms that can be useful in crop improvement. Each technique has advantages and disadvantages from both the conceptual and the methodological perspectives. Numerous methods are being used to study plants proteomics and these include electrophoresis and/or chromatography combined with chemical or metabolic labeling and mass spectroscopy. The most common method in proteomics is the use of two-dimensional (2D) gel electrophoresis for separating the proteins followed by excision of individual spots, proteolytic in gel digestion of the isolated protein, purification of the resulting peptides followed by the fragments analyzed by mass spectrometry and finally computer analysis of the resulting mass fingerprint of the peptides is then compared with protein-sequence databases. Several software tools are available which allow extreme selectivity and specificity for the protein identification process.

In case of plants, although database resources are less, several efforts are being undertaken to identify the corresponding proteomes by using specific sub-cellular fractions. In Arabidopsis, large-scale proteomic efforts have been carried out, for example to determine the proteome of cell wall and organelles such as the chloroplasts, mitochondria and vacuoles. These researches have shown that many bioinformatics predictions of protein sub-cellular localization are incorrect and they must be confirmed experimentally. These findings also demonstrated that cellular trafficking is more multifaceted. In rice, sub-cellular protein mixtures are analyzed using a two-dimensional liquid-chromatography system ending with online digestion directly coupled to a nanospray mass spectrometer based on the available genome data and EST sequences. This technique allows complete automation of the process, a higher sensitivity than 2-D electrophoresis and thus a high throughput. Combination of mass spectrometry analysis and Edmann sequencing has allowed the identification of 61 proteins. Plant-specific proteomes have been demonstrated by the construction of proteome maps of the maritime pine wood forming tissue, maize, wheat, barley and soybean. In plants, comparative analysis of 2-D gel protein patterns is an essential tool for studying the genetic and physiological differences between variants of cultivated species. In principle, variations between protein patterns arising from post-translational modifications change the structural properties of the protein of interest. Relative quantification of proteins or peptides using mass spectrometry analyses has high accuracy, resolution and sensitivity but the sample preparation seems to be laborious. High-throughput methods are available for the preparation of protein digests and mass spectrometric analyses. The next technological challenge will be to speed up and standardize the separation steps, at first through liquid chromatography processes and, in the near future, through protein-chip technology. The usage of techniques such as matrix-assisted laser desorption ionization time-of-flight (MALDI-TOF) MS and electrospray ionization (ESI) MS with tandem MS (MS/MS) has overcome the slower and less sensitive methods The most commonly used pipeline for the identification of proteins involves comparing the MS/MS spectra to the reference databases. Although SEQUEST, MASCOT and X!Tandem are the widely used database search programmes for protein identification, MassMatrix is also suited for tandem MS data.

Metabolome

The metabolome refers recent high-throughput approaches in the field of metabolic genomics that aim to identify gene function, the full complement of metabolites of a biological system. Numerous analytical techniques such as nuclear magnetic resonance (NMR), GC-MS, LC-MS, capillary electrophoresis–MS (CE-MS) and Fourier transform infrared (FT-IR) spectroscopy have been reported for analyzing the data from metabolite profiling (Minai-Tehrani *et al.*, 2016). The use of metabolic profiling has been limited in various crops but this approach has been successfully demonstrated in model plants. One of the first studies of this kind was examined to find out the mechanism of action of herbicides like acetyl CoA carboxylase and acetolactate synthase inhibitors, using GC–MS. Another technology used in metabolomics is the Fourier transform-ion cyclotron MS that separates metabolites based on differences in their isotopic masses. In Medicago untargeted quantitative, MS approach was used to profile metabolites treated with rhizobial Nod factors to find the metabolic alterations between the symbionts. By profiling the chemical constituents of the defined cell, it is feasible to achieve a full non-biased description of the metabolome status of plant system under defined conditions, at a given time in a given genetic background. Technology related to study the metabolome has the potential to visualize relationships that occur via regulation at the metabolic level. Unlike transcriptomic and proteomic approaches analyzing variations in mRNA and protein levels, this technology allows to get closer to the specific role that a certain gene performs in a biological process. High-throughput metabolic screening methods involve automated multiparallel analyses of numerous compounds by one single approach. Classically, gene function prediction was made based on studies at the mRNA or protein levels. However, metabolite profiling is currently being transformed into a diagnostic tool in agriculture to link genotype to phenotype (Bino *et al.*, 2004; Beale and Sussman *et al.*, 2018).

Fiehn and co-workers developed a novel tool for plant functional genomics. Metabolite profiles are typically recorded by using techniques such as gas chromatography coupled with mass spectrometry (GC/MS) and high performance liquid chromatography (HPLC). Apart from GC/MS and HPLC, nuclear magnetic resonance (NMR) spectroscopy-based techniques have also been applied for the profiling of small compounds. The key to the applicability of these techniques for profiling methods is the development of fast and simple extraction and separation protocols. The feasibility of metabolome studies for plant functional genomics has, for instance, been established by recording distinct metabolic profiles from plant extracts (Hong *et al.*, 2016). Different Arabidopsis genotypes were compared (two homozygous ecotypes and a mutant of each ecotype) and distinctive metabolite profiles of each genotype were reported. Also, data analysis tools enabled the assignment of 'metabolic phenotypes'. These study mirror the metabolic flexibility of plants in general – a phenomenon that this approach has to cope with. Metabolite profiling findings have been demonstrated in a diverse array of plant species including Arabidopsis, tomato, oilseed rape, potato, rice, wheat, strawberry and eucalyptus. Plant metabolome will allow the comprehensive analysis of the metabolic consequences of genetic modifications and as a result will help towards a better understanding of biochemical networks and their regulation – a prerequisite for improving the metabolic engineering of crop plants. Metabolic profiling is also being used in studies of environmental perturbations in order to understand the complex shifts under various stresses. The application of forward and reverse genetics in conjunction with metabolite profiling plays a vital role in gene annotation, characterization of biochemical pathways and the identification of new genes for their use in biotechnological applications. Remarkably the ongoing researches elucidating the metabolic responses to various stresses involved indicate that metabolomics-directed breeding could also be helpful in obtaining stresses resistant crops. The significant application of metabolomics in crop improvement will become increasingly evident in the future.

Phenomics

The term phenomics stands for the large-scale phenotypic analysis of mutants. Classical genetics is largely based on the understanding gene function by the study of the phenotypes caused by mutations of the gene. In general, mutations in genes that result in similar phenotypes often imply that the genes function in the same process. This hypothesis has led to the development of large-scale initiatives to illustrate phenotypes caused by the systematic inactivation of genes in the genome of model plants. Currently, with the establishment of non-invasive methods for plant phenotyping and their large-scale high-throughput application, phenomic technologies are applied in basic plant research and crop plant breeding (Furbank and Tester, 2011). Even though high-throughput phenotyping techniques were originally developed for controlled environments, their application under natural conditions is receiving cumulative attention. The plant phenotype is the set of structural, physiological, and performance-related traits of a genotype at specific time points during its life cycle. Understanding results on a certain set of morphological and developmental traits allows researchers to put together a comprehensive picture of a mutant's phenome (Yang *et al.*, 2013). Such a mutant phenotype profile can then be compared to the profile of the wild-type plant, which is verified simultaneously. Such profiling methodologies provide huge data which serve as the basis for systemically understanding the relationship between a mutant genotype and its respective phenotype. The model of phenomics is nicely demonstrated by the automated high-throughput phenotyping platform, the Phenovator, capable of screening 1440 Arabidopsis plants multiple times per day for photosynthesis, growth and spectral reflectance at eight wavelengths. Using this unprecedented phenotyping capacity, it has been possible to identify significant genetic differences between Arabidopsis accessions for all traits measured both on temporal and environmental scales. Further, large-scale phenomic studies with thousands of mutant plants are currently being applied by researchers. Once a desirable mutant is identified, the integration of phenomics data with data sets from transcriptome, proteome and metabolome analyses helps in achieving a clear image of mutant gene function. The taxonomy of Omics science is illustrated through a simple diagramme in Fig. 22.1.

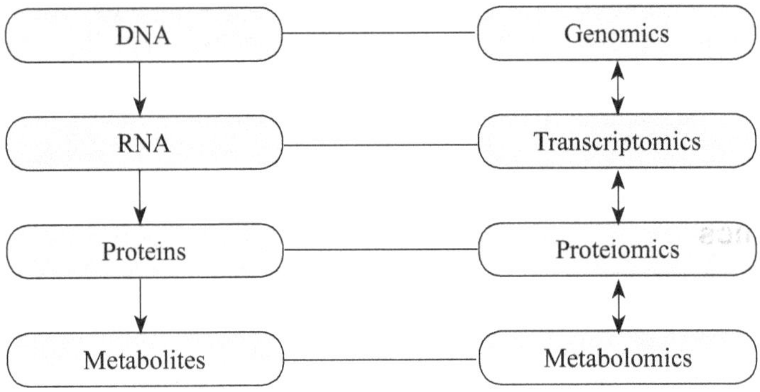

Fig 22.1. The taxonomy of 'Omics'sciences.

Advantages of the CRISPR/Cas9 System and its Applications

Crop genome editing is among the most promising techniques to accelerate plant breeding by allowing the introduction of predictable modifications directly in an elite background. The CRISPR/Cas9 system is useful not only for genetic improvement of agricultural crops but also to functional characterization of specific plant genes for fundamental research purposes. Earlier, the dominant genome editing tools were zinc finger nucleases (ZFNs) and transcription activator-like effector nucleases (TALENs). ZFNs and TALENs are artificial fusion proteins comprising an engineered DNA-binding domain fused to the nonspecific nuclease domain of the restriction enzyme FokI, and they have been applied successfully in various plants. The latest ground-breaking technology for genome editing is the type II clustered regularly interspaced short palindromic repeat (CRISPR)/Cas9 (CRISPR-associated) system based on RNA-guided engineered nucleases, which already hold great promise due to their simplicity, accessibility, cost and versatility (Belhaj *et al.*, 2015). CRISPR/Cas systems have been characterized in bacteria and archaea, protecting them against invading nucleic acids by cleaving the foreign DNA in a sequence-dependent manner, as an easily programmed tool for facile genome editing. The immunity is acquired by the integration of short fragments of the invading DNA known as spacers between two adjacent repeats at the proximal end of a CRISPR locus. Its natural state, a spacer-containing crRNA hybridises (approximately 40 nt in length) with a non-coding, trans-activating crRNA (tracrRNA) that facilitates DNA cleavage as shown in Fig.22.2. The action mode of CRISPR/Cas9 biological system is based on the participation of two distinct elements: the Cas9 protein (CRISPR-associated protein 9) and sgRNA (single guide RNA). Cas9 protein has two representative domains: the first one is known as nuclease-associated proteins, which is responsible for cleaving and regrouping the complementary strand of crRNA and the second domain RuvC-like (nuclease domain that cleaves complementary DNA strands) has the purpose of carrying out the cleavage of the complementary strand of dsDNA (double stranded DNA). Also, a fusion of these RNA moieties into a 'single guide' RNA (sgRNA) was initially demonstrated to be functional and subsequently used to introduce double-strand breaks into specified regions of genomes (Fig. 22.2, Cong *et al.*, 2013).

The CRISPR/Cas9 system does not require any protein engineering steps, making it much simpler to evaluate multiple gRNAs for each target gene. Also, only small gRNA sequence (20 nt) need to be altered to confer a different target specificity, that means that cloning is not required. Thus, the CRISPR/Cas9 system can be used for high-throughput functional genomics applications by permitting to develop the inexpensive assembly of large gRNA libraries. The main convenient advantage of CRISPR/Cas9 is the ease of multiplex editing that simply requires the monomeric Cas9 protein and any number of different sequence-specific gRNAs. In plants, the CRISPR/Cas9 technology is more versatile for genome editing in monocots with high genomic GC content such as rice. The feasibility of CRISPR/Cas9 system for genetic engineering of crops has been demonstrated in ***A. thaliana***, ***Nicotiana benthamiana***, ***Glycine max***, ***Zea mays***, ***Vitis vinifera***, ***Oryza sativa*** and ***Triticum aestivum***. Other crops in which the CRISPR/Cas9 technology has been optimized include barley (*Hordeum vulgare*) and ***Brassica oleracea***, watermelon (***Citrullus lanatus***), as well as the non-herbaceous sweet orange (***C. sinensis*** cultivar *valencia*) and poplar (***Populus tomentosa***) as reported by Ricroch ***et al.*** (2017).

Lastly, the open access policy of the CRISPR research community has endorsed the widespread use of this technology. The research community provides access to plasmids (e.g., via the non-profit repository Addgene), web tools for selecting gRNA sequences and predicting specificity in plant genomes (http://cbi.hzau.edu.cn/cgibin/CRISPR; http://www.genome.arizona.edu/crispr/; and http://www.rgenome.net/cas-offinder, http://www.e-crisp.org/E-CRISP/index.html) and hosts active discussion groups (https://groups.

google.com/forum/#!forum/crispr). These all services have encouraged new researchers to implement the technology and contributed to the rapid improvement of crop traits in understanding of the plant system and its practical applications in plant functional genomics studies. Future application of CRISPR/Cas9 is the targeted insertion of specifically modifying genomic loci in the fields of metabolic engineering and molecular farming, where plant cells are used as units for the production of specific metabolites or proteins.

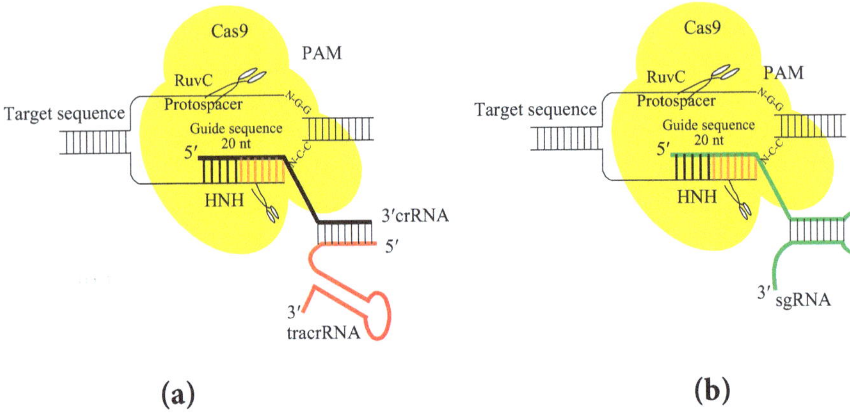

Figure 22.2. Schematic representation of RNA-guided gene editing by Cas9. (a) The Cas9 protein (in yellow) is guided by the CRISPR RNA (crRNA, in black), that contains a 20nt guide sequence, and a trans-activating CRISPR RNA (tracrRNA, in red), which stabilizes the structure and activates CAS9 to cleave the target sequence (protospacer). The presence of a protospacer-adjacent motif (PAM), i.e. NGG sequence directly downstream from the target sequence, is a prerequisite for DNA cleaavage by Cas9. (b) The sgRNA (single guide RNA, in green) can identify the target gene, and then the two domains of Cas9 (in yellow) cleave the target sequence (Bortesi and Fischer, 2015).

Challenges Ahead

The revolutionizing advances made in the past more than three decades in plant genomics and its sub-disciplines provided a mass of novel opportunities with easy-solution applications and high throughput, cost-effective, and time-effective technologies. Plant genomics era increased understanding of the basis of complex life processes/traits in plants and crop species, and it paved a way for effective improvement of plants to fulfill the dietary and other needs of mankind. However, it also piled up challenging grand tasks ahead for current genomics and post-genomics era. Tremendous achievements have been made toward sequencing more than hundreds of plant genomes including major crop species and specialty, model/non-model, wild, vascular, flowering, and polyploid plants. There are ongoing and fascinating consortia projects of sequencing "1001 genotypes of Arabidopsis" and "1000 various plant species". However, the challenge ahead is to extend such large-scale, multiple accession genome sequencing initiatives for each priority agricultural and specialty crop species including their wild relatives and ancestor-like genome representatives.

This is to effectively utilize all variations existing among plant/crop germplasm resources and its ecotypic populations and to design efficient GWAS analysis and consequent genomic selections as well as tools/software programmes for better and in-depth understanding of plant genomes and improving genome assembly issues. In this context, the polyploid crops pose a major challenge because the sequencing of many polyploids and their sub-genomes would be increasingly difficult in the understanding of the complexity of polyploidy, gene silencing, epigenetics, and biased retention and expression of genes after polyploidization. Furthermore, it also helps to discover all natural variations and lost genes during crop domestication that should be useful to restore the key agriculturally important traits in the future.

Sequencing the entire genome of 1KP samples, rather concentrating on only transcriptome/ exome, is also the necessary task ahead that would elucidate many important non-coding sequences from these plant species. Results would be useful for plant evolutionary, speciation and taxonomy studies. Although high-throughput DNA sequencing instrumentation exists and keeps evolving to better versions year-to-year, the consequent task is still to improve the sequence length that would solve many incorrect sequence sites and genome assembly challenges that plant genomics faces currently. A consequent grand task is the handling, organizing, systematizing, and visualizing a huge amount of plant genome sequencing ("Big Data") data that require urgent attention, effort, collaborative work, and investment. There is an urgent need to develop more efficient bioinformatics platforms to handle plant genome data due to challenges, specificities, complexities, and sizes of currently available and future sequenced plant genomes mentioned herein. Furthermore, the most important current and future post-genomics grand task ahead is to link the sequence variation(s) with phenotype(s), trait expression, and epigenetic and adaptive features of plants to their living environment and extreme conditions. The successful completion of this task will require the combined approaches of genomics with bioinformatics, proteomics, metabolomics, phenomics, genomic selections, genetical genomics, reverse genomics, system biology, etc. as summarized by Abdurakhmonov (2016). In other words, there is a need to make sequenced genomes "functional" and biologically meaningful. This also requires the integration of all available genomic and phenotypic data to identify key networks that also require downstream effort of integration of specific networks to networks of other systems in order to connect heterogeneous data.

There is a need to use molecular phenotyping (i.e., using molecular process such as protein-RNA interactions, translation rates, etc.) in QTL mapping to precisely link the sequence variation(s) to its phenotype(s). The concept "personalized agriculture" needs an attention as an unexplored area in crops with the availability of sequenced genomes and high-throughput genotype, proteome, metabolome, and phenotype profiling platforms and rapid crop line development tools such as genomic selection and new-generation genome-editing tools. All these will help in accelerated and precision crop breeding. These particular grand tasks further highlight a need for extended effort and work on the development of inexpensive high throughput plant phenotyping and plant proteome and metabolome profiling tools and instrumentation by utilizing small amount single-cell-derived samples. A parallel grand challenge to the above-outlined needs is to have concentrated efforts on the timely application of novel transgenomics and genome-editing tools for all types of plants and to optimize it for routine large- and short-scale biotechnology industry usage. There are key important tasks to (i) utilize the complex effects of plant developmental genes (e.g., core microRNA/RNAi machinery) to simultaneously improve the key traits and overcome negative trait correlations, (ii) to optimize and better design novel transgenomics and genome editing technologies for the key priority crops and plant by-product production, (iii) to identify the appropriate choice of plant tissues for genome editing, (iv) to reduce or eliminate side effects and off-target toxicity and mutagenesis of application of novel genome modification technologies, and (v) to develop reliable screens for the detection of edited genome samples. The revolutionizing effects of these novel genome-editing/manipulation technologies and genome-edited organisms (GEOs) as well as their safer nature compared to conventional transgenesis are evident. However, without objective or proper regulatory policies in place, "these technologies may not live up to their full potential" if they are regulated as genetically modified organisms bearing foreign genes as this will delay the commercialization of the novel products. Finally, the grandest task is a preparation of well-qualified next-generation scientists capable of continuing plant genomics tasks highlighted herein with the understanding of conventional plant biology, ecology, plant breeding, evolution, taxonomy, modern "omics" disciplines, and cross-related scientific disciplines (e.g., mathematics, computing, and modeling) as highlighted by Abdurakhmonov (2016). The breeders also need to have abilities to handle, work, and utilize the sequenced genomes, high-throughput genotyping, and phenotyping platforms.

Conclusions

In the past more than three decades, plant genomics has evolved from the enrichment and advances made in conventional genetics and breeding, molecular biology, molecular genetics, molecular breeding, and molecular biotechnology in the land of high-throughput DNA sequencing technologies powering the plant research community to sequence and understand the genetic compositions, structures, architectures, and functions of full plant genomes. An urgent need has been realized to increase agricultural production and productivity to feed ever increasing human population which is projected to touch 9 billion by 2050. Climate change is affecting the crop productivity adversely. All these emphasize the need of strengthening of accelerated breeding efforts to evolve superior crop varieties with precision. This is where plant genomics is getting increasing attention and application in plant biology where technological developments led to sequence and assemble entire plant genomes including very complex polyploid plants, annotate gene functions, link the sequence variation(s) to the phenotype(s), and exploit sequence variation(s) in plant/crop improvement in genome-wide scale or through targeted native modification of plant genomes in a highly sequence-specific manner. To date, more than 100 plant genomes including a large number of crops as well as flowering, non-flowering, crop wild relative, model and non-model, and specialty plants have been fully sequenced. As a result, the knowledge has expanded in understanding of many aspects of plant biology, genetics, breeding, and crop evolution and domestication, which contributed to the development of analytical and breeding tools, resulting in accelerated crop improvement programs. To look even deeper scales, more than 1100 Arabidopsis accessions from various eco-geographic origin and experimental populations have been fully sequenced, which will equip plant researchers with better analysis tools and help in tagging and exploiting the biologically meaningful variations. Furthermore, transcriptome profiling of 1000 distinct plant species with agricultural, medicinal, biochemical, and evolutionary utilization has a great value and will be "a gold mining" opportunity for plant biology to explain the evolution of tree of life and Plant Kingdom speciation. All of these successes are expected to significantly accelerate crop improvement using novel genomic selections and new-generation genome-editing and manipulation technologies. There are many tasks ahead for the plant genomics community, which require more collaboration, integrated approaches, better computing capacity and analytical tools, accelerated training and education of well-qualified researchers, and larger investments and a regulatory policy frame-work which should facilitate the commercialization of the products rather than creating hurdles and delaying the exploitation of the novel products.

23

GM Crops

Biotech crops or genetically modified (GM) crops or the transgenic crops are those crop plants into which certain genes have been introduced from unrelated sources beyond the boundary of sexual compatibility using techniques which bypass the normal sexual cycle. The transgenic production technology uses recombinant DNA technology where a recombinant DNA molecule is a vector into which the desired DNA fragment has been inserted to enable its cloning in an appropriate host. The recombinant DNA technology has opened up the flood gate of plant biotechnology to produce plants which are equipped with novel genes to produce products never heard of that a plant will do. Now transgenics in several crops have been commercialized globally. The importance and potential of GM crops can be realized from the fact that acreage under such crops at global level has reached to 189.8 million ha covering 24 countries in 2017. One can have a feel of this huge acreage under commercialized GM crops globally if it is compared with about total cultivated area of 143 million ha in India.

Over the last 30 years field of agricultural biotechnology has expanded rapidly due to greater understanding of DNA as the chemical double helix code from which genes are made. Genetic engineering is one of the modern biotechnology tools that are based on recombinant DNA technology. The term genetic engineering, often interchanged with terms such as gene technology, genetic modification, or gene manipulation, is used to describe the process by which the genetic makeup of an organism can be altered using "recombinant DNA technology." This involves the use of laboratory tools and specific enzymes to cut out, insert, and alter pieces of DNA that contain one or more genes of interest. The ability to manipulate individual genes and to transfer genes between species that would not readily cross is what distinguishes genetic engineering from traditional plant breeding and opens up a huge opportunity to create crop plants equipped with desirable genes from the sources which cannot be sexually crossed. With conventional plant breeding, there is little or no guarantee of obtaining any particular gene combination from the millions of crosses generated. Undesirable genes can be transferred along with desirable genes or while one desirable gene is gained, another is lost because the genes of both parents are mixed together and re-assorted more or less randomly in the offspring. These are the major limitations of conventional crop breeding.

In contrast, genetic engineering allows the direct transfer of one or just a few genes, between either closely or distantly related organisms. Not all genetic engineering techniques involve inserting DNA from other organisms. Plants may also be modified by removing or switching off particular genes and genetic controls (promoters). Genetic engineering techniques are only used when: (i) the trait to be introduced is not present in the germplasm of the crop; (ii) the trait is very difficult to improve by conventional breeding methods; and (iii) it will take a very long time to introduce and/or improve such trait in the crop by conventional breeding methods. Modern plant breeding is a multi-disciplinary and coordinated process where a large number of tools and technologies of conventional breeding techniques, bioinformatics, biochemistry, molecular genetics, molecular biology and genetic engineering are utilized in an integrated manner.

Development of Transgenic Crops

The basic principles of genetic engineering are reasonably simple. It is however, very important to know the biochemical and physiological mechanisms of action, regulation of gene expression and safety of gene and gene product to be utilized. The general process of genetic engineering requires the successful completion of a series of six steps Crop Biotech Update (2014).

Step 1. Nucleic Acid (DNA/RNA) Extraction

Nucleic acid extraction, either DNA or ribonucleic acid (RNA) is the first step in the genetic engineering process. In any isolation procedure, the initial step is the disruption of the cell of the desired organism, which may be viral, bacterial or plant cells, in order to extract the nucleic acid. After a series of chemical and biochemical steps, the extracted nucleic acid can be precipitated to form thread-like pellets of DNA/RNA.

Step 2. Gene Cloning

There are basically four stages in any cloning experiment: generation of DNA fragments, joining to a vector, propagation in a host cell, and selection of the required sequence. In DNA extraction, all DNA from the desired organism is extracted. This genomic DNA is treated with specific enzymes called restriction enzymes cutting it into smaller fragments with defined ends to allow it to be cloned into bacterial vectors. Copies of the vector will then harbor many different inserts of the genome. These vectors are transformed into bacterial cells and thousands of copies are produced. Using information relating to specific molecular marker sequences and the desired phenotype, the vector harboring the desired sequence is detected, selected, isolated and clones are produced. Restriction enzymes are again utilized to determine if the desired gene insert was cloned completely and correctly.

Step 3. Gene Design and Packaging

Once the gene of interest has been cloned, it has to be linked to pieces of DNA that will control its expression inside the plant cell. These pieces of DNA will switch on (promoter) and off (terminator) the expression of the gene inserted. Gene designing/packaging can be done by replacing an existing promoter with a new one, incorporating a selectable marker gene and reporter gene, adding gene enhancer fragments, introns, and organelle-localizing sequences, among others.

Promoters

Promoters allow differential expression of genes. For instance some promoters cause the inserted genes to be expressed all the time, in all parts of the plant (constitutive) whereas others allow expression only at certain stages of plant growth, in certain plant tissues, or in response to external environmental signals. The amount of the gene product to be expressed is also controlled by the promoter. Some promoters are weak, whereas others are strong. Controlling the gene expression is an advantage in developing GM plants.

Selectable Marker Genes

Selectable marker genes are usually linked to the gene of interest to facilitate its detection once inside the plant tissues. This enables the selection of cells that have been successfully incorporated with the gene of interest, thus saving considerable expense and effort. Genetic engineers use antibiotic resistance and herbicide resistance marker genes to detect cells that contain the inserted gene. Cells that survive the addition of marker agents to the growth medium indicate the presence of the inserted gene. Although increase in antibiotic resistance in humans and animals is unlikely to occur using antibiotic resistance marker, genes coding for resistance to non-medically important antibiotics are preferred. In addition, alternative types of marker genes have been developed which are related to plant metabolism such as phosphomannose isomerase, xylose isomerase and others.

Reporter Genes

Reporter genes are cloned into the vector in close proximity to the gene of interest, to facilitate the identification of transformed cells as well as to determine the correct expression of the inserted gene. Reporter genes that have been used include: the beta glucuronidase gene (gus A gene) which acts on a particular substrate producing a blue product, hence making the transformed cells blue; the green fluorescent protein (gfp) which allows transformed cells to glow under a green light; and luciferase gene that allows cells to glow in the dark, among others.

Enhancers

Several genetic sequences can also be cloned in front of the promoter sequences (enhancers) or within the genetic sequence itself (introns, or non-coding sequences) to promote gene expression. An example is the cloning of the cauliflower mosaic virus promoter enhancers in front of the plant promoter. Once the gene of interest is packaged together with the promoter, reporter and the marker gene, it is then introduced into a bacterium to allow for the creation of many copies of the gene package. The DNA isolated from the bacterial clones can then be used for plant cell transformation using particle bombardment. If however the use of bacteria *Agrobacterium tumefaciens* is preferred in the plant transformation, the whole gene package should be cloned in between two border sequences (left and right border) of a binary vector plasmid. This will allow processing of the *Agrobacterium* so that only the transfer DNA (T-DNA) will be incorporated into the plant genome.

Step 4. Transformation

The most common methods used to introduce the gene package into the plant cells in a process called transformation or gene insertion, include biolistic transformation using the gene gun and *Agrobacterium*-mediated transformation.

Particle Bombardment

Particle bombardment is a mechanical method of introducing the desired gene. The desired genetic sequence is cloned into a plant DNA vector and introduced into the plant using the gene gun or particle gun. As in the common gun, the gene gun uses minute particles of tungsten or gold as the bullet. These particles are coated with the DNA solution and fired to the plant cells through the force of the Helium gas inside a vacuum-filled chamber. The DNA and the tungsten/ gold particles get inside the cell, and within 12 hours, the inserted DNA gets inside the nucleus and integrated with the plant DNA. The tungsten/ gold particles are sequestered to the vacuole and eliminated later. Transformed cells are cultured in vitro and induced to form small plants (regeneration) that express the inserted gene.

Step 5. *Agrobacterium tumefaciens*-Mediated Transformation

The "sharing" of DNA among living forms is well documented as a natural phenomenon. For thousands of years, genes have moved from one organism to another. For example, *Agrobacterium tumefaciens*, a soil bacterium known as 'nature's own genetic engineer', has the natural ability to genetically engineer plants. It causes crown gall disease in a wide range of broad-leaved plants, such as apple, pear, peach, cherry, almond, raspberry and roses. The disease gains its name from the large tumor-like swellings (galls) that typically occur at the crown of the plant, just above soil level. Basically, the bacterium transfers part of its DNA to the plant, and this DNA integrates into the plant's genome, causing the production of tumors and associated changes in plant metabolism. Molecular biologists have utilized this biological mechanism to add desired gene into crops. The genes that cause the galls are removed and replaced with genes coding for desirable traits. Plant cells infected with the bacterium will not form galls but produce cells containing the desired gene, which when cultured in a special medium will regenerate into plants and manifest the desired trait. The main goal in any transformation procedure is to introduce the gene of interest into the nucleus of the cell without affecting the cell's ability to survive. If the introduced gene is functional, and the gene product is synthesized, then the plant is said to be transformed. Once the inserted gene is stable, inherited and expressed in subsequent generations, then the plant is considered a transgenic.

Step 6. Detection of Inserted Genes

Molecular detection methods have been developed to determine the integrity of the trans-gene (introduced gene) into the plant cell. Polymerase chain reaction or PCR is a quick test to determine if the regenerated transgenic cells or plants contain the introduced gene. It uses a set of primers (DNA fragments) – forward and backward primers, whose nucleotide sequences are based on the sequence of the inserted gene. The primers and single nucleotides are incubated with the single stranded genomic DNA and several cycles of DNA amplification is conducted in a PCR machine. Analysis of the PCR products in agarose gel will show if the plants are really transformed when DNA fragments equivalent in size with the inserted gene is present and amplified.

Southern Blot Analysis

This technique determines the integrity of the inserted gene: whether the gene is complete and not fragmented, at the correct orientation, and with one copy number. The DNA coding sequence is the probe binding to the single stranded genomic DNA of the transgenic plant which is implanted on a nitrocellulose paper. Autoradiography will reveal the transgenic status of the plant. One copy of the trans-gene is desired for optimum expression.

Northern Blot Analysis

This technique determines whether the transcript or the messenger RNA (mRNA) of the introduced DNA is present and is correctly transcribed in the transgenic plant. The messenger RNA of the transgenic plants are isolated and processed to bind to the nitrocellulose membrane. Labeled DNA is used to bind to the mRNA and can be visualized through autoradiography.

Western Blot Analysis or Protein Immune-Blotting

This is an analytical technique used to detect whether the transgenic plants produce the specific protein product of the introduced gene. Protein samples are extracted from the transgenic plants, processed into denatured proteins and transferred to a nitrocellulose membrane. The protein is then probed or detected using the antibodies specific to the target protein.

Step 6. Backcross Breeding (If Needed)

Genetic transformation is usually conducted in elite or commercial varieties which already possess the desired agronomic traits but lack the important trait of the trans-gene. Thus, once successfully transformed and regenerated, the genetically modified plant will be easily recommended for commercialization, if it shows stability in several generations and upon successfully passing and fulfilling varietal evaluation and release requirements. However, some plant transformations may have been performed in plant varieties which are amenable to genetic transformation but are neither adapted nor important in the target country. There may also be sterility problems in the transgenic plant. In such cases, conventional plant breeding is performed where the transgenic plant becomes the pollen source in the breeding programme and the elite lines or commercial varieties as the recurrent parent. Backcross breeding enables the combination of the desired traits of the recurrent parent and the transgenic line in the off-springs. The length of time in developing transgenic plant depends upon the gene, crop species, available resources and regulatory approval. It varies from 6 to 15 years before a new transgenic plant or hybrid is ready for commercial release.

A few simple schematic diagrammes on above mentioned processes are given in Figs. 23.1, 23.2, 23.3 and 23.4.

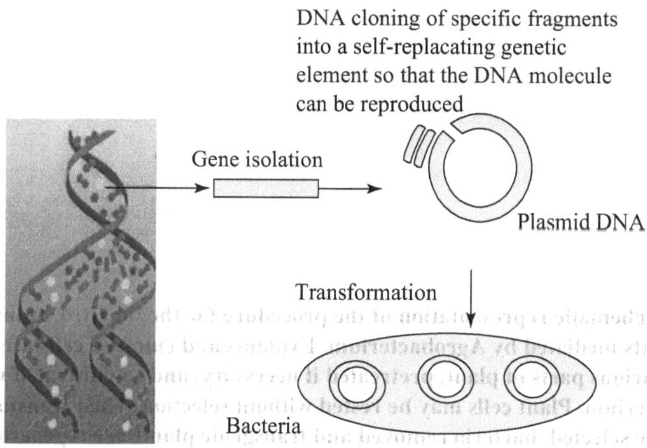

Fig. 23.1. Gene cloning (Source: Tabien, R. 2000).

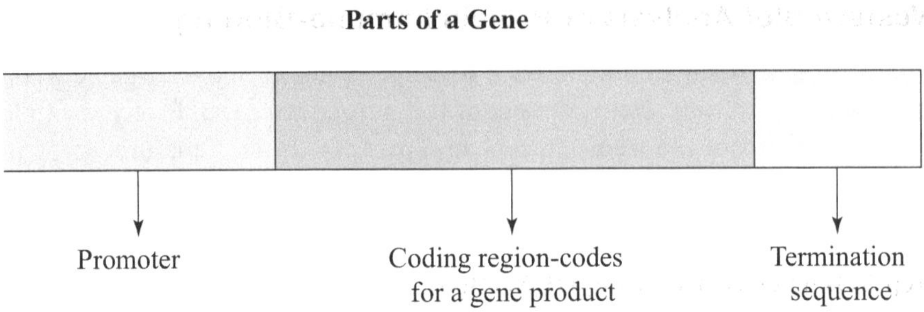

Fig. 23.2. Parts of a gene (Source: Alfonso, A. 2007).

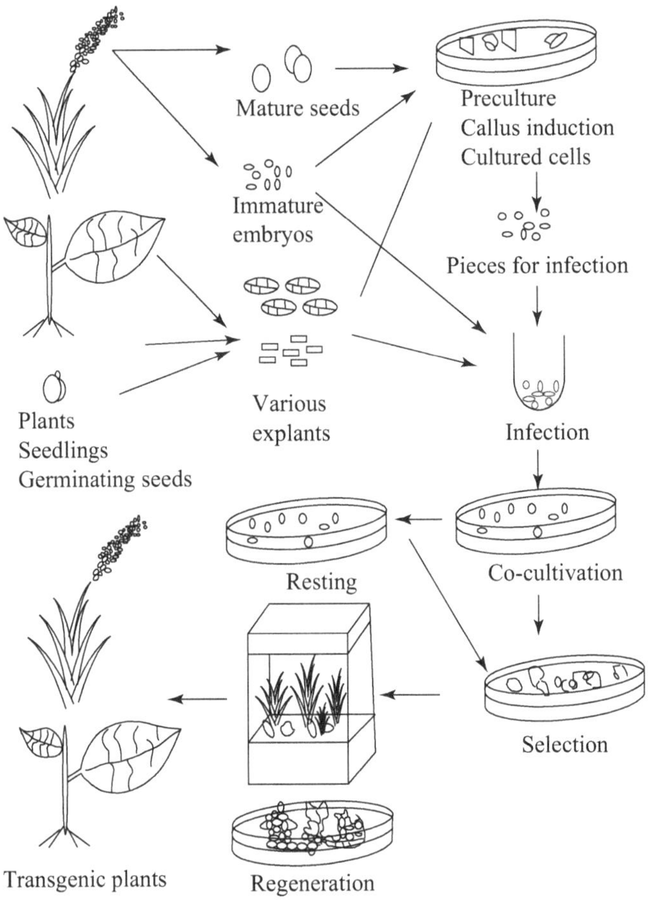

Fig. 23.3. A schematic representation of the procedure for the 'in vitro' transformation of higher plants mediated by Agrobacterium. Explants and cultured cells are prepared from various parts of plant, pretreated if necessary, and co-cultivated with Agrobacterium. Plant cells may be rested without selection. Then transformed cells are selected, bacteria removed and transgenic plants are regenerated (Source: Komari, *et al.*, 2004).

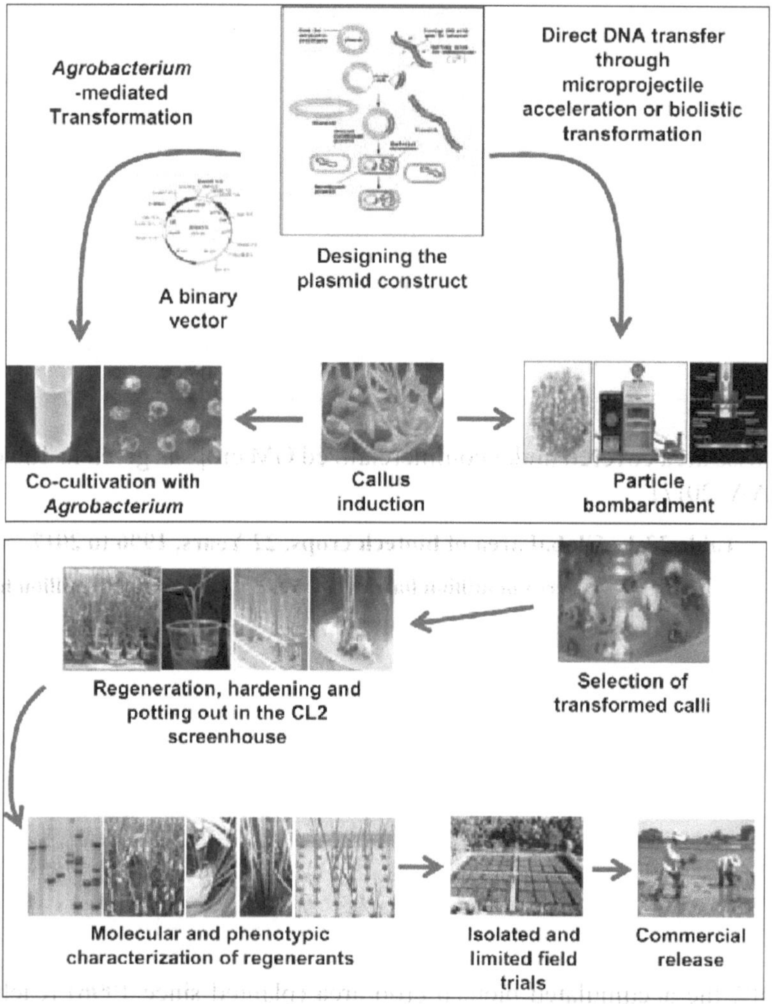

Fig. 23.4. Genetic transformation methods (biolistic or gene gun and *Agrobacterium tumefaciens* mediated transformation methods). (Source: Alfanso, A. 2007).

Global Status of Biotech/GM Crops Commercialized

Global Acreage

The first approval of GM tomato (FlavrSavr) occurred in USA in 1994 but the commercialization of the same happened in 1996 in USA when Calgene first marketed the unsuccessful FlavrSavr delayed-ripening tomato. Since then most plant transgenics/genetic modifications have primarily focused on cash crops in high demand by farmers such as soybean, corn, canola and cotton. GM crop technology is the fastest adopted crop technology in the world. Acreage increased from 1.70 million hectares in 1996 to 185.1 million hectares in 2016 and 189.8 million hectares in 2017. The more than 18 million farmers (up to 90% were small/poor farmers) in up to 30 countries who have planted biotech crops attest to the multiple benefits they derived in the last 21 years as follows:

- Increased productivity that contributes to global food, feed, and fiber security;
- Self-sufficiency on a nation's arable land;
- Conserving biodiversity, precluding deforestation and protecting biodiversity sanctuaries;
- Mitigating the challenges associated with climate change; and
- Improving economic, health, and social benefits.

Year-wise area covered under commercialized GM crops is given in Table 23.1 (ISAAA, 2017).

Table 23.1. Global area of biotech crops, 22 Years, 1996 to 2017.

Year	Area in million ha	Year	Area in million ha
1996	1.7	1997	11.0
1998	27.8	1999	39.9
2000	44.2	2001	52.6
2002	58.7	2003	67.7
2004	81.0	2005	90.0
2006	102.0	2007	114.3
2008	125.0	2009	134.0
2010	148.0	2011	160.0
2012	170.3	2013	175.2
2014	181.5	2015	179.7
2016	185.1	2017	189.8

In 2017, the accumulated biotech crop area (planted since 1996) reached to a record 2.3 billion hectares or 5.8 billion acres. Of the total number of 24 countries planting biotech crops in 2017, 19 were developing countries and 5 industrialized countries (Table 23.2). To put the 2017 global area of biotech crops into context, 189.8 million hectares of biotech crops is equivalent to almost 20% of the total land area of China (956 million hectares) or the USA (937 million hectares) and more than 7 times the land area of the United Kingdom (24.4 million hectares).

Global Area of Biotech Crops by Country

Globally 26 countries planted biotech crops in 2016 and 24 countries in 2017 (Table 23.2, ISAAA, 2017).

Table 23.2. Global area of biotech crops in 2016 and 2017: by country (million hectares).**

S. No.	Country	2016	%	2017	%	+/–	%
1	USA*	72.9	39	75.0	40	2.1	3
2	Brazil*	49.1	27	50.2	26	1.1	2
3	Argentina*	23.8	13	23.6	12	–0.2	–1
4	Canada*	11.1	6	13.1	7	1.5	18
5	India*	10.8	6	11.4	6	0.6	6
6	Paraguay*	3.6	2	3.0	2	–0.6	–18
7	Pakistan*	2.9	2	3.0	2%	0.1	3%
8	China*	2.8	2	2.8	1%	0.0	0
9	South Africa*	2.7	1	2.7	1%	0.0	0
10	Bolivia*	1.2	1	1.3	1%	0.1	7%
11	Uruguay*	1.3	1	1.1	1%	–0.2	–12%
12	Australia*	0.9	<1	0.9	<1	0.1	0%
13	Philippines*	0.8	<1	0.6	<1	–0.2	–21%
14	Myanmar*	0.3	<1	0.3	<1	0.0	0
15	Sudan*	0.1	<1	0.2	<1	0.1	59
16	Spain*	0.1	<1	0.1	<1	0.0	0
17	Mexico*	0.1	<1	0.1	<1	0.1	9
18	Colombia*	0.1	<1	0.1	<1	<0.1	7
19	Vietnam	<0.1	<1	<0.1	<1	<0.1	29
20	Honduras	<0.1	<1	<0.1	<1	<0.1	3
21	Chile	<0.1	<1	<0.1	<1	<0.1	23
22	Portugal	<0.1	<1	<0.1	<1	<0.1	0
23	Bangladesh	<0.1	<1	<0.1	<1	<0.1	242
24	Costa Rica	<0.1	<1	<0.1	<1	<0.1	0
25	Slovakia	<0.1	<1	—	—	—	—
26	Czech Republic	<0.1	<1	—	—	—	—
	Total	**185.1**	**100**	**189.8**	**100%**	**5.4**	**+3**

* Biotech mega-countries growing 50,000 hectares or more.
** Rounded-off to the nearest hundred thousand or more.
Source: ISAAA, 2017.

Country-wise Distribution of Commercialized GM Crops

In 2015, GM crops were grown in 28 countries and on 179.7 million hectares – that is over 10% of the world's arable land and equivalent to seven times the land area of the UK. The USA, Brazil, Argentina, Canada and India are the leading producers of GM crops. The GM crops grown commercially include: potato (USA), squash/pumpkin (USA) alfalfa (USA), aubergine/eggplant (Bangladesh), sugar beet (USA, Canada), papaya (USA and China), oilseed rape (4 countries), maize/corn (17 countries), soybean (11 countries) and cotton (15 countries).

The top GM crop grown in 2015 was soybean (92.1 million ha), followed by maize (53.6 million ha), then cotton (24 million ha) and oilseed rape/canola (8.5 million ha). This represents 83% of the world production of soybean, and 75% of production of cotton. GM crops made up 29% of the world's maize produce, and almost a quarter of the world's oilseed rape that year. The crops which accounted for substantial acreage under biotech crops in 2016 and 2017 are summarized in Table 23.3.

Table 23.3. Global area of biotech crops, 2016 and 2017: by crop (million hectares).

Crops	2016	%	2017	%	+/–	%
Soybeans	91.4	50	94.1	50	+2.7	3
Maize	60.6	33	59.7	31	–0.9	–1
Cotton	22.3	12	24.1	13	+1.8	8
Canola	8.6	5	10.2	5	+1.6	19
Alfalfa	1.2	<1	1.2	<1	+<1	<1
Sugar beet	0.5	<1	0.5	<1	–<1	<1
Papaya	<1	<1	<1	<1	–<1	<1
Others*	<1	<1	<1	<1	+ <1	<1
Total	**185.1**	**100**	**189.8**	**100**	**4.7**	**+3%**

*Others include biotech squash, potato, eggplant and apples. Source: ISAAA, 2017

Among the countries growing GM crops, as per 2017 production figures, the USA (75.0 m ha), Brazil (50.2 m ha), Argentina (23.6 m ha) Canada (13.1 m ha) and India (11.4 m ha) are the largest users. Within Europe, five EU countries, namely, Spain, Portugal, Czech Republic, Romania and Slovakia grow GM maize. In Africa, GM crops are grown in South Africa, Burkino Faso and Sudan. Commercialized GM crops being grown in specific countries are shown in Table 23.4.

Table 23.4. Country-wise distribution of GM crops.

Rank	Country	GM crops commercialized
1	USA	Maize, soybean, cotton, canola, sugar beet, alfalfa, papaya, squash, potato, apple
2	Brazil	Soybean, maize, cotton
3	Argentina	Soybean, maize, cotton
4	Canada	Canola, maize, soybean, sugar beet, potato
5	India	Cotton
6	Paraguay	Soybean, maize, cotton
7	Pakistan	Cotton
8	China	Cotton, papaya, poplar, tomato, sweet pepper
9	South Africa	Maize, soybean, cotton

Rank	Country	GM crops commercialized
10	Bolivia	Soybean
11	Uruguay	Soybean, maize
12	Australia	Cotton, canola
13	Philippines	Maize
14	Burkina Faso	Cotton
15	Myanmar	Cotton
16	Spain	Maize
17	Mexico	Cotton, soybean
18	Colombia	Cotton, maize
19	Sudan	Cotton
20	Chile	Maize, soybean, canola
21	Honduras	Maize
22	Portugal	Maize
23	Costa Rica	Cotton, soybean
24	Romania	Maize
25	Bangladesh	Brinjal
26	Slovakia	Maize
27	Cuba	Maize
28	Czech Republic	Maize

Distribution of Biotech Crops, by Trait

During the 22-year period of 1996 to 2017, herbicide tolerance has consistently been the dominant trait of biotech crops planted (Table 23.5), but is slowly declining through the years with the increasing prominence of the stacked traits. In 2017, stacked IR/HT traits deployed in soybean, maize, and cotton had the highest increase in year-on-year increment of 3% and occupy 41% of the global area. In 2017, there were percentage increases in area planted to stacked IR/HT cotton (22%), soybeans (4%), and maize (1%). Stacked trait products were preferred by farmers due to its cost saving technology, especially in soybean and cotton. Various IR/HT products were approved for food/feed and commercialization in 2017 for soybeans, maize, and cotton. Stacking is a very important feature of the technology with SmartStax™ comprising eight genes coding for three traits, launched in the USA and Canada in 2010, as well as the Innate™ potato generation 2 which was approved for cultivation in 2015 in the USA and in Canada in 2016. The deployment of stacked traits of different Bt genes and herbicide tolerance is becoming increasingly important and is most prevalent in the USA and Brazil. Distribution of trait-wise GM crops is given in Table 23.5 (ISAAA, 2017).

Table 23.5. Distribution of biotech crops by traits.

Traits	2016	%	2017	%	+/–	%
Herbicide Tolerance	86.6	47	88.7	47	2.1	2%
Stacked Traits	75.4	41	77.7	41	2.3	3%
Insect Resistance	23.1	12	23.3	12	0.2	1%
Virus Resistance/						
Other	<1	<1	<1	<1	<1	<1
Total	**185.1**	**100**	**189.8**	**100**	**4.7**	**+3%**

GM Crops in India

The only crop where GM versions have been commercialized in India is cotton imparting resistance to cotton boll worm. The first event MON531(Bollgard I, BG I) featuring the *cry1Ac* developed by Maharashtra Hybrid Seed Company Ltd (Mahyco) was commercialized in 2002 followed by commercialization of the event MON15985 featuring two genes, namely, *cry1Ac* and *cry2Ab* known as Bollgard II, again developed by Maharashtra Hybrid Seed Company Ltd (Mahyco). Several companies outsourced these genes from MMB (Monsanto-Mahyco-Biotech) and converted their non-Bt cotton hybrids into Bt cotton hybrids (mostly Bollgard II) and started selling seeds as there has been huge demand for such Bt cotton seeds. In 2017, the adoption of Bt cotton in India reached 93% planted by 7.5 million farmers among 10 major cotton growing states. In 2017, India planted 11.4 million ha of IR (Bt) cotton (insect resistant Bt cotton) recording an increase of 600, 000 ha from 10.8 million ha in 2016. Since 2015, India has been the world's top cotton producing country, achieving cotton production of 37.7 million bales in 2017. India produces more than a quarter of world's cotton.

Bt brinjal developed by Mahyco using *cry1Ac*, was recommended for commercial release by GEAC in 2009. However, in view of divergent views of proponents and opponents of this recommendation, Minister of Environment and Forest, imposed a moratorium on its release on 9 Feb, 2010 till safety and other issues were resolved. The moratorium on insect resistant (IR) Bt brinjal by MOEF&CC in 2010 has not yielded any outcome in last nine years and thus a careful consideration of the recommendation of the regulatory agency of the GM crops by MOEF&CC is immensely needed at this time. The same Bt event in brinjal has been commercialized in Bangladesh on 30 October, 2013.

In 2017, the Genetic Engineering Appraisal Committee (GEAC) of Ministry of Environment, Forest and Climate Change thoroughly assessed the safety and performance of GM mustard and recommended the environmental release of transgenic mustard hybrid Dhara Mustard Hybrid 11 (DMH 11) and parental lines containing events bn 3.6 and modbs 2.99 developed using *barnase*, *barstar* and *bar* genes developed by the Centre for Genetic Manipulations of Crop Plants (CGMCP) of the university of Delhi on May 11, 2017. GM mustard is the India's first state-of-art farm innovation that is expected to allow increase in mustard productivity. However, on October 26, 2017, the MOEF and CC decided to keep the release of transgenic mustard pending further review based on receipt of various representations from different stakeholders.

World Food Prize's Recognition Of Biotechnology's Contribution To Food, Feed And Fiber Seccurity

The World Food Prize (WFP) is the foremost International Foundation that recognizes accomplishments of individuals who have advanced human development by improving the quality, quantity, or availability of food in the world. The 2013 Laureates are three biotechnologists who have independently discovered molecular techniques for genetically engineering improved crops. As the founder of the World Food Prize and a strong advocate of biotech/GM crops, Norman Borlaug, Nobel Peace Prize Laureate in 1970 had expressed his views to the WFP Foundation that biotechnologists should not be excluded from consideration as World Food Prize Laureates because of the controversy surrounding GM crops. He contended that they should be considered on their own merit and judged by their contribution to global food security and the alleviation of poverty.

Borlaug would have been pleased with the decision to award the 2013 World Food Prize to three internationally recognized biotechnologists, whom he knew personally and respected: Marc Van Montagu, Mary-Dell Chilton and Robert Fraley, who have all made important contributions in their respective areas of crop biotechnology. "The three Laureates have in their own unique ways established the science behind the transfer of genes from other species to the target crops through *Agrobacterium tumefaciens* in the late 1970's. Marc Van Montagu and colleague Jeff Schell were the first to discover, in 1974, that the bacteria carry a Ti-plasmid (plant tumor-inducing plasmid). They did a thorough study on its structure and function which led to the stable transfer of foreign genes into plants. Mary- Dell Chilton and her research team discovered that there is a segment in this plasmid, the Transfer- DNA (T-DNA) that is processed and transferred into the genome of the infected plant cell. Her work provided evidence that plant genomes could be manipulated more precisely than in conventional plant breeding. Robert Fraley and his team's research works were built on the advances made by Van Montagu and Chilton. The team was able to isolate a bacterial marker gene, which was expressed in plant cells. This became the scientific basis of the development of Roundup

Ready soybeans. "The work of the three Laureates became the foundation of plant cell transformation technologies that enabled the development of a host of genetically-enhanced crops with improved yields; resistance to insects and diseases; and tolerance against extreme variations in climate. Their combined achievements have contributed significantly to increasing the quantity and availability of food, and can play a critical role as we face the global challenges of the 21st century of producing more food in a sustainable way, while confronting an increasingly volatile climate."

It is noteworthy that the 2013 World Food Prize served as a unique global forum to stimulate and encourage professional debate, and to increase the awareness of both the scientific community and the public about the formidable challenge of food security and the current and future contributions that biotechnology can make to help feed the world of tomorrow with a projected population of 9 billion in 2050.The three 2013 Laureates were of the unanimous view that sharing knowledge and communicating with the public on biotech crops was the top priority. ISAAA is of the same view and initiated its extensive global knowledge-sharing activities with the public more than ten years ago in 2000. ISAAA's flagship publication, the Annual Brief on the *Global Status of Commercialized Biotech/GM Crops*, authored for the last 17 years by Dr. Clive James, is the most quoted publication on biotech crops globally. The major messages from the Brief typically reach up to an unprecedented 3 billion people in about 50 countries and languages. Knowledge-sharing is achieved through multi-media channels, thereby reaching a remarkably large number and broad range of stakeholders from global society at large. Other ISAAA complementary activities organized by the Global Knowledge Center (KC) in knowledge-sharing include its active user-friendly website with various educational/learning materials, including, videos, and info-graphics as well as its weekly newsletter *Crop Biotech Update* distributed to subscribers in 140 countries. In addition, ISAAA organizes a continuing series of workshops in developing countries to meet the multiple and changing needs of policy makers, regulators and other stakeholders in crop biotechnology. ISAAA, like the three Laureates, believes that knowledge-sharing is the key to increasing biotech crop understanding, acceptance and adoption globally. The 2013 World Food Prize and the Borlaug dialogue have contributed in a unique and significant way towards an increased measure of consensus by the scientific community and the public about major issues that have been debated for over a decade or more. For example, there has been a marked shift in public sentiment and an increased trust in science-based assessments that confirm that foods from biotech products are safe and that significant productivity and environmental benefits have accrued to both producers and consumers.

Similarly, there is a favorable shift in public opinion on Golden Rice (vitamin A rich transgenic).

Looking Ahead

There is no denying the fact that biotech/GM crops have been adopted by the farmers globally reflected in continuous increase of acreage under biotech crops from 1.70 million ha in 1996 (the first year of biotech crop commercialization) to 189.8 million ha in 2017. This growth has been unprecedented. The crops covered so far are maize, soybean, cotton, canola, sugar beet, alfalfa, papaya, squash, potato, apple, papaya, poplar, tomato, sweet pepper and eggplant. The top five countries accounting for major GM crop acreage are USA, Brazil, Argentina, Canada and India. The major crops accounting for substantial acreage under GM crops are soybean, maize, cotton and canola all jointly representing almost 99% of GM crop acreage. The major traits introduced are herbicide tolerance, stacked traits and insect resistance. It is noteworthy; however, that productivity gained in the last 21 years (1996-2017) through biotech crops proved that conventional crop technology alone cannot allow us to feed the immense increase in population, but neither is biotechnology a panacea. The global scientific community adheres to the option of a balanced, safe, and sustainable approach using the best of conventional crop technology such as the well-adapted and agronomically desirable and high-yielding germplasm, and the best of biotechnology (GM and non-GM traits), to achieve sustainable intensification crop productivity on the 1.5 billion hectares of cropland globally. GM crops help in increasing productivity by suppressing insects and weeds, conserve biodiversity precluding deforestation and protecting biodiversity sanctuaries, mitigating the challenges associated with climate change; and improving economic, health, and social benefits. Several exciting developments in GM crops, namely, Golden Rice, biotech-DMH 11 mustard and Bt. brinjal for India; Golden Rice for Philippines, Bean Golden Mosaic Virus Resistant Bean for Brazil and drought tolerant biotech-maize in Africa are waiting approval and commercial release. However, there are concerns in some NGOs and people pursuing and encouraging traditional and organic farming regarding safety of these products and their environmental impacts and these need to be addressed to through dialogue and discussion but the farmers who wish to plant the GM crops based on their own experiences and expectations should not be denied at least the option of planting/not planting the GM crops and the same holds true for the consumers.

24

International Research Centres on Plant Breeding

There have been lots of international efforts to establish crop based research centres primarily in the developing/under-developed countries to carry out concerted breeding and plant genetic resources management activities pooling germplasm from all over the world and also engaging scientists from different countries to do breeding work on the certain mandate crops of the particular region using all kinds of germplasm and technologies with strong financial support from a central agency known as Consultative Group on International Agricultural Research (CGIAR) for doing the stipulated work as per programme and deliver the products which can be used by different national plant breeding programmes either to use them directly as cultivars or to use them in their breeding programmes in various ways including parents in the crossing programme.

CGIAR (Consultative Group on International Agricultural Research)

CGIAR System Organization headquartered in Montpellier, France is a global agricultural research partnership for a food secure future that was founded in the 1970s in response to the Green Revolution. There are presently 15 CGIAR agricultural research centres employing more than 8,500 researchers and support staff worldwide, with an annual budget of US$ 800 million.

CGIAR is a global research partnership for a food-secure future. CGIAR science is dedicated to reducing poverty, enhancing food and nutrition security, and improving natural resources and ecosystem services. Its research is carried out by 15 CGIAR Research Centers in close collaboration with hundreds of partners, including national and regional research institutes, civil society organizations, academia, development organizations and the private sector. The CGIAR Fund was established as a Financial Intermediary Fund (FIF) in 2010. It was subsequently closed and replaced by a new FIF in 2017, following the CGIAR System governance reorganization.

The 15 CGIAR Research Centers are independent, non-profit research organizations, conducting innovative research. Home to more than 8,000 scientists, researchers, technicians, and staff, CGIAR research works to create a better future for the world's poor. Each Center has its own charter, board of trustees, director general, and staff. CGIAR Research Centers are responsible for hands-on research programs and operations guided by policies and research directions set by the System Management Board.

The CGIAR Excellence in Breeding Platform is the newest Platform arrival, starting up in August 2017 with the recruitment of the director. It aims to modernize breeding programs, targeting the developing world for greater impact on food and nutrition security, climate change adaptation, and development. The Platform draws from innovations in the public and private sectors to provide access to cutting-edge tools, services and best practices, application-oriented training and practical advice, with the aim of increasing the effectiveness and efficiency of breeding.

In its first few months, the Platform:

- Forged partnerships across CGIAR breeding programs, as well as with external bodies such as Monsanto (now merged with Bayer), Diversity Arrays Technology, Cornell University, Corteva Agriscience, the University of Queensland, Hiphen, Institut National de la Recherche Agronomique -Avignon, CSIRO, and Kansas State University.
- Held a successful series of workshops to bring breeders together to address practical problems of product development and breeding program management, including overcoming bottlenecks and identifying lower-cost and more rapid approaches to common breeding processes.
- Conducted two surveys of member programs to assess needs for genotyping and breeding informatics support for CGIAR breeding programs.
- Worked closely with the Breeding Program Assessment Tool team to identify priority improvements for participating breeding programs.

The CGIAR Genebank Platform supports the core activities of the CGIAR genebanks to conserve collections of plant genetic resources for food and agriculture. Through the Platform, CGIAR fulfills its legal obligation to conserve and make available the 35 crop and tree collections under its management and works towards meeting international standards, improving efficiency, and ensuring more effective use of collections within a supportive policy environment. The work directly contributes to indicator 2.5.1 of SDG Target 2.5: "maintain the genetic diversity of seeds, cultivated plants and farmed and domesticated animals and their related wild species, including through soundly managed and diversified seed and plant banks and promote access to and fair and equitable sharing of benefits arising from the utilization of genetic resources and associated traditional knowledge".

CGIAR genebanks presently manage 768,576 accessions, including 25,301 *in vitro* accessions and 28,063 accessions held as plants or trees in the field. Taken together, CGIAR genebanks represent the largest and most widely used collections of crop diversity in the world.

In 2017, 79% of total accessions are immediately available for international distribution, a significant improvement from 66% in 2012. Of the seed accessions, 55% are secured in safety duplication at two levels and 73% are duplicated at the Svalbard Global Seed Vault. 79% of clonal crop collections are 'safety duplicated' in the form of cryopreserved samples or *in vitro* cultures (https://www.cgiar.org/.)

1. AfricaRice (Africa Rice Centre)

AfricaRice (Africa Rice Centre) headquartered in Côte d'Ivoire and also in Africa Rice Research Stations in Benin, Liberia, Madagascar, Nigeria, and Senegal (www.AfricaRice.org) is a CGIAR Research Center – part of a global research partnership for a food-secure future. It is also an intergovernmental association of African member countries.

The Center was created in 1971 by 11 African countries. Today its membership comprises 27 countries, covering West, Central, East and North African regions, namely Benin, Burkina Faso, Cameroon, Central African Republic, Chad, Côte d'Ivoire, Democratic Republic of Congo, Egypt, Ethiopia, Gabon, the Gambia, Ghana, Guinea, Guinea Bissau, Liberia, Madagascar, Mali, Mauritania, Mozambique, Niger, Nigeria, Republic of Congo, Rwanda, Senegal, Sierra Leone, Togo and Uganda. The centre is engaged among others on genetic diversity and improvement on rice.

Rice yields in Africa are affected by a large number of abiotic and biotic stresses. The major abiotic stresses addressed through this program are drought, acidity, salinity, iron toxicity and extreme temperatures. The major biotic stresses that are considered are: RYMV, BLB, blast, African rice gall midge, stem borers and termites. In this program both conventional breeding and biotechnology will be used to exploit the rich reservoir of genetic resources present in the indigenous germplasm pool of African rice *Oryza glaberrima*, its wild relatives – *O. barthii* and *O. longistaminata* – and *O. sativa* landraces.

2. Biodiversity International

Biodiversity International headquartered at Rome is a global research-for-development organization. The centre has a vision – that agricultural biodiversity nourishes people and sustains the planet. It delivers scientific evidence, management practices and policy options to use and safeguard agricultural and tree biodiversity to attain sustainable global food and nutrition security. It works with partners in low-income countries in different regions where agricultural and tree biodiversity can contribute to improved nutrition, resilience, productivity and climate change adaptation. Bioversity International contributes to the following CGIAR Research Programs:

- Agriculture for nutrition and health
- Big Data
- Climate change, agriculture and food security
- Excellence in breeding
- Forests, trees and agroforestry
- Genebank Platform
- Policies, institutions, and markets
- Roots, tubers and bananas
- Water, land and ecosystem

The international status of Bioversity International is conferred under an Establishment Agreement that had been signed by the Governments of: Algeria, Australia, Belgium, Benin, Bolivia, Brazil, Burkina Faso, Burundi, Cameroon, Chile, China, Congo, Costa Rica, Côte d'Ivoire, Cyprus, Cuba, Czech Republic, Denmark, Ecuador, Egypt, Ethiopia, Ghana, Greece, Guinea, Hungary, India, Indonesia, Iran, Israel, Italy, Jordan, Kenya, Malaysia, Mali, Mauritania, Mauritius, Morocco, Nepal, Norway, Oman, Pakistan, Panama, Peru, Poland, Portugal, Romania, Russia, Senegal, Slovakia, Sudan, Switzerland, Syria, Tunisia, Turkey, Uganda and Ukraine.

3. CIFOR (Centre for International Forestry Research)

CIFOR is headquartered in Bogor, Indonesia, and maintains hubs in Nairobi Kenya, Yaounde, Cameroon, and Lima Peru. Research in the six thematic work areas is underway in more than 50 countries. Occupying less than a third of the world's land mass, forests provide $250 billion in income, are essential to the livelihoods of 1 out of every 4 people on the planet, contain 80% of Earth's terrestrial biodiversity, and absorb up to a third of all carbon emissions. Since 1990, more than 300 million hectares of forests have been destroyed. Deforestation contributes 10–15% of the world's greenhouse gas emissions, reduces agricultural productivity, and undermines food security globally. Forest loss endangers both biodiversity and the livelihoods of a quarter of the world's population. CIFOR is proud to work with local and international partners. CIFOR is a CGIAR Center and leads the CGIAR Research Program on Forests, Trees and Agro-forestry.

CIFOR envisions a more equitable world where forestry and landscapes enhance the environment and well-being for all. The Center for International Forestry Research is a non-profit, scientific facility that conducts research on the most pressing challenges of forest and landscape management around the world. Using a global, multidisciplinary approach, it aims to improve human well-being, protect the environment, and increase equity. To do so it helps policymakers, practitioners and communities make decisions based on solid science about how they use and manage their forests and landscapes.

CIFOR's work is based on three pillars, all of which are essential for achieving the targeted mission:

- Research for impact
- Capacity development
- Outreach and engagement

Research focuses on the following six thematic areas:

- Forests and human well-being
- Sustainable landscapes and food
- Equal opportunities, gender, justice and tenure
- Climate change, energy and low-carbon development
- Value chains, finance and investment
- Forest management and restoration

4. ICARDA (International Center for Agricultural Research in the Dry Areas)

ICARDA headquartered at Beirut, Lebanon works with partners worldwide, developing innovations to improve food security and the livelihoods of the rural poor. ICARDA's work targets non-tropical dry areas in developing countries, and also produces international public goods with potential for global application. Research covers crops (wheat, barley, faba bean, lentil, chickpea and forage legumes), the management of natural resources (water, land, biodiversity), small ruminant production (sheep and goats), farming systems (intensification, diversification, integration between farming system components), and socio-economics and policy research – on how policies can be more relevant to the situation of low income countries.

The estimated benefits of ICARDA's crop improvement research and production of new varieties over the past 3 decades has been estimated at US$ 850 million per year. Over 900 improved varieties of wheat, developed from ICARDA material, have been released for cultivation worldwide.

The new varieties offer higher yields; better tolerance to drought, heat, cold and salinity; and improved resistance to diseases, weeds and insect pests.

Other innovations include:

- Technology and policy packages that countries can apply to increase the productivity and income of smallholder farmers.
- Water harvesting, supplemental irrigation and other low-cost methods to increase water productivity and bring water to rural communities in dry areas.
- Conservation agriculture methods to reduce production costs and improve sustainability.
- Diversification of production systems to improve farmer incomes through high-value crops – horticulture, herbal and medicinal plants.
- Integrated crop/rangeland/livestock production systems including non-traditional sources of livestock feed.
- Policy and institutional options to improve rural livelihoods.
- Poverty mapping and other tools to guide investment targeting.

Plant Genetic Resources at ICARDA

ICARDA's genebank holds over 135,000 accessions from over 110 countries: traditional landraces, improved germplasm, and a unique set of wild crop relatives. These include wheat, barley, oats and other cereals; food legumes such as faba bean, chickpea, lentil and field pea, forage crops, rangeland plants, and wild relatives of each of these species.

Capacity development – through training programs, support for graduate education, internships, farmer field schools and other opportunities – is a vital part of ICARDA's work and has benefitted over 17,000 young developing-country researchers, and many more farmers and extension staff.

5. CIAT (International Center for Tropical Agriculture)

CIAT headquartered at Palmira, Colombia working in collaboration with hundreds of partners across the developing world, is dedicated to developing technologies, innovative methods, and new knowledge that better enable farmers, mainly smallholders, to improve their crop production, incomes, and management of natural resources. Within CGIAR, CIAT has global responsibility for the improvement of beans, cassava, and tropical forages – crops that have historically been neglected by research despite their vital importance for food and nutrition security. It also conducts research on rice and tropical fruits for Latin America and the Caribbean.

All of the Center's work on agricultural biodiversity – encompassing diverse food groups – employs advanced biotechnology to discover useful knowledge and accelerate crop improvement. Progress in CIAT's crop improvement research also depends on unique collections of genetic resources (67,700 crop samples in all) which it holds in trust for humanity.

CIAT conducts research on two major issues that cut across tropical crops and production niches: (1) sustainable management of tropical soils, and (2) decisions and policies that are critical for coping with challenges such as climate change, environmental degradation, gender inequities, and weak links between farmers and markets.

CIAT safeguards the world's largest and most diverse collection of bean germplasm. The collection consists of around 36,000 samples of cultivated materials mostly from the crop's Mesoamerican and Andean centers of origin together with wild species related to these materials. The CIAT collection constitutes a valuable resource for bean improvement worldwide. CIAT bean researchers are developing a wide range of user-friendly molecular markers (especially SSRs) and have implemented marker-assisted selection for disease resistance. They also aim to combine tolerance to various abiotic stresses (drought and poor soils) through the use of gene discovery, marker-assisted selection, and farmers participatory methods.

Cassava is another most important crop for CIAT after beans. Cassava is the third most important food crop in the tropics, after rice and maize. Esteemed by smallholder farmers for its tolerance to drought and infertile soils, the crop is inherently eco-efficient, offering a reliable source of food and income.

CIAT scientists work with partners to boost yields by at least another 30% through continued genetic improvement and better agronomy as well as pest and disease management. CIAT seeks to develop and scale up cassava varieties that are agronomically competitive and more nutritious than varieties currently grown. Particular emphasis is placed on breeding for high content of provitamin A carotenoids. The Center's cassava team also develops novel starch quality traits such as the waxy-starch and small-granule mutations which promise to strengthen the crop's appeal to industrial markets.

To help counter the threat of pests and diseases, CIAT scientists devise diagnostic kits for disease detection and use advanced spatial analysis to study changes in the distribution of key pests and diseases. In search of solutions, researchers identify biological control agents and develop sources of genetic resistance. CIAT aims at developing cassava that is:

- Higher yielding
- Better starch quality

- Nutritionally improved (especially provitamin A carotenoids)
- Resistant to pests and diseases

CIAT conserves in vitro the world's most important collection of cassava gemplasm. The cassava collection held in trust at CIAT includes a total of 6,155 accessions from 28 countries, represented in 5,690 clones of (*Manihot esculenta*) and 465 genotypes of wild species conserved using in vitro techniques. Over 37% of the cassava diversity held in CIAT's genebank originates from Colombia, with another 24% coming from Brazil. Collections from other South American countries (21%), as well as Central America and the Caribbean (7%), and Asia (7%) are also conserved at CIAT.

Cassava improvement at CIAT relies on biotechnology tools to speed progress and solve recalcitrant problems. Center researchers develop molecular markers for traits such as whitefly resistance, quantitative trait loci (QTLs) in populations derived from heterozygous parent materials, and protocols for rapid multiplication of disease-free planting materials through tissue culture.

Through the Cassava Genome Hub, an online platform that produces and stores more than 15 terabytes of genetic data on cassava, CIAT is pioneering a new approach to big data management and analysis. Launched publically in late 2015, and accessible from anywhere in the world, the Hub allows researchers to manage and mine this huge amount of data themselves, using graphical and analytical tools to conduct complex analysis in a user-friendly way

6. ICRISAT (The International Crops Research Institute for the Semi-Arid Tropics)

ICRISAT headquartered at Patancheru, Hyderabad, India is a non-profit, non-political organization that conducts agricultural research for development in the dry-lands of Asia and sub-Saharan Africa. Covering 6.5 million square kilometers of land in 55 countries, the semi-arid or dry-land tropics has over 2 billion people, and 644 million of these are the poorest of the poor. With two regional hubs (Nairobi, Kenya and Bamako, Mali) and country offices in Niger, Nigeria, Zimbabwe, Malawi, Ethiopia and Mozambique, ICRISAT conducts research on five highly nutritious drought-tolerant crops: chickpea, pigeonpea, pearl millet, sorghum, and groundnut.

Chickpea

In a scientific advance involving scientists from more than 20 research organizations, sequencing of the chickpea genome was recently completed for 90 chickpea genotypes, including several wild species. The researchers responsible for the work, led by ICRISAT, succeeded in identifying more than

28,000 genes and several genetic markers that scientists expect will lead to the development of superior varieties.

The new research will benefit the millions of developing country farmers who grow chickpea as a source of much needed income, as well as for its ability to add nitrogen to the soil in which it grows. A close relative of peas, chickpeas are rich in protein, but low in fat. Consumed in many different forms, the crop is probably best known as the central ingredient in hummus, the Arabic word for chickpea.

Production is growing rapidly across the developing world, especially in West Asia where production has grown four-fold over the past 30 years. India is by far the world largest producer but is also the largest importer. Modern early maturing chickpea lines have had a considerable impact in short season environments across South and Southeast Asia transforming what was once a subsistence crop to into significant income generator and export commodity.

Currently, efforts are being made to enhance tolerance to terminal drought and heat stresses. In addition, researchers are working to boost the crop's natural nutritional advantages by increasing its protein, iron and zinc content. Plans also call for developing varieties that are resistant to herbicides and can be harvested with small machinery to reduce costs and overcome labor shortages that arise at weeding and harvest time. These efforts should help to reduce production shortages, stabilize prices for producers and consumers, and help to drive the development of overseas markets.

Pigeonpea

Pigeonpea is grown by millions of resource-poor farmers on marginal land across the semi-arid regions of Asia and Africa. Producing small seeds similar in appearance to soybean, pigeonpea is valued by farmers and consumers as food and forage crop. It is also a major contributor to food security in areas facing the early effects of global climate change.

Long considered an under-researched 'orphan crop', its standing has grown considerably in recent years as researchers have improved the crop's potential and resiliency. For example, early- and medium-duration varieties, including hybrids that mature in 4 to 6 months, have been developed and are proving highly successful in farmers' fields. The new cultivars are attractive to farmers and commercial seed producers who recognize their potential for income generation and international trade.

Crop protection specialists estimate that the pests and diseases that attack pigeonpea are responsible for US$ 1.1 billion in annual losses across the semi-arid tropics. Principal plant diseases include Fusarium wilt and sterility mosaic; major insect pests are pod borers, mainly *Helicoverpa, Maruca* and pod fly.

Eighty five (85) cultivars based on germplasm and improved breeding lines developed by ICRISAT have been released in India (23), Kenya (9), Tanzania (7), Myanmar (7), Malawi (6), China (6), Mozambique (5), USA (3), Australia (3), Philippines (3), Venezuela (3), Nepal (2), Uganda (2), Fiji (2), Indonesia (1), Sudan (1), Sri Lanka (1), and Thailand (1). Short- and medium-duration cultivars have had a significant impact in Asian countries. Long- and medium-duration varieties with Fusarium wilt resistance and consumer-preferred large, cream-colored seeds have been well received in eastern and southern Africa.

Hybrids: A major progress was achieved with the development of the world's first commercial pigeonpea hybrid. This hybrid technology was developed after intensive research efforts spanning more than three decades and has provided the opportunity to break a decades-old yield plateau in pigeonpea. This system, in combination with natural out-crossing of the crop, was successfully used to develop economically viable hybrid seed production technology. It has been refined to suit different agro-ecologies, and the hybrids ICPH 2671 and ICPH 2740 were released in India for general cultivation by farmers.

Genome Sequencing: Pigeonpea is the first non-industrialized commodity, and only the second food legume, with a published genome sequence. Genome analysis identified more than 48,000 genes, including several hundred that are unique to the crop and are closely associated with drought tolerance. The availability of the sequence opens up new avenues for crop improvement and could help plant breeders raise yields from an average of 700 kg/ha closer to the crop's full potential, an estimated 3.5 tons/ha. In time it could also lead to the production of short-duration lines that are both temperature and photoperiod insensitive, thereby expanding the geographical range of the crop.

Pearl Millet

Pearl millet is a dependable nutritious source of food for millions in marginal agricultural areas. It is the world's hardiest warm-season cereal crop. It can survive and produce grain even on the least fertile soils in the driest regions, on highly acidic and saline soils, and in the hottest climates.

Since 1982, the year when WC-C75 – the first ICRISAT-bred open-pollinated variety (OPV) – was released for cultivation in India, 67 cultivars based on ICRISAT-bred germplasm (14 OPVs and 53 hybrids) have been released and notified in India. Five of the OPVs were developed by ICRISAT, and nine were produced by national agricultural research partners. Five of the hybrids were developed by ICRISAT, and 31 by national partners using ICRISAT-bred male-sterile lines. Seventeen hybrids developed by the private sector are also based on ICRISAT-bred male-sterile lines, or on selections made within these lines. Many other hybrids from the private sector, and several others marketed as truthfully labeled seeds, also involve some degree of ICRISAT-bred materials – both seed parents and pollen parents.

Main Threats to Production

- Such bacterial diseases as bacterial spot (*Pseudomonas syringae*) and bacterial leaf streak (*Xanthomonas campestris* pv. *pennamericanum*);
- Various fungal and pseudo-fungal diseases, especially downy mildew (caused by *Sclerospora graminicola* and *Plasmopara penniseti*), blast (caused by *Pyricularia grisea*), smut (caused by *Moesziomyces penicillariae*) , ergot (caused by *Claviceps fusiformis*) and rust (caused by *Puccinia substriata* var. *penicillariae*);
- The parasitic weed, *Striga hermonthica*, which can significantly reduce productivity;
- Insect pests, including millet head miner and stem borers;
- Parasitic nematodes; and
- Abiotic stresses such as drought, soil acidity, soil salinity, and high temperatures during the time when seedlings are just starting to grow and when the plants are flowering.

Current and Future Research

- ICRISAT scientists have conducted extensive research on resistance to the major insect pests and diseases that limit pearl millet production, as well as on improving the crop's natural tolerance to abiotic stresses, including drought, poor soils and high temperatures;
- Considerable effort has gone into the development of high yielding, early maturing pearl millet hybrids, especially for the sufficiently moist rain-fed areas (600-800 mm of annual precipitation) and irrigated areas of the semi-arid tropics of the Indian Subcontinent;
- Future research will build on the growing interest in higher yielding hybrids for Sub-Saharan Africa, building on past successes in India and on the initial hybrid vigor classification of pearl millet landraces from West Africa;
- Research will focus on exploiting the high levels of mineral micronutrients (iron and zinc) found in pearl millet grain, the crop's high optimum temperature for growth, and its increased use for alternative food products, feed, and fodder; and

- Work will also continue to develop new genetic and genomic tools for identifying and deploying favorable genes that can significantly improve grain and stover yield potential, biotic stress resistances and abiotic stress tolerances, as well as the nutritional value of pearl millet grain, green fodder and stover.

Sorghum

Sorghum is a highly reliable crop that grows well in hot, dry environments. It is "climate change-ready", and provides food security and income for millions of poor farmers living in such locations. Over half a billion people rely on sorghum as a dietary mainstay and, given its diversity of uses, as an important source of income. The grain is used mainly for food, prepared in the form of flat breads and porridges of different kinds. Sorghum stover is a vital source of fodder for livestock. Sorghum is also used for a wide range of industrial purposes, including the production of sweet syrups, as a source of starch for fermentation, and for producing biofuel. Sorghum is an extraordinarily robust and reliable crop. Given that sorghum requires less water, it is usually grown instead of maize in the hotter and drier areas of Africa, South Asia, and Central America.

Main Threats to Production

- Shoot fly, stemborer, head bug and aphid insect pests;
- Grain mold, anthracnose diseases, and leaf blight;
- Weed competition and (in Africa) the parasitic plant *Striga* spp.; and
- Abiotic stresses such as drought (especially terminal drought), high temperatures, acid soils (which are often associated with toxic levels of aluminum saturation) and low soil fertility (in terms of both macronutrients such as nitrogen and phosphorus, as well as micronutrients such as iron and zinc).

Varietal Releases: Over the past four decades, a total of 242 cultivars have been released using germplasm and ICRISAT-bred lines and hybrid parents in all regions (Asia, Eastern and Southern Africa, West and Central Africa, and Latin America). The number of cultivar releases has been highest in Asia (75), including 35 in India, closely followed by Eastern and Southern Africa (74), West and Central Africa (58), and Latin America (35). While released cultivars include both hybrids and varieties in Asia, in Africa releases were composed mostly of varieties (an exception being a hybrid released in Sudan).

Current and Future Research: Current and on-going research by ICRISAT is aimed at increasing yield potential, disease and insect pest resistance, and tolerance to such abiotic stresses as drought, heat, and various soil-related limitations. All this research is done within the context of the Institute's Inclusive Market-Oriented Development framework (IMOD) which seeks to harness markets for the rural poor. Looking ahead, research opportunities include:

- Creation of hybrids to increase yields for a wider range of production systems in Africa, building on successes in India, Mali and elsewhere (see Guinea-race Sorghum Hybrids);
- Development of improved "dual purpose" plant types for grain, as well as feed and fodder, that can increase the value of the crop and strengthen the integration of animal husbandry with crop production, resulting in higher and more stable incomes while improving soil health through increased organic matter cycling; and
- Efficient use of the crop's rich genetic diversity for the improvement of sorghum and other cereals, made possible by the availability of the full genome sequence and other genetic and genomic tools that will enable identification and transfer of favorable alleles for stress tolerance (such as phosphorus deficiency, aluminum toxicity and terminal drought), product quality (micronutrient content, digestibility and industrial qualities) and superior agronomic performance.

Groundnut

Rich in protein and edible oil, groundnut (*Arachis hypogaea)* is central to the financial and nutritional well-being of hundreds of millions of farmers and consumers across the semi-arid tropics. Nutrition experts point out that groundnut provides over 30 essential nutrients and is an excellent source of niacin, fiber and vitamin E. Rich in anti-oxidants, the crop is naturally free of trans-fats and contains about 25% protein, a higher proportion than any other true nut.

Increasingly, however, its greatest value is as an income generator. At many locations in West Africa, groundnuts are largely grown by women for the benefit of their families and are used to pay school fees and purchase household essentials. Many of the best improved varieties, some of which are twice as productive as the lines they replaced, were selected by women farmers in Mali and Niger working with national and international plant breeders.

Threats to Production: Low yields in the semi-arid tropics are attributable mainly to the fact that groundnut is grown in marginal area under rain-fed conditions and is subject to periodic drought. The crop is highly susceptible to contamination by some 20 different mycotoxins, including aflatoxin caused by a fungus *Aspergillus flavus.* Aflatoxin is extremely hazardous to human health and is especially harmful to the physical and mental development of young children. Over time, exposure to alflatoxin-infected foods can lead to hepatitis, immune system suppression and liver cancer.

The risks associated with aflatoxin contamination have led industrialized countries to establish rigorous quality standards that often deny developing country farmers the opportunity to export. In West Africa, where groundnut is largely produced by women, export prohibitions have important implications for family well-being and frequently prevent farmers from purchasing resources that might otherwise be used to increase productivity. Researchers have developed a simple, inexpensive testing kit to identify the presence of aflatoxin on stored grain. The kits use reagents that are commonly available in developing countries and generate results comparable to the more expensive techniques used in industrialized country laboratories. They are currently being used in India, Kenya, Malawi, Mali and Mozambique in an effort to improve consumer safety and help farmers re-enter the European market.

The kits also support national plant breeding programs and play a central role in the development of new varieties that are now beginning to displace the older cultivars that have long dominated the Indian market. One such drought-tolerant variety named after the Hindu goddess Devi (ICGV 91114) ranks among India's most popular groundnut varieties. Economic projections indicate that annual benefits in just one district where the new varieties are grown should surpass $ 500 million by 2020.

7. IFPRI

The International Food Policy Research Institute (IFPRI) headquartered at Washington DC, established in 1975, provides research-based policy solutions to sustainably reduce poverty and end hunger and malnutrition. IFPRI's strategic research aims to foster a climate-resilient and sustainable food supply; promote healthy diets and nutrition for all; build inclusive and efficient markets, trade systems, and food industries; transform agricultural and rural economies; and strengthen institutions and governance. Gender is integrated in the entire Institute's work. Partnerships, communications, capacity strengthening, and data and knowledge management are essential components to translate IFPRI's research from action to impact. The Institute's regional and country programs play a critical role in responding to demand for food policy research and in delivering holistic support for country-led development. IFPRI collaborates with partners around the world.

8. IITA

The International Institute of Tropical Agriculture (IITA), headquartered at Ibadan, Nigeria is one of the world's leading research partners in finding solutions for hunger, malnutrition, and poverty. Its award-winning agricultural research for development (R4D) addresses the needs of the poor and vulnerable in the tropics. It works with public and private sector partners to enhance crop quality and productivity, reduce risk to producers and consumers, and generate wealth from agriculture.

The Institute's R4D covers the themes, biotechnology and genetic improvement, natural resource management, plant production and plant health, and social science and agribusiness. For the last 45 years, IITA's agricultural R4D has delivered over 70 per cent of CGIAR impact in sub-Saharan Africa. It has achieved this impact by focusing on key tropical food crops, such as banana and plantain, maize, cassava, soybean, cowpea, tree crops, and yam.

IITA is an award-winning, research-for-development (R4D) organization, providing solutions to hunger, poverty, and the degradation of natural resources in Africa. Since 1967, IITA has worked with international and national partners to improve livelihoods, enhance food and nutrition security, increase employment, and preserve natural resource integrity.

To help address poverty and the high number of unemployed young men and women, IITA has initiated a Youth Agripreneurs program to create jobs by making agriculture and agribusiness appealing to the youth. The program is integrated into the Business Incubation Platform (BIP), the technology delivery arm of IITA, which serves as a model to stimulate product development and to provide opportunities for market expansion.

Genebank: Preserving biodiversity outside of its natural environment is done in the Genebank. IITA's Genebank holds over 28,000 accessions of plant material or germplasm, of major African food crops. This germplasm is held in trust on behalf of humanity under the sponsorship of the United Nations. It is distributed without restriction for use in research for food and agriculture. Started in the mid-seventies the Genebank helps in crop improvement and also provides "seeds of hope" for people affected by flood, fire, wars, and other disasters.

The main crops stored in the Genebank are cowpea, cassava, plantain and banana, yam, soybean, bambara groundnut, and maize. In addition, substantial collections of wild cowpea relatives and miscellaneous legumes have been collated over the past 30 years. More recently a small collection of African yam bean, an underused legume, has been put together.

By far the most important crop in the Genebank is the cowpea. The IITA Genebank holds the world's largest and most diverse collection of cowpeas with 15,122 unique samples from 88 countries, representing 70% of African cultivars and nearly half of the global diversity. This incredible collection makes IITA integral in the protection of the cowpea species.

9. ILRI

The International Livestock Research Institute (ILRI), co-hosted by Kenya and Ethiopia is headquartered at Nairobi, Kenya and Addis Ababa, Ethiopia. It works with partners worldwide to enhance the roles that livestock play in food security and poverty alleviation, principally in Africa and Asia. The outcomes of these research partnerships help people in developing countries keep their farm animals alive and productive, increase their livestock and farm productivity in sustainable ways, find profitable markets for their animal products, and reduce the risk of livestock-related human diseases. It works through a network of regional and country offices and projects in East, South and Southeast Asia, Central, East, Southern and West Africa.

ILRI is organized in the following major research programs:

- Animal and Human Health
- Feed and Forage Development
- Livestock Genetics (LiveGene)
- Policies, Institutions and Livelihoods
- Sustainable Livestock Systems
- Impact at Scale
- Biosciences Eastern and Central Africa (BecA)-ILRI Hub

Cross-cutting gender, research methods, communications, knowledge management, capacity development and partnership programs support these research programs.

ILRI leads the CGIAR Research Program on Livestock, ILRI leads one flagship and co-leads another of the CGIAR Research Program on Agriculture for Nutrition and Health, and ILRI contributes substantially to the CGIAR Research Program on Climate Change, Agriculture and Food Security and several other CGIAR research programs and platforms.

10. CIMMYT

The International Maize and Wheat Improvement Center, known by its Spanish acronym, CIMMYT, headquartered near Mexico City is a non-profit research and training organization with more than 400 partners in over 100 countries. CIMMYT is the global leader in publicly-funded maize and wheat research and related farming systems. The Center works with its partners throughout the developing world to sustainably increase the productivity of maize and wheat cropping systems, thus improving global food security and reducing poverty. For this it applies the best of agronomy practices and breeding, socioeconomics, agricultural extension, and capacity building to create sustainable solutions with lasting impact and a strong focus on climate change, hunger and nutrition, rural community development, and the environment. CIMMYT's germplasm bank is home to humanity's largest collection of maize and wheat varieties made freely available to scientists, researchers and farmers around the world.

CIMMYT works throughout the developing world to improve livelihoods and foster more productive, sustainable maize and wheat farming systems. Its portfolio squarely targets critical challenges, including food insecurity and malnutrition, climate change and environmental degradation. Through collaborative research, partnerships, and training, the center helps to build and strengthen a new generation of national agricultural research and extension services in maize- and wheat-growing nations. As a member of the CGIAR System composed of 15 agricultural research centers, CIMMYT leads the CGIAR Research Programs on Maize and Wheat, which align and add value to the efforts of more than 500 partners.

Turning Research into Impact

- CIMMYT generates benefits of $ 3.5-4 billion annually.
- CIMMYT alumni include a Nobel Peace Prize laureate and three World Food Prize winners.
- CIMMYT's success depends on the longstanding partnerships and trust of public agricultural research systems, private companies, advanced research institutes and academia, and non-governmental and farmer organizations.
- More than 70 percent of the wheat grown in developing countries and more than 50 percent of improved maize varieties derive from CIMMYT breeding materials.
- More than 10,000 scientists have trained at CIMMYT and gone on to become leaders in their own countries. The center empowers thousands of students, extension workers and farmers through courses, workshops and field days.
- CIMMYT crop-breeding research begins with its Germplasm Bank, a remarkable living catalog of genetic diversity comprising over 28,000 unique seed collections of maize and over 140,000 of wheat. From its breeding

programs, each year CIMMYT sends half a million seed packages to 600 partners in 100 countries. CIMMYT's germplasm bank is home to humanity's largest collection of maize and wheat varieties made freely available to scientists, researchers and farmers around the world. With researchers and farmers, the center also develops and promotes more productive and precise maize and wheat farming methods and tools that save money and resources such as soil, water, and fertilizer.

11. CIP

The International Potato Center, known by its Spanish acronym CIP, headquartered at Lima, Peru was founded in 1971 as a root and tuber research-for-development institution to deliver sustainable solutions to the pressing world problems of hunger, poverty, and the degradation of natural resources. CIP is truly a global center, with headquarters in Lima, Peru and offices in 18 countries across Asia, Africa, and Latin America. Working closely with partners worldwide, CIP seeks to achieve food security, increased well-being, and gender equity for poor people in root and tuber farming, and food systems in the developing world. CIP furthers its mission through rigorous research, innovation in science and technology, and capacity strengthening among farmers and partners.

CIP and its partners work across the agricultural spectrum. From the laboratory to the market place, CIP biologists, entomologists, agronomists, nutritionists, and social scientists conduct research and carry out projects on the Potato and other roots and tubers.

CIP's research originally focused on the potato, which originated and was first domesticated in the Andean Highlands. Produced in over 100 countries, the potato is the world's third most important food crop after rice and wheat. The potato is a resilient tuber that grows underground making it more robust to climate changes. Together with roots, tubers play a critical role in the global food system, especially in the developing world. They are among the most commonly consumed food staples and provide one of the cheapest sources of energy and vital nutrients. Additionally, they grow in marginal conditions with relatively few inputs and simple techniques – making them ideal "climate-smart" crops.

Sweet-potato is a cheap, nutritious solution for developing countries needing to grow more food on less land for rapidly multiplying populations. The orange-fleshed varieties of sweet potato can play a key role in alleviating vitamin A deficiency, which is rampant among children in Asia and sub-Saharan Africa. Roots and tubers share important scientific similarities. They are vegetatively-propagated, meaning that they are not grown from seed but from cuttings or clones. As these crops are often locally grown and consumed, they are less affected than grains by food price fluctuations.

Programs

CIP's programs focus on six strategic objectives. Three programs are in the research and development space and designed to deliver shorter term solutions to food security in the target commodities and geographies by going to scale with flagship technologies. Two programs address more upstream research for development with the intention of delivering future research outputs, through the discovery flagships, representing longer term solutions for development. A sixth program, on biodiversity conservation and use, underlines continuing commitment to protect and utilize the world potato and sweet potato collections.

The Center's global priorities include sustaining root and tuber biodiversity; breeding more nutritious, adaptable, pest-and-disease-resistant varieties; and building resilient agro-economic-social systems for marginal populations in developing countries.

CIP's six programs include:

1. Resilient Nutritious Sweet potato
2. Agile Potato for Asia
3. Potato Seed for Africa
4. Game Changing Solutions
5. Resilient Food Systems
6. Conserving Biodiversity for the Future: The Genebank

Targeted Regions

Using the "Pro-Poor Research and Development" model, CIP completed a rigorous targeting exercise to identify regional priorities. The first step was defining the agro-ecological regions where potato and other root and tuber cultivation are widespread among poor people, and where increasing productivity is most likely to enhance their livelihoods. These data were then combined with an analysis of livelihood indicators (income per capita, nutritional status, child mortality rates, maternal mortality, etc.).

CIP's targeted regions include:

1. Latin America and the Caribbean
2. Sub-Saharan Africa
3. Asia

Building partnerships, promoting gender equity, and developing climate smart solutions are themes common to all CIP research.

12. The International Rice Research Institute (IRRI)

IRRI headquartered in Los Baños, Philippines is the world's premier research organization dedicated to reducing poverty and hunger through rice science; improving the health and welfare of rice farmers and consumers; and protecting the rice-growing environment for future generations. IRRI is an independent, nonprofit, research and educational institute, founded in 1960 by the Ford and Rockefeller foundations with support from the Philippine government. The institute, headquartered in, Manila has offices in 17 rice-growing countries in Asia and Africa, and over 1,000 staff. Working with in-country partners, IRRI develops advanced rice varieties that yield more grain and better withstand pests and diseases as well as flooding, drought, and other harmful effects of climate change.

Working with in-country partners, IRRI develops advanced rice varieties that yield more grain and better withstand biotic stress as well as flooding, drought, and other harmful effects of climate change. More than half of the rice area in Asia is planted to IRRI-bred varieties or their progenies. The institute develops new and improved methods and technologies that enable farmers to manage their farms profitably and sustainably, and recommends rice varieties and agricultural practices suitable to particular farm conditions as well as consumer preferences. IRRI assists national agricultural research and extension systems in formulating and implementing country rice sector strategies.

IRRI works worldwide with all research institutions that share its goal. In particular, IRRI works with the national agricultural research and extension systems of the countries where its target beneficiaries live. Rice research has no political boundaries. IRRI searches for new solutions to both old and emerging problems through personal and institutional efforts and through partnerships with farming communities and other institutions, both public and private.

IRRI is a member of the CGIAR Consortium, a global research partnership committed to a food-secure future. The Institute is the lead center for the CGIAR Research Program on RICE (RICE CRP).

IRRI is also a member of the following CGIAR research programs:

CGIAR Research Program on Maize

CGIAR Research Program on Climate Change, Agriculture and Food Security (CCAFS)

Key Research Areas

- Breeding for marginal environments: Developing rice varieties resilient and adaptive to biotic and abiotic stresses for non-ideal rice growing areas.
- Breeding for favorable environments: Optimizing rice varieties and cropping systems for intensive rice-growing areas.
- Quantitative genetics: Studying rice measurable phenotypes that depend on the cumulative actions of many genes and the environment.
- Seed and delivery systems, and germplasm evaluation: Continuous improvement of systems for evaluation and delivery of rice seeds of new varieties to farmers.
- Host-resistance: Discovering and applying the resistance of rice plants to various environmental stresses.

Research Network

- C4 Rice Project
- Central Systems Initiative for South Asia (CSISA)
- CGIAR Research Program on Climate Change, Agriculture and Food Security (CCAFS)
- CGIAR Research Program on Rice (Rice CRP)
- Climate Change Adaptation in Rain-fed Areas (CCARA)
- Climate Change Affecting Land Use in Mekong Delta (CLUES)
- Closing Rice Yield Gaps in Asia With Reduced Environmental Footprint (CORIGAP)
- Consortium for Unfavourable Rice Environment (CURE)
- Council for Partnerships on Rice Research in Asia (CORRA)
- Direct Seeded Rice Consortium (DSRC)
- Hybrid Rice Development Consortium (HRDC)
- International Network for Genetic Evaluation in Rice (INGER)
- International Network for Quality Rice (INQR)
- Rice and Climate Change Research
- Rice Straw Project
- Stress Tolerant Rice for Africa and South Asia (STRASA)
- Sustainable Intensification Innovation Lab (SIIL)
- Sustainable Rice Platform (SRP)
- Temperate Rice Research Consortium (TRRC)

Rice Genebank

The T.T. Chang Genetic Resources Center at IRRI in The Philippines is the largest collection of rice diversity in the world, with more than 130,000 accessions, including genetic stocks, landraces and wild relatives. IRRI has a strong focus on seed conservation research, largely directed to greater longevity of seeds in storage and improved precision of predicted seed longevity. The T.T. Chang Genetic Resources Center is actively working to improve efficiency, for example by testing an automated seed sorter that will allow seed processing to take place overnight and thus free staff for other routine operations.

The genetic resources center also has a molecular genetics team working on genetic diversity and biosystematics, and using genetic markers to find beneficial versions of genes. Data are managed, maintained and made available through the International Rice Genebank Collection Information System (IRGCIS). The Germplasm Health Unit is well established and is a recognized component of the Philippines quarantine and phytosanitary certification system with responsibilities to the Philippines Government as well as to IRRI.

13. International Water Management Institute (IWMI)

IWMI headquartered at Colombo-Sri Lanka is a non-profit, scientific research organization focusing on the sustainable use of water and land resources in developing countries. IWMI works in partnership with governments, civil society and the private sector to develop scalable agricultural water management solutions that have a real impact on poverty reduction, food security and ecosystem health. Headquartered in Colombo, Sri Lanka, with regional offices across Asia and Africa, IWMI is a CGIAR Research Center and leads the CGIAR Research Program on Water, Land and Ecosystems (WLE). The International Water Management Institute (IWMI) is a non-profit, scientific research organization focusing on the sustainable use of water and land resources in developing countries.

14. The World Agro-forestry (ICRAF)

International Centre for Research on Agro-Forestry headquartered at Nairobi, Kenya is a centre of scientific excellence that harnesses the benefits of trees for people and the environment. Leveraging the world's largest repository of agro-forestry science and information, it develops knowledge practices, from farmers' fields to the global sphere, to ensure food security and environmental sustainability.

ICRAF is the only institution that does globally significant agro-forestry research in and for all of the developing tropics. Knowledge produced by ICRAF enables governments, development agencies and farmers to utilize the power of trees to make farming and livelihoods more environmentally, socially and economically sustainable at scales.

Trees play a crucial role in almost all terrestrial ecosystems and provide a range of products and services to rural and urban people. As natural vegetation is cleared for agriculture and other types of development, the benefits that trees provide are best sustained by integrating trees into agriculturally productive landscapes — a practice known as agro-forestry.

15. World-Fish

World-Fish headquartered at Penang, Malaysia is an international, non-profit research organization that strives to strengthen livelihoods and enhance food and nutrition security by improving fisheries and aquaculture.

Fish is vital source of food, nutrition and income for millions around the world. Around 800 million people depend on fishing and aquaculture for their livelihoods. More than one billion poor people obtain most of their animal protein from fish, which are rich in micronutrients and essential fatty acids.

However, climate change, water and land scarcity, and ecosystem decline pose real threats to the sustainable production of fish. The very poor, who often rely on fishing as a primary source of income, are particularly vulnerable when fish stocks decline. Postharvest losses are a widespread challenge, with 27%-39% of fish caught going to waste globally.

Recognizing the power of aquaculture and fisheries, and the sectors' challenges, WorldFish focuses its research on sustainable aquaculture, resilient small-scale fisheries and enhancing the contributions of fish to nutrition of the poor. Using a multidisciplinary approach, it strives to help those who can most benefit – poor producers and consumers, women and children.

With a 40-year track record, WorldFish uses its expertise to develop technological innovations, strengthen institutions and policies, and deliver transformational impacts. This is done in partnership with communities, research innovators, entrepreneurs and investors.

WorldFish is a member of CGIAR and leads the CGIAR FISH Research Program. WorldFish is headquartered in Penang, Malaysia and it operates regional offices in Bangladesh, Cambodia, Egypt, Myanmar, Solomon Islands and Zambia. Its research is currently underway in more than 15 countries around the developing world in Africa, Asia and the Pacific.

Focal Areas for Research

- Sustainable aquaculture: Enable sustainable increases in livelihoods from aquaculture production without creating adverse socio-economic or environmental impacts.
- Value chains and nutrition: Increase the availability, access and consumption of nutrient-rich, safe fish, especially for women of reproductive age, infants and young children.
- Resilient small-scale fisheries: Secure and enhance the contribution of small-scale fisheries to poverty reduction and food security in priority geographies.

Cross-Cutting Research Questions

- Entrepreneurship: How can youth be engaged in fisheries and aquaculture value chains, and how can they help diffuse innovative ideas and knowledge within households?
- Gender equity: How can enabling women's access to productive assets and resources help boost agricultural production and reduce poverty and hunger?
- Climate change: How will climate change affect fisheries and aquaculture in developing countries, and how can adaptive capacity be built?

Recognizing the power of aquaculture and fisheries, and the sectors' challenges, WorldFish focuses its research on sustainable aquaculture, resilient small-scale fisheries and enhancing the contributions of fish to nutrition of the poor. Using a multidisciplinary approach, it strives to help those who can most benefit – poor producers and consumers, women and children.

With a 40-year track record, WorldFish uses its expertise to develop technological innovations, strengthen institutions and policies, and deliver transformational impacts. This is done in partnership with communities, research innovators, entrepreneurs and investors. WorldFish is a member of CGIAR and leads the CGIAR FISH Research Program.

25

Statistical Tools in Plant Breeding

Statistics

Statistics may be defined as the mathematical analysis, based on the theory of probability, of experimental data in an attempt to summarize or to describe them so that conclusions can be drawn with respect to the phenomena that supply the data. It is the branch of mathematics concerned with collection, classification, analysis, and interpretation of numerical facts, for drawing inferences on the basis of their quantifiable likelihood (probability). Statistics can interpret aggregates of data too large to be intelligible by ordinary observation because such data (unlike individual quantities) tend to behave in regular, predictable manner. It is subdivided into descriptive statistics and inferential statistics. Statistics is also defined as a discipline that allows researchers to evaluate conclusions derived from sample data. Plural of statistic is also statistics. In practice, statistics refers to a scientific approach used to:

- Collect data.
- Interpret and analyze data.
- Assess the reliability of conclusions based on sample data.

The development of statistics arose out of a need to assist researchers in those areas where the laws of cause and effect are not apparent to the observer and an objective approach is needed. For example, a researcher may ask which medicine is more effective - Drug A or Drug B. The researcher could use the tools of statistics to design a study to compare the two drugs. Using statistical methods, the researcher might conclude with 95% confidence that one medicine was superior to the other. Plant breeders use statistics in designing studies, analyzing results and drawing conclusions while dealing with the plant materials grown in variable environments and where environment also affects the phenotypic expression of a character. The role of statistics in plant breeding may be as follows regarding its application:

- To obtain a descriptive summary of the sample.
- To provide a means of statistical inference.
- To provide a mechanism to compare certain quantitative values.

Sample

The objective of statistical methods is to discover facts about a given population relative to one or more variables. In statistics, a sample refers to a set of observations drawn from a population. Often, it is necessary to use samples for research, because it is impractical to study the whole population. For example, suppose we wanted to know the average height of 12-year-old Indian boys. We cannot afford to measure the height of all the Indian boys of the 12-year-old to get the right answer. However, we can draw a representative sample of the 12 years old Indian boys and measure the height. The height based on the sample (sample mean) will be quite close to the height of population mean. A study of the sample reveals information about the population which it represents.

Population

In statistics, population refers to the total set of observations that can be made. For example, if we are studying the weight of adult men in a particular country, the population is the set of weights of all the men in that particular country. If we are studying the grade point average (GPA) of students at a particular college, the population is the set of GPA's of all the students at that particular college.

Statistic

A statistic is a characteristic of a sample. Generally, a statistic is used to estimate the value of a population parameter. For instance, suppose we selected a random sample of 100 students from a school with 1000 students. The average height of the sampled students would be an example of a statistic. In fact, any measurable characteristic of the sample would be an example of a statistic.

Parameter

A parameter is a measurable characteristic of a population, such as a mean or a standard deviation.

Population vs. Sample

A statistical population is the totality of the units or individuals of the interest to the researcher. A population can be small or infinitely large. A small population can be measured in its entirety but a huge population cannot be measured in its entirety. Plant breeders often deal with large population of plants and taking data on the plants of the huge population is often impractical and under such situations, the researchers obtain measurements/data from a subset of the target population. This representative subset of the population is called a sample. The data from the samples are used to infer or estimate the values one would expect to find if it were possible to take the data on all the individuals of the population. In order to draw accurate and valid conclusions about the population, the sample must be representative of the population. To obtain a representative sample, random sampling is used in which all the individuals of the designated population have an equal chance of being included in the sample. A number that describes a characteristic of a population is called a sample statistic or simply a statistic. A number that describes a population is called a population parameter or simply a parameter. The size of sample, the number of observations afforded by the sample, is represented by n. Samples are sometimes classified as large or small. Samples of size (n) less than 100 arbitrarily may be considered as small, whereas those with n = 100 or more are considered large.

Simple Statistics Commonly Used in Plant Breeding

In present days soft-wares are available to handle all kinds of statistical analysis to handle data in plant breeding programmes and also to derive several kinds of statistic to have an idea about the population parameters. However, in day to day handling of data by the plant breeders, they often encounter difficulties in interpreting the data and at times are unable to calculate required statistic using hand calculators while they have small amount of data. Sometimes, the plant breeders have to gloss over the available data and the statistical inferences drawn in a superficial manner without in depth understanding of the statistical techniques. To equip the students and plant breeders with basic operational knowledge of day to day required statistic, a few statistics and experimental designs have been explained with demonstration of calculation using small set of data for which major information has been drawn from LeClerg *et al.* (1962) and these are as follows:

Frequency Table

Once the data are tabulated, they can be arranged in a frequency table where frequencies are entered to correspond to their class values as shown in the following table (Table 25.1).

Graphical Representation of Frequency Table

Histogram is commonly used to represent frequency table. The character measured is represented along the horizontal axis (abscissa) while the frequencies are represented vertically (ordinate) to correspond to each class. The frequency table on plant height as given in Table 25.1 is plotted as given in Fig. 25.1.

Table 25.1. Frequency table for plant height in wheat.

Class value (cm)	Class frequencies	
40	1	An inspection of the frequency distribution affords a much better idea of the wheat plant height than does the original mass of the unorganized data. However, it should be evident that the classification of data into a frequency distribution has distorted them into from their original form.
45	1	
50	3	
55	4	
62	8	
70	12	
89	17	
94	15	
102	12	
112	10	
120	7	
129	6	
135	3	
140	1	
	$n = 100$	

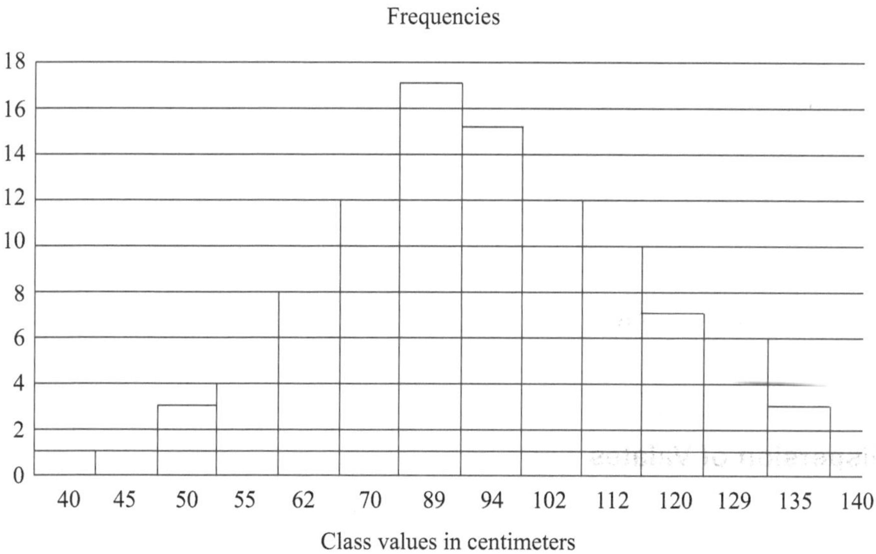

Fig. 25.1. Histogram based on data in Table 25.1

Arithmetic Mean

The arithmetic mean, commonly represented by $\bar{x}$ is equal to the sum of the individual measurements divided by their number. The simple equation follows:

$$\bar{x} = \frac{x_1 + x_2 + x_3 + \cdots + x_n}{n} \quad \text{or} \quad \bar{x} = \frac{S(x)}{n}$$

where $S(x)$ = the sum of all the variables, n = the total number of variates, $\bar{x}$ = the arithmetic mean, and $x_1, x_2, \ldots\ldots x_n$, the individual variates.

For example, if we have yield of five varieties of wheat as 40, 45, 50, 55, 60 q/ha, then the arithmetic mean would be:

$$\bar{x} = \frac{40 + 45 + 50 + 55 + 60}{5} = 50 \text{ q/ha}$$

The arithmetic mean from frequency table as given in Table 25.1 can be calculated as follows:

x	f	fx	
40	1	40	Each class value is multiplied by its frequency (f) to give fx. These values are added to arrive on $S(fx)$. $S(fx)$ is divided by the total number (n) = 100.
45	1	45	
50	3	150	
55	4	220	
62	8	496	
70	12	840	
89	17	1513	
94	15	1410	
102	12	1224	
112	10	1120	
120	7	840	
129	6	774	
135	3	405	
140	1	140	
	n = 100	$S(fx)$ = 9217	

$$\bar{x} = \frac{s(fx)}{n} = \frac{9217}{100} = 92.17$$

Dispersion of Vaiates

Samples exhibit variability or dispersion with respect to the character (variable) under study. Statistical measures or statistics commonly used to describe dispersion of the variates are the standard deviation, variance and coefficient of variability.

Standard Deviation

Standard deviation is the statistic most commonly used to measure dispersion of sample. It is defined as the square root of the mean of the squares of deviations of the variates from the arithmetic mean. This definition provides a measure of dispersion that is useful for descriptive purpose. Standard deviation is represented as follows:

$$\text{Standard deviation} = \sqrt{S(x-\overline{x})^2/n}$$

The standard deviation, as merely a measure of dispersion of the sample, is of no significant value. The primary interest for the research investigators is to obtain an estimate of the standard deviation of the population from which the sample was drawn. An unbiased estimate is obtained when the sum of deviations squared $s(x-\overline{x})^2$ is divided by $n - 1$, instead of n (number of samples). The use of $n - 1$ is explained by the fact that only $n - 1$ variates are free to vary. One value is fixed by the mean ($\overline{x}$) calculated from the sample which in turn is used to estimate the standard deviation. This situation is described by a statement that there are $n - 1$ degrees of freedom associated with the estimate. For example, in a sample of $n = 10$, the degrees of freedom are $n - 1 = 9$. The formula for the standard deviation of the sample as an uniased estimate of the population standard deviation is as follows:

$$S = \sqrt{\frac{S(x-\overline{x})^2}{n-1}}$$

The square of the standard deviation (s) is variance (s^2). Variance is a term widely used in statistical analysis.

Calculation of standard deviation is done using the following formula suitable for hand calculator calculation:

$$S = \sqrt{\frac{S(x)^2 - \{S(x)\}^2/n}{n-1}}$$

As an example, the standard deviation is calculated using direct computation procedure applicable on hand calculator for grain yield of 10 plants of wheat as shown below:

Yield per plant in grams (x)	x^2	Calculation
62	3844	
68	4624	$S = \sqrt{\frac{S(x)^2 - \{S(x)\}^2/n}{n-1}}$
58	3364	
57	3249	
67	4489	$= \sqrt{\left[\frac{42701 - \{(651)(651)\}/10}{9}\right]}$
71	5041	

68	4624	$= \sqrt{\frac{42701 - 42380}{9}}$
76	5776	
63	3969	
61	3721	$= \sqrt{\frac{321}{9}}$
$S(x) = 651$	$S(x^2) = 42701$	
		$= \sqrt{35.66}$ $= 5.97$

Standard deviation has an important application in normally distributed populations which are supposed to have a mean of 0 and variance of 1.0. Further, a range of ± 1 SD from the mean will cover 68.26% of the observations and a range of ± 2 SD will include 95.45% of the observations and ± 3 SD will cover almost the entire observations (99.70% of the observations). This is illustrated in Fig. 25.2.

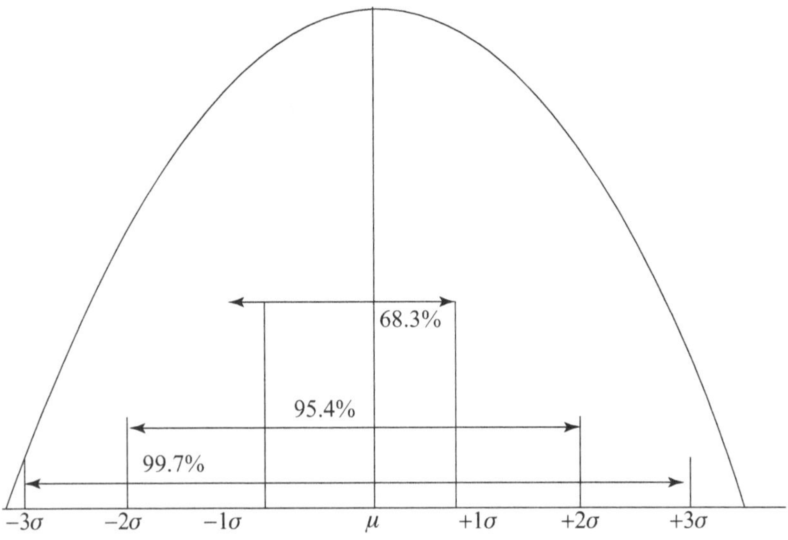

Fig. 25.2. Normal distribution curve showing population mean and the range of observations covered by mean ± 1 σ, ± 2 σ and ± 3 σ.

Coefficient of Variability (CV)

Coefficient of variability is defined as the ratio of standard deviation of a sample to the mean of the sample expressed as a percentage. This tells a relative measure of dispersion so that the variation may be compared in features expressed in different units of measurement. The formula is as follows:

$$CV = \frac{s}{\bar{x}} \times 100$$

For wheat yield data given earlier, the mean is 65.1 g and the standard deviation is 5.97. Thus in this case the CV will be:

$$CV = \frac{100s}{\overline{x}} = \frac{100 \times 5.97}{65.1} = 9.17$$

CV is useful to compare the variance for two characters which are measured in two different units.. It is generally seen that the measurements in larger units have more variance than those with the smaller units in the same population. Under such situations, the coefficient of variation facilitates the comparison because it is unit free. A CV of 10% or less is generally desirable in biological experiments.

Standard Error of Mean (SE)

Simply stated, the standard deviation of the sample means is commonly termed as standard error. For a variety of populations, the means of the samples are distributed near normal. This agreement between the distribution of many sample means and a normal distribution becomes more perfect as the sample size increases. In practice, a sample mean ($\overline{x}$) may be regarded as a random normal variable. Mathematically, it has been shown that the standard deviation of the population mean equals:

$$\sigma\overline{x} = \frac{\sigma}{\sqrt{n}}$$

Since the population standard deviation is unknown, it is estimated by the standard deviation derived from the sample. Standard deviation derived from samples means is termed as standard error of means and the square of standard error is termed as sampling variance of the mean. Samples are drawn from the population because entire population cannot be subjected to measurements due to practical difficulties. Therefore, the standard error of the theoretical distribution of means of a large number of samples is estimated as follows:

$$s\overline{x} = \frac{s}{\sqrt{n}}$$

It is obvious that the standard error of variable mean keeps on reducing as '*n*' the number of observations on which '*s*' is based becomes larger. For a hypothetical case, let us say that 80 plants of wheat in a sample had a mean plant height of 130 cm and standard deviation of 90, then the standard error (SE) will be equal to:

$$\frac{90}{\sqrt{80}} = \frac{90}{8.9} = 10.1 \text{ cm}$$

The standard error of mean indicates how precisely the population parameter has been estimated. It is attached to the mean in the presentation of results in the annual reports/publications, for example, 130 ± 10.1 as per above example.

Test of Significance with 't' Distribution

The distribution of mean divided by standard error of mean, $\frac{\bar{x}}{s/\sqrt{n}}$ is not sufficiently close to normal for small samples. The nature of distribution of $\bar{x}$ was published by "Student" in 1908. He prepared a series of tables based on distribution of s', where $s' = \sqrt{\frac{s(x-\bar{x})^2}{n-1}}$, which he designated as "Z". Later "Student" prepared tables for the distribution $\bar{x}\sqrt{n}/s$, which was designated as "*t*" by Fisher (LeClerg *et al.*, 1962). In the "*t*" table devised by R. A. Fisher in 1954, a probability of 0.05 is the 5 per cent level of significance for which odds are 19: 1. A probability of 0.01 is the 1 per cent level of significance for which the odds are 99:1 that an error of first type is avoided (the assertion of a significant difference when none exists is called an error of the first kind or false rejection-Type I and the assertion of a non-significant difference when one exists is called an error of second kind or false acceptance-Type II). In a strict sense, the test for significance of "*t*" can be applied only to the ratio of the mean difference to the standard deviation of the sample considered.

"t" Test for Means of Paired Samples

The "t" test of significance may be applied to variables which are paired. When there are n pairs of variates, the number of degrees of freedom is *n*–1. The formulae for computation of the pooled variance and "t" for paired values are as follows:

$$S^2 = \frac{\dfrac{s(d^2) - [s(d)]^2}{n}}{n-1}$$

$$t = \frac{\bar{d}}{\sqrt{s^2/n}}$$

Where n = number of pairs, d = the difference between paired values, i.e., $x1 - x2$, $\bar{d}$ = mean difference, and s^2 = the estimated variance. The statistical hypothesis is that the mean difference between populations is zero. A significant "t" value would mean the rejection of hypothesis meaning thereby that the difference between the means of the two populations is significant. The method of computation can be demonstrated with hypothetical data as follows (Table 25.2).

Table 25.2. Plant height of 9 wheat varieties in cm.

Locations	Variety V1 (cm)= x1	Variety V2 (cm) =x2	Difference between means (d)	d^2
L1	71	55	16	256
L2	74	61	13	169
L3	49	45	4	16
L4	79	71	8	64
L5	43	41	2	4
L6	48	45	3	9
L7	63	53	10	100
L8	48	41	7	49
L9	48	44	4	16
	S(x1) = 523	S(x2) = 456	S(d) = 67	S(d^2) = 683

$n = 9,\ \bar{d} = 67/9 = 7.4$

$S(d^2) = 683,\ [S(d)]^2 = (67)^2 = 4489$

$$S^2 = \frac{\frac{s(d^2)-[s(d)]^2}{n}}{n-1} = \frac{\left[\frac{683-(67)^2}{9}\right]}{8} = \frac{683-498.7}{8} = \frac{184.3}{3} = 23.0$$

$$t = \frac{\bar{d}}{\sqrt{s^2/n}} = \frac{67/9}{\sqrt{s^2/n}} = \frac{7.44}{\sqrt{23.0/9}} = \frac{7.44}{1.59} = 4.67$$

"*t*" at n – 1 (9 – 1) = 8 degrees of freedom at $P = 0.05$ is 2.306. Since calculated "*t*" value is greater than the table value of "*t*" and hence null hypothesis is rejected and it is concluded that the difference between the plant heights of two wheat varieties is significant i.e., cultivar *V*1 is significantly taller than the variety *V*2.

"t" Test for Unpaired Observations with Equal Number of Observations

This is based on LeClerg *et al.* (1962). Here number of observations in two samples are equal ($n1 = n2$) and there is no basis for pairing the variables. The formula for the estimated variance of a single observation is as follows:

$$S^2 = \frac{S(x1-\bar{x}1)^2 + (x2-\bar{x}2)^2}{2(n-1)}$$

Where $n - 1$ is the degree of freedom in a sample. The value of "*t*" is calculated as follows:

$$t = \frac{\bar{x}_1 - \bar{x}_2}{s}\sqrt{\frac{n}{2}}$$

The computation will be illustrated with the following example:

Year	Yield of variety A (q/ha) x1	Yield of variety B (q/ha) x2
1.	71.0	54.7
2.	73.9	60.6
3.	48.9	45.1
4.	78.9	71.0
5.	43,5	40.9
6.	47.9	45.4
7.	63.0	53.4
8.	48.4	41.2`
9.	48.1	44.8
	S(x1) = 523.6	S(x2) = 457.1

$\overline{x1} = 523.6/9 = 58.18$ $\overline{x2} = 457.1/9 = 50.79$

$S(x1^2) = (71.0)^2 + (73.9)^2 \ldots\ldots\ldots\ldots + (48.1)^2 = 31930.46$

$S(x2)^2 = (54.7)^2 + (60.6)^2 \ldots\ldots\ldots\ldots + (44.8)^2 = 24029.47$

$[S(x1)]^2 = (523.6)^2/9 = 30461.88$ $[S(x2)]^2 = (457.1)^2/9 = 23215.60$

$$s^2 = \frac{31930.46 - 30461.88 + 24029.47 - 23215.60}{16} = 142.65$$

$$s = \sqrt{142.65} = 11.94$$

$$t = \frac{\overline{x}1 - \overline{x}2}{s}\sqrt{\frac{n}{2}} = \frac{58.18 - 50.79}{11.94} \times \sqrt{\frac{9}{2}} = 1.31$$

The table value of "t" at $2(n - 1)$ degrees of freedom at 0.05 P is 2.120. Since the calculated value of "t" is less than the table value of "t", the null hypothesis is accepted and it is concluded that there is no significant difference between the yields of two varieties.

"t" Test for Unpaired Observations with Unequal Number of Observations

When it is assumed that the standard deviations of the populations are same, or that the samples have been drawn from the same population, the best estimate of the population variance (σ^2) is as follows:

$$\frac{S(x1 - \overline{x}1)^2 + (x2 - \overline{x}2)^2}{(n1 - 1) + (n2 - 1)}$$

where $n1$ and $n2$ are number of observations in the 2 samples respectively. The formula for "t" is as follows:

$$t = \frac{\overline{x}2 - \overline{x}2}{s}\sqrt{\frac{n1n2}{n1+n2}}$$

The method of computation can be illustrated with the following data where the yield of two wheat varieties V1 and V2 is given in q/ha.

Variety V1 (x1): 38, 40, 40, 42, 39, 35,32, 28,42, 44

Variety V2 (x2): 37, 37, 40, 40, 32, 30, 31

Variety V1 (x1)	Variety V2 (x2)
n1 = 10	n2 = 7
S(x1) = 380	S(x2) = 247
$\overline{x1} = 38$	$\overline{x2} = 35.29$
$S(x1)^2 = (38)^2 + (40)^2 \ldots\ldots + (44)^2 = 14662$	$S(x_2^2) = (37)^2 + (37)^2 + \cdots + (31)^2 = 8823$
$[S(x1)]^2 = (380)^2/10 = 14440$	$[S(x2)]^2 = (247)^2/7 = 8716$

$$S^2 = \frac{S(x1-\overline{x1})^2 + S(x2-\overline{x2})^2}{(n1-1)(n2-1)}$$

$$= \frac{(14662-14440)+(8823-8716)}{9+6} = \frac{222+107}{15} = \frac{329}{15} = 21.93$$

$$s = \sqrt{21.93} = 4.68$$

$$t = \frac{\overline{x}1 - \overline{x}2}{2}\sqrt{\frac{n1n2}{n1+n2}} = \frac{38.00-35.29}{4.68}\sqrt{\frac{70}{17}} = 1.18$$

"t" value at (n1 − 1) + (n2 − 1) = 15 degrees of freedom at 0.05 P is 2.131. Thus calculated value of "t" is less than the table value of "t" and hence null hypothesis is not rejected concluding that the varieties V1 and V2 do not differ significantly in yield.

Confidence Interval

Confidence Interval of a Mean

Use of sample mean is common to estimate the population mean. However, it does not necessarily mean that the sample mean is equal to the population mean. In most cases, the two are different and also in most cases, the population mean remains unknown. The confidence interval determined by two points, one above and one below the estimate, asserts a degree of confidence i.e., a certain probability that the population mean (μ) lies within these limits as determined by the sample. The probability that the parameter lies within these limits has been termed as the confidence coefficient. It is an arbitrarily selected probability that the confidence interval contains the population mean. The level of confidence may be represented by a probability, say 0.95 which is commonly used. If someone desires to be more confident of the inferences, 0.99 level of confidence may be chosen.

It should be clear that the confidence interval is a way of drawing an inference about a population mean. At 0.95 level of confidence, one can say that the chances are 95 in 100 that population mean falls within the confidence limit. But this conclusion is wrong as population mean is constant and has no probability distribution. Probabilities are only associated with variables. Therefore, it would be correct to state that an interval will probably contain the population mean but not the probability that the mean lies within the interval.

The method of calculation off confidence limits of a population mean is simple. The procedure requires the sample mean ($\bar{x}$), the sample variance (s^2) and the value of '$t_{0.025}$' which is available in t-table. The calculation is illustrated as follows:

($\bar{x}$) = 48.8, n = 5, s^2 = 149.7.

One can calculate $s_{\bar{x}} = \sqrt{\dfrac{149.7}{5}} = 5.472, t_{0.025} = 2.776$

The formula for limits of the confidence interval is:

$$\bar{x} \pm t_{0.025} \times S_{\bar{x}} = 48.8 \pm (2.776)(5.472) = 48.8 \pm 15.2$$

Thus, the confidence interval is 48.8 – 15.2 = 33.6 and 48.8 + 15.2 = 64.0.

This indicates that it is quite likely that the population mean is between these limits (33.6 and 64.0). It also means that one can expect to be correct 95 times in 100 and wrong 5 times in 100.

Confidence Interval of Mean Differences

The confidence interval of the difference between two population means may be calculated by a procedure similar to that for the determination of the confidence interval of a single population mean. In this procedure, the information required is as follows:

1. Number of observations
2. The degrees of freedom
3. Value of "t" at the desired confidence coefficient
4. Standard error of the difference

The data from Table 22.2 is used to illustrate the computation of the 95 per cent confidence interval for the difference between two population means. The standard error of the difference (s_d) is 2.33, $n = 9$, $t_{0.025} = 2.306$, $\overline{} = 58.18$ and $\overline{x2} = 50.79$. The difference $d = 7.39$.

The limits of the confidence interval for the difference between two population means would be:

$\overline{x1} - \overline{x2} \pm t_{0.025}\, s_d$, by substituting the values as above, one gets as:

7.39 ± (2.306) (2.33) = 7.39 ± 5.37 giving rise to confidence interval limits of 2.02 and 9.41. In other words, the difference between the two population means, μ_1-μ_2 is between 2.02 and 9.41. Since the confidence limits do not include 0, the mean difference is significant which is similar to the conclusions drawn under Table 25.2.

Regression

Simultaneous variation of two (or more) variables is understood by regression. The regression relationship represents a mathematical function which expresses the mean value expected for one variable for any given value of the correlated variable. The variable chosen as independent is customarily designated with the letter *x* and the dependent variable with the letter *y*. Thus, the study of effect of plant stand (*x*) on the yield (*y*) of any crop would be called as regression of *y* on *x*. This functional relationship gives the rate of change of yield as per unit change of plant stand. Such a relationship has been termed as "regression".

The relation between *x* and *y* from standpoint of linear regression is expressed as follows:

$\hat{y} = a + b(x - \overline{x})$ where

$\hat{y}$ = the expected value obtained for any given value of *x*.

a = *a* constant which fixes the position of the regression line and

b = the regression coefficient of *x* on *y*, which represents the slope of the regression line.

The *a* and *b* are calculated as follows:

$$a = \hat{y} - b\overline{x} \text{ and}$$

$$b = \frac{s[(x-\overline{x})(y-\overline{y})]}{s(x-\overline{x})^2}$$

The quantity $S\left[(x-\overline{x})(y-\overline{y})\right]$ represents the sum of products of deviations of *x* and *y* from their respective means.

Computation of Regression Coefficient

Regression coefficient of y on x is determined from the following formula:

$$b_{yx} = \frac{S[(x-\overline{x})(y-\overline{y})]}{s(x-\overline{x})^2}$$

Thus, the linear regression coefficient of b_{yx} is the sum of the products of the deviations of the two variables from their respective means divided by the sum of squares of the deviations of the independent variable (*x*). For calculation purposes, it is preferable to use the following formula:

$$\frac{S(xy)-s(x)\overline{y}}{s(x)^2-[S(x)]^2/n}$$

As an illustration, following data is used (Table 25.3):

Table 25.3. Number of plants and fruit yield (kg) in tomato and calculation of regression of yield on number of plants.

Number of plants (x)	Yield in kg (y)	x2	y2	xy
12	15	144	225	180
14	11	196	121	154
14	12	196	144	168
15	8	225	64	120
17	6	289	36	102
10	5	100	25	50
11	6	121	36	66
8	10	64	100	80
8	9	64	81	72
12	4	144	16	48
7	4	49	16	28
5	7	25	49	35
5	6	25	36	30
8	3	64	9	24
7	4	49	16	28
Totals 153	110	1755	974	1185

The sum of products of deviations is obtained as follows:

$$S[(x-\overline{x})(y-\overline{y})] = S(xy) - S(x)\overline{y} = 1185 - 153\left(\frac{110}{15}\right)$$

$$= 1185.00 - 1121.99 = 63.01$$

The sum of squares of deviations of the independent variable $(x-\overline{x})^2$ is calculated as follows:

$$S(x-\bar{x})^2 = \frac{S(x)^2 - [S(x)]^2}{n}$$

= 1755 – $(153)^2/15$ = 1755.00 – 1560.60 = 194.40 and thus the regression coefficient (b) = 63.01/194.40 = 0.3241.

This regression coefficient (0.3241) shows that for each increase of one tomato plant, the fruit yield increases on an average by 0.3241 kg.

After regression coefficient (b) is known, it is possible to calculate the expected values of $\hat{y}$, which corresponds to a given value of x on the basis of the following formula:

$$\begin{aligned}\hat{y} &= \bar{y} + b(x-\bar{x}) \\ &= \bar{y} - b\bar{x} + bx \\ &= 7.3333 - (0.3241)(10.2) + 0.3241x \\ &= 7.333 - 3.058 + 0.3241x \\ &= 4.0275 + 0.3241x\end{aligned}$$

For drawing regression line, three values of $\hat{y}$ may be calculated corresponding to any three values of x and these values, if calculated correctly, will fall on a straight line drawn on a graph paper. For example, the three $\hat{y}$s corresponding to x- values of 15, 10 and 5 will be as follows:

$\hat{y}$ = 4.0275 + (0.3241) (15) = 8.889

$\hat{y}$ = 4.0275 + (0.3241) (10) = 7.268

$\hat{y}$ = 4.0275 + (0.3241) (5) = 5.648

Significance of the Regression Coefficient

Significance of regression coefficient against zero can be tested by the *t*-test for which it is necessary to compute $S(y-\hat{y})^2$ which is follows:

$$S(y-\hat{y})^2 = S(y-\bar{y})^2 - \frac{[S(x-\bar{x})(y-\bar{y})]^2}{S(x-\bar{x})^2}$$

$$= (974) - (110 \times 110)/15 = 974 - 806.6667 = 167.3333$$

$$= 167.3333 - (63.01)^2/194.40 = 167.3333 - 20.4231 = 146.9102$$

The standard error of the estimate is determined as follows:

$$s\hat{y} = \sqrt{\frac{S(y-\hat{y})^2}{n-2}} = \sqrt{\frac{146.9102}{13}} = 3.36$$

Standard error of regression coefficient is as follows:

$$s_{byx} = \frac{s\hat{y}}{\sqrt{S(x-\bar{x})^2}} = \frac{3.36}{\sqrt{S(194.40)^2}} = \frac{3.36}{13.94} = 0.2410$$

The calculated value of "*t*" is $t = b_{yx}/s_{byx} = 0.3241/0.2410 = 1.34$

The table value of "*t*" at $n - 2$ i.e., 13 degrees of freedom at 5 percent level is = 2.16

Since the calculated value of "*t*" is less than the table value of "t", the regression coefficient is not significantly greater than zero.

Correlation

The degree of association between two variables is measured by correlation coefficient. It is a statistic which measures the extent of a co-relationship between two variables and is symbolized by "*r*". It ranges from –1 to +1. Mathematically correlation coefficient between the two variables is the ratio of their covariance or the sum of products to the geometric mean of their sum of squares of deviations. Thus,

$$\frac{[S(x-\bar{x})(y-\bar{y})]}{\sqrt{[S(x\quad \bar{x})\ ][s(y\quad \bar{y})\ ]}} \qquad (1)$$

$$= \frac{(S(xy) - S(x)\bar{y})}{\sqrt{[S(x)^2 - S(x)(\bar{x})][S(y^2) - S(y)(\bar{y})]}} \qquad (2)$$

The covariance or sum of products of *x* and *y*, $S[(x-\bar{x})(y-\bar{y})]$ may be obtained directly by multiplication of the corresponding mean deviations. However, the computations are much simplified through use of the following formula:

$$[S(x-\bar{x})(y-\bar{y})] = S(xy) - S(x)\bar{y}$$

Simple correlation coefficient (*r*), also known as Pearson correlation coefficient as applicable to samples unlike population correlation coefficient represented by Greek lower case letter Rho (ρ) is also given by:

$$r = \frac{Cov_{xy}}{\sqrt{Var_x Var_y}} = \frac{\dfrac{\Sigma xy - [(\Sigma x)(\Sigma y)]}{n}}{\sqrt{\dfrac{\Sigma x^2 - (\Sigma x)^2}{n}}\sqrt{\dfrac{\Sigma y^2 - (\Sigma y)^2}{n}}}$$

(correlation can be computed using directly *x* and *y*) (3)

The calculation of correlation coefficient is demonstrated with the data in Table 25.4 (LeClerg *et al.*, 1962).

Table 25.4. Numerical illustration of computation of correlation coefficient.

x	y	xy	$(x-\bar{x})$	$(y-\bar{y})$	$(x-\bar{x})^2$	$(y-\bar{y})^2$	$(x-\bar{x})(y-\bar{y})$
2.5	3.5	8.75	–4.0	–5.5	16.00	30.25	22.00
2.5	3.0	7.50	–4.0	–6.0	16.00	36.00	24.00
3.0	5.0	15.00	–3.5	–4.0	12.25	16.00	14.00
4.0	7.0	28.00	–2.5	–2.0	6.25	4.00	5.00
4.5	5.5	24.75	–2.0	–3.5	4.00	12.25	7.00
5.0	8.0	40.00	–1.5	–1.0	2.25	1.00	1.50
5.5	8.0	44.00	–1.0	–1.0	1.00	1.00	1.00
6.0	10.0	60.00	–0.5	1.0	0.25	1.00	–0.05
6.0	7.0	42.00	–0.5	–2.0	0.25	4.00	1.00
6.5	10.5	68.25	0.0	1.5	0.00	2.25	0.00
6.5	6.5	42.25	0.0	–2.5	0.00	6.25	0.00
7.5	10.0	75.00	1.0	1.0	1.00	1.00	1.00
8.0	8.0	64.00	1.5	–1.0	2.25	1.00	–1.50
8.0	10.0	80.00	1.5	1.0	2.25	1.00	1.50
8.0	12.0	96.00	1.5	3.0	2.25	9.00	4.50
8.5	13.0	110.50	2.0	4.0	4.00	16.00	8.00
9.0	12.0	108.00	2.5	3.0	6.25	9.00	7.50
9.0	14.0	126.00	2.5	5.0	6.25	25.00	12.50
9.5	13.0	123.5	3.0	4.0	9.00	16.00	12.00
10.5	14.00	147.00	4.0	5.0	16.00	25.00	20.00
S(x) = 130.00	S(y) = 180.00	S(xy) = 1310.50	0	0	107.50	217.00	140.95
$\bar{x}=\frac{130}{20}$ = 6.5	$\bar{y}=\frac{180}{20}$ = 9						

Substituting the values in correlation formula-1, we get the following:

$$r = \frac{+140.95}{\sqrt{(107.50)(217.00)}} = \frac{140.95}{(10.37 \times 14.73)} = \frac{140.95}{152.75} = 0.92$$

For machine/calculator calculation, '*r*' is easily obtained using the correlation formula-2.

$$r = \frac{1310.50 - 130.00\,(9.0)}{\sqrt{[952.50 - 130.0\,(6.5)][1837.0 - 180.0\,(9.0)]}}$$

$$= \frac{140.5}{\sqrt{107.5 \times 217.0}} = \frac{140.5}{10.37 \times 14.73} = \frac{140.5}{152.75} = 0.92$$

The following hypothetical data is used to compute correlation coefficient using equation-3 where direct values of x and y are used.

Individuals

	x	y	x^2	y^2	xy
1	3	5	9	25	15
2	4	6	16	36	24
3	5	8	25	64	40
4	6	9	36	81	54
5	7	10	49	100	70
Total	25	38	135	306	203

$$\Sigma(x-\bar{x})(y-\bar{y}) = 203 - \frac{25 \times 38}{5} = 203 - 190 = 13$$

$$\Sigma(x-\bar{x})^2 = 135 - \frac{(25)^2}{5} = 135 - 125 = 10$$

$$\Sigma(y-\bar{y})^2 = 306 - \frac{(38)^2}{5} = 306 - 288.8 = 17.2$$

$$r = \frac{13}{\sqrt{10 \times 17.2}} = \frac{13}{13.11} = 0.99$$

Test of Significance of Correlation Coefficient

The correlation coefficient is calculated by comparing calculated value of "*t*" with the table value of "*t*" at *n*-2 degrees of freedom at 5 percent level of significance. The *t*-value is calculated as follows:

$$t = \frac{r\sqrt{n-2}}{\sqrt{1-r^2}}$$

Based on data and the calculations in Table 25.4, we get the following:

$t = \frac{0.92\sqrt{20-2}}{\sqrt{1-(0.92)^2}} = \frac{3.90}{0.3919} = 9.95$ ("*t*' value at 0.05 p at 20 – 2 = 18 degrees of freedom = 2.101). Since, calculated value of "*t*" is more than the table value of "*t*", the correlation is considered to be significant.

26

Field Plot Technique and Simple Experimental Designs in Plant Breeding

Plant breeding activities basically involve development of new and superior crop varieties to replace the old ones and this naturally needs comparison of several lines/cultivars for agronomically important traits which are often quantitative in nature. Therefore, lots of environmental influences are there on the mean performance and mere numerical comparison of mean data is not sufficient to draw valid conclusions regarding the superiority of one variety over the other. Great advances have been made during the last about 100 years in the area of experimental designs and analytical procedures to facilitate such comparisons. Most commonly used designs by the plant breeders are included here for better understanding by the students and the breeders and to handle the experimental data on their own when confronted with such issues. Before applying statistical procedures on the experimental designs, it must be well understood that proper use of field plot technique is of great value to have meaningful data for further statistical analysis and use.

Field Plot Technique

Uniformity of Test Area: The area under the experiment/trial should be uniform and free from heterogeneity as far as possible. Pre-cropping the test area uniformly to a single variety of a crop will often eliminate some variability and will provide an opportunity to understand the pattern of remaining variability so that some remedial measures may be taken further to reduce the variability within the test site.

Crop Growing: Best cultural practices covering seed bed preparation, time and amount of fertilizer and irrigation application, seed rate, weed control and management of diseases and insects, etc should be adopted to raise a good crop. Protection against rodents and birds may also be necessary to have valid yield data. Crop should be carefully harvested and threshed without loss of grains due to shattering. Care should be taken that no grains are left after manual threshing/machine threshing of the harvested produce as small amount of grains left after the threshing affect the actual yield drastically.

Size and Shape of Plots: It is better to have small plots and more number of replications. Small plots are used while comparing large number of entries and relatively large plots are used while comparing small number of entries. For cereals, three row plots of 5-6 m length are common. Sometimes, number of rows could be increased to five-six. Outside rows are discarded at harvest to eliminate border effects and competition between test cultivars. Large plots could be 40 to 400 square meters. The actual size of plots will depend upon the available land and number of test genotypes and other facilities at the experiment station.

Spacing: The row to row and plant to plant spacing within rows should be as normally followed in commercial crop raising.

Check Variety: It is always necessary to include successful commercial cultivars as checks in the experimental trials. Wherever feasible, same cultivar may be used for several years to monitor the progress of a plant breeding programme. If a check variety becomes susceptible to any particular disease or pest, it should be replaced with appropriate checks. In some projects, it is usual to have an old established variety as a long term/national check and newly released cultivar as the local check for a particular location.

Basic Principles of Experimental Design

The basic principles of experimental design as presently understood were developed by R. A. Fisher and his associates at the Rothamsted Experimental Station in England. The three basic principles randomization, replication and what he termed as "local control", are in use.

Randomization: Randomization of experimental variables is essential to have a valid estimate of experimental error and minimization of bias in the results. Randomization also reduces the effect of soil heterogeneity to some extent.

Replication: Soil heterogeneity is the main source of error in conduct of the field experiments. Soil heterogeneity can be overcome to some extent by replication. Replication functions to provide an estimate of the magnitude of error to which comparisons are likely to be subjected and to diminish the experimental error. The diminution of error may be carried to any required degree of precision, provided a sufficient number of replications are used with local control. The variance of any treatment comparison decreases directly with an increase in the number of replications. The standard error of mean is proportional to the square root of the number of replications or observations as:

$$S_{\bar{x}} = \frac{s}{\sqrt{n}}$$

It is obvious that 4 replications will reduce standard error to 1/2, 9 replications to 1/3 and 16 replications to 1/4. Thus, increase in replications beyond 5 or 6, reduces the standard error very little. It is impossible to determine exact number of replications needed for any experiment. The number of replications required depends on such factors such as cost, labour, variability of the test material, probable size of the mean differences and level of desired accuracy. Generally, it is unwise to use less than four and more than six replications in field experiments. A rule-of-thumb is useful to provide for not less than 10 degrees of freedom for the error component.

Local Control: Besides, randomization and replication, a further principle of experimental design is called as local control. Experimental error is more adequately controlled and measured through division of the experimental site into a number of blocks of equal size. In simple experiments, each block or replication contains the same number of plots to which all the varieties or treatments being compared are distributed in a random manner. With this arrangement, the experimental error is minimized due to the fact that the variation between plot yields over the whole experiment can be partly attributed to a measurable amount caused by replication differences. The variability between replications is irrelevant to the experimental test when each variety occurs only once in a replication. Therefore, the variance due to replications is generally removed from the experimental error. As a result, the precision of the experiment becomes greater when a large amount of the total variability is removed in this manner.

Randomized Complete Block Design

This is probably the most commonly and widely used design in plant breeding trials. In this design, the experimental area is divided into a number of "blocks" of land which is as the number of replications. Each replication consists of as many plots as the number of varieties. The varieties are distributed in the replication/block randomly with the restriction that each variety should occur only once in each replication. The improved efficiency of a randomized complete block design over a completely randomized design is due to the fact that plots within a replication tend to be more similar than plots scattered over an entire area of the experimental plot. Replications are placed across the fertility gradient, if it is known. The computational illustration is shown based on hypothetical data as given in Table 26.1.

Table 26.1. Yield of 5 varieties (q/ha) in a completely randomized block design with four replications.

Variety	R1	R2	R3	R4	Total	Mean
A	10	9	9	8	36	9
B	11	10	10	9	40	10
C	18	17	19	18	72	18
D	14	15	13	14	56	14
E	14	13	12	13	52	13
Total	67	64	63	62	256	

Calculations are as follows:

$$\text{Correction factor (CF)} = \frac{(Sx)^2}{n} = \frac{(256)^2}{20} = \frac{65536}{20} = 3276.80$$

Total sum of squares (S. S.) = Sum of squares of individual cell values – CF.

$= [(10)^2 + (11)^2 \ldots\ldots\ldots + (13)^2] - \text{CF} = 3490.00 - 3276.80 = 213.20$

Replication S. S. = Sum of square of total of each replication/number of observations on which each total is based – CF=

$= [(67)^2 + (64)^2 + (63)^2 + (62)^2]/5 - \text{CF}$

$= (16398/5) - \text{CF} = 3279.60\text{-}3276.80 = 2.80$

Variety sum of square (S. S.) = Sum of square of total of each variety/number of observations on which each total is based-

$\text{CF} = [(36)^2 + (40)^2 + (72)^2 + (56)^2 + (52)^2]\ /4 - \text{CF}$

$= 13920/4 - \text{CF} = 3480.00\text{-}3276.80 = 203.20$

Error sum of square (Error S. S.) = Total sum of square – (replication S. S + variety S. S.)

$= 213.20 - (2.80 + 203.20) = 213.20 - 206.00 = 7.2$

Based on above sum of squares, the analysis of variance table is as given in Table 26.2.

Table 26.2. Analysis of variance of data given in Table 26.1

Source of variation	Degrees of freedom		Sum of squares	Mean squares	Variances	F
	Symbolic	Numerical				
Rep (R)	$r - 1$	3	2.80	0.93		
Variety (V)	$v - 1$	4	203.20	50.80	$\sigma^2 + r(V)$	84.66**
Error	$(r - 1)(v - 1)$	12	7.2	0.60	σ^2	
Total	$rv - 1$	19	213.20	52.33		

** Calculated value of F is more than the table value of F at 4 and 12 degrees of freedom and hence, the differences among the varieties are treated as highly significant.

Statisticians agree that where the observed F value for varieties/treatments exceeds the tabular F value at the 5 percent probability point at f1 = 4 and f2 = 12 degrees of freedom, then the varieties means differ significantly. Similarly, when the observed F values for varieties exceeds the F value at 1 percent probability, the treatment mean differences are considered as highly significant.

Standard error of means $\sqrt{\frac{EMS}{r}} = \sqrt{\frac{0.60}{4}} = \sqrt{0.15} = 0.39$

Standard error of difference between two treatment means ($SE_{\bar{d}}$) = $\sqrt{\frac{2EMS}{r}}$ (r = number of replications) $= \sqrt{\frac{(2 \times 0.60)}{4}} = \sqrt{0.30} = 0.55$

Least significant difference (LSD)/critical difference (CD) = Standard error of difference between two treatment means x "t" value at error degrees of freedom at 5 percent or 1 percent level of significance. The least-significant-difference test should not be applied unless the F-test indicates the existence of significant differences among the means. A non-significant F-test should preclude any attempt to make interpretations of differences among a set of means. Under such circumstances, the investigator should look upon widely separated means as worthy of future experiments. Based on analysis of variance (Table 26.1), the least significant difference (LSD) or critical difference (CD) is as follows:

LSD/CD = 0.55 x t value at 12 degrees of freedom at 0.05 P =0.55 × 2.179 = 1.20

LSD/CD = 0.55 × *t* value at 12 degrees of freedom at 0.01 P = 0.55 × 3.055 = 1.68.

All varieties are compared in Table 26.3.

Table 26.3. Difference in mean yields of varieties in Table 26.1.

	A	B	C	D	E
B	1				
C	9**	8**			
D	5**	4**	4**		
E	4**	3**	5**	1	

Any difference which is more than the LSD or CD, will be considered as significant or highly significant depending upon the use of "t" at 0.05/0.01 P. But for the difference between mean yield of D and E, all the differences were found to be highly significant in this example.

Dunkan Range Test

Frequently, it is desirable to make any possible combination of mean comparisons. The range test, devised by D. B. Dunkan in 1955 is appropriate for this type of evaluation of mean differences. To use this test, the means in Table 26.1 are arranged in order from the lowest to the highest as follows:

$$\frac{A}{9} \quad \frac{B}{10} \quad \frac{E}{13} \quad \frac{D}{14} \quad \frac{C}{18}$$

The following information is required to apply this test:

1. The means
2. The standard error of the mean ($S_{\bar{x}}$)
3. The degree of freedom on which the error mean square is based

The standard error of mean is calculated as follows:

$$S_{\bar{x}} = \sqrt{\frac{\text{Error mean square}}{r}} \quad (r = \text{number of replications})$$

In the experiment under consideration, there are five means and standard error of mean is as:

$$\sqrt{\frac{0.60}{4}} = 0.39 \text{ and error D. F.} = 12$$

In addition to the above information, special tables of Significant Studentized Ranges for 5 and 1 percent levels of significance are required. These are available in Standard Books on Statistics. For this experiment, appropriate "significant student ranges" (r_p) for the various size of range (p) at 5 percent level of significance are taken from the table. The various "significant studentized ranges" are each multiplied by the standard error of the mean to get what Dunkan has termed as the "shortest significant ranges" (R_p).

The above-mentioned "significant studentized range" (r_p) values (taken from the table) and the "shortest significant range" (R_p) (calculated by multiplication of standard error of mean and the r_p) are as follows at 0.05 P:

P	2	3	4	5
r_p at 5% level	3.08	3.23	3.33	3.37
R_p at 5% level	1.20	1.26	1.30	1.31

"p" is the number of variables in the range. The mean differences are tested in order as follows: The largest minus the smallest, the largest minus the second smallest up to the largest minus the second largest, then second largest minus smallest, the second largest minus the second smallest and so on up to the second smallest minus the smallest. The mean differences in the present example are as follows:

$C - A = 9^*, D - A = 5^*, E - A = 4, B - A = 1$

$C - B = 8^*, D - B = 4^*, E - B = 3^*$
$C - E = 5^*$, D - E = 1
C - D = 4*

Latin Square Design (LSD)

The Latin square design is one with double restrictions, i.e. variables (varieties or treatments) are grouped in replications ways. The variables are so arranged that each cultivar or treatment occurs only ones in the same row or column, but at random within these restrictions. Thus, the number of replications is same as the number of variables. In this design, variability due to soil heterogeneity is removed in two directions i.e., for rows and columns. In contrast, the randomized complete block design removed soil variability in only one direction as replication. The replications run linearly in two directions which are referred to as rows and columns. This design is an excellent design for (i) small number of entries and (ii) for situations where the soil variability is unknown or known to run in two directions. Normally the number of test entries does not exceed eight. For illustration of computations, hypothetical data is given in Table 26.4.

Table 26.4. Randomization and yield of 5 varieties (q/ha) in 5 × 5 Latin square design (LSD).

Row	Column 1	Column 2	Column 3	Column 4	Column 5	Row total	Row mean
1	D, 9	A, 12	B, 12	C, 10	E, 7	50	10.0
2	C, 8	E, 8	A, 14	B, 12	D, 8	50	10.0
3	A, 10	B, 10	E, 7	D, 9	C, 9	45	9.0
4	B, 11	D, 7	C, 9	E, 6	A, 12	45	9.0
5	E, 8	C, 10	D, 7	A, 12	B, 13	50	10.0
Column total	46	47	49	49	49	240	
Column mean	9.2	7.4	9.8	9.8	9.8		
	Varieties						
	A	**B**	**C**	**D**	**E**		
Variety Total	60	58	46	40	36		
Variety Mean	12.0	11.6	9.2	8.0	7.2		

The analysis of variance proceeds in the same manner as for randomized complete block design.

Correction Factor (C. F.) = $(240)^2/25 = 2304$

Total sum of squares = $[(9)^2 + (12)^2 \ldots\ldots + (13)^2] - 11520 = 2418 - 2304 = 114.0$

Row sum of squares = $[(50)^2 + (50)^2 + (45)^2 + (45)^2 + (50)^2]/\, 5 - CF$
$= 11550/5 - CF = 2310 - CF = 2310 - 2304 = 6.0$

Column sum of squares = $[(46)^2 + (47)^2 + (49)^2 + ((49)^2 + ((49)^2]/5 - CF$
$= 11528/5 - CF = 2305.6 - 2304 = 1.6$

Variety sum of square $= [(60)^2 + (58)^2 + (46)^2 + (40)^2 + (36)^2]/5 - CF$
$= 11976/5 - CF = 2395.2 - 2304 = 91.2$

Error sum of square = Total sum of square – (row sum of square + column sum of square + variety sum of square)
$= 114.0 - (6.0 + 1.6 + 91.2) = 114.0 - 98.8 = 15.2$

Based on above sum of squares, the analysis of variance is as follows (Table 26.5).

Table 26.5. ANOVA for LSD.

Source of variation	Symbolic DF	Numerical DF	Sum of squares	Mean squares	Variances	F
Row	r – 1	4	6.0	1.5		
Column	r – 1	4	1.6	0.4		
Varieties	v – 1	4	91.2	22.8	$\sigma^2 + r(V)$	17.54**
Error	(v – 1)(r – 2)	12	15.2	1.3	σ^2	

Since calculated value of F is greater than the table value of F at 4 and 12 degrees of freedom at 1 percent level of significance, it is concluded that highly significant differences exist between some of the varieties.

Standard error of means $= \sqrt{\frac{EMS}{r}} = \sqrt{\frac{1.3}{5}} = \sqrt{0.26} = 0.50$

Standard error of difference between two treatment means

$$(SE_{\bar{d}}) = \sqrt{\frac{2EMS}{r}} \quad (r = \text{number of replications}) = \sqrt{2 \times \frac{1.3}{5}} = \sqrt{0.52} = 0.72$$

LSD = t at error degrees of freedom × 0.72 = 2.179 × 0.72 = 1.5

Differences between any two varietal means exceeding 1.5 will be considered as significant.

Split Plot Design

This design is used when two factors are to be compared where the second factor is assessed with greater degree of precision than that of the first factor. The randomization is restricted within blocks in such a manner that several levels of second factor are assigned to contiguous plots with a common level of the first factor instead of scattering them over the entire replication. This is achieved by first dividing each block into main plots to be assigned to the levels of the first factor at random and next sub-dividing each main plot into sub-plots to be allotted randomly to the levels of the second factor. Such a layout in which one set of treatments are assigned to larger plots called main plots into which a replicate is divided and another set of treatments to subdivisions of the main plots called sub-plots, is termed a split-plot design.

This design should not be used when all comparisons must be made with equal precision because this design provides a more precise comparison of the sub-plot treatments and a less precise comparison of main plot treatments. Consequently, less important treatment comparisons should be allotted to the main plots. The main plot treatments are arranged as per randomized complete block design within a replication.

Statistical Analysis

This is illustrated taking 6 crosses as main plots and 3 generations of each cross as sub-plots and 2 replications. The hypothetical data is on grain yield per plant in wheat as given in Table 26.6.

Table 26.6. Wheat grain yield (g/plant).

Replication I Cross Number							
	5	3	6	2	4	1	Total
	F1 (50)	F2 (52)	F3 (20)	F2 (55)	F1 (40)	F1 (52)	
	F3 (35)	F1 (62)	F2 (25)	F3 (35)	F2 (30)	F3 (38)	
	F2 (40)	F3 (46)	F1 (28)	F1 (65)	F3 (25)	F2 (43)	
Total	125	160	73	155	95	133	741
Replication II Cross Number							
	1	3	4	5	2	6	
	F2 (40)	F2 (52)	F1 (38)	F3 (36)	F2 (51)	F3 (24)	
	F1(53)	F3 (48)	F2 (32)	F2 (38)	F3 (30)	F1 (27)	
	F3 (40)	F1 (60)	F3 (20)	F1 (52)	F1 (66)	F2 (23)	
Total	133	160	90	126	147	74	730

Let us prepare a two-way table between replications and main-plots (crosses) from the data of Table 26.6 as follows:

Table 26.7. Two-way table of replications and crosses in split-plot design (each figure is a total of 3 sub-plots).

Rep	**Crosses**						**Total**
	1	2	3	4	5	6	
I	133	155	160	95	125	73	741
II	133	147	160	90	126	74	730
	266	302	320	185	251	147	1471

Grand total = 741 + 730 = 1471

Correction factor = $(1471)^2$/Total number of observations on which the total is based
= 2163841/36 = 60106.69

Total sum of square based on Table 23.7 = $[(133)^2 + (155)^2 \ldots\ldots + (74)^2]/3 - CF$
= 191643/3 – CF = 63881.00 – 60106.69 = 3774.31

Replication sum of squares = $[(741)^2 + (730)^2]/18 - CF$ = 1081981/18 – 60106.69
= 60110.05 – 60106.69 = 3.36

S. S. due to crosses = $[(266)^2 + (302)^2 \ldots\ldots + (147)^2]/6 - CF$
= 383195/6 – CF = 63865.83 – 60106.69 = 3759.14

S. S. for main plot error (Error-a) = 3774.31 – 3.36 – 3759.14 = 11.81

In order to obtain the sum of squares for the crosses and the generations and interaction between them, it is necessary to prepare a two-way table with yields of the two replications of each factors combined as given in Table 26.8.

Table 26.8. Total yields for the crosses and the generations.

Generations	Crosses						Total
	1	2	3	4	5	6	
F1	105	131	122	78	102	55	593
F2	83	106	104	62	78	48	481
F3	78	65	94	45	71	44	397
Total	266	302	320	185	251	147	1471

Generations sum of squares = $[(593)^2 + (481)^2 + (397)^2]/12 - CF$
= 740619/12 – CF
= 61718.25 – 60106.69
= 1611.56 for 2 degrees of freedom.

Crosses x generations interaction sum of squares = $[(105)^2 + (131)^2 \ldots\ldots + (147)^2]/2$ –CF – (crosses SS + generation SS)
= 131903/2 – CF = 65951.50 – 60106.69
= 5844.81 for 17 DF – (3759.14+ 1611.56)
= 5844.81 – 5370.70 = 474.11 at 10 DF.

Also from the entire Table 23.6 of all the 36 plots,

Total S. S. (35 degrees of freedom) = $(50)^2 + (52)^2 \ldots\ldots + (23)^2 - CF$ = 66015.00 – 60106.69 =5908.31

Error-b sum of squares = Total SS based on individual data – replication SS – crosses SS – error-a SS-generation SS – crosses x generation interaction SS = 5908.31-3.36-3759.14-11.81-1611.56-474.11 = 5908.31- 5859.98 = 48.33

The above sums of squares are entered in Table 26.9 with their appropriate degrees of freedom.

Table 26.9. ANOVA for data in Table 26.6.

Sources of variation	DF		SS	MS
Replications (R)	r–1	1	3.36	1.68
Crosses (C)	c–1	5	3759.14	751.83**
Error (a) = R x C	(r–1) (c–1)	5	11.81	2.37
Generations	g–1	2	1611.56	805.78**
Crosses x generations	(c–1) (g–1)	10	474.11	47.41
Error (b) = R x C x G	c(r–1) (g–1)	12	48.33	4.03
Total	rcg–1	35	5908.31	

Main plot variance is tested against error-a variance and sub-plot variance against error-b variance.

Standard error of the average difference between two levels of main plot treatment = $\sqrt{\dfrac{2 \text{ error} - \text{‘}a\text{’ mean square}}{rk}}$ where r = number of replications and k = number of sub-plots within a main plot. In the given case, error-a mean square = 2.37, number of replications = 2 and number of sub-plots = 3.

Standard error of the average difference between two levels of main plot treatment $= \sqrt{\dfrac{2(2.37)}{6}} = \sqrt{0.79} = 0.88$

Standard error of difference between any two levels of sub-plots depends solely on the sub-plot error mean square (error-b mean square) and is:

$\sqrt{\dfrac{2 \text{ error} - \text{‘}b\text{’ mean square}}{rm}}$ where r = number of replications and m = number of main plot treatments. In the present case this $= \sqrt{\dfrac{2(4.03)}{12}} = \sqrt{0.67} = 0.82$

The LSD are calculated by multiplying the standard error of differences with the table value of "t" at error-a degrees of freedom for comparing main plot means and that of "t" at error-b degrees of freedom for comparing sub-plot means.

27

Varietal Release and Notification in India

Varietal release and notification comes under the purview of the Seeds Act which is intended to provide for regulating the quality of certain notified seeds for sale and for marketing and for the matters connected therewith. Released and notified varieties are offered for seed certification by the seed certification agencies which ensure that the varieties under certification meet certain prescribed field and seed standards failing which they stand to be rejected during the process of seed certification.

Purpose of Release

The purpose of release of cultivars is to introduce the newly evolved varieties to the public for general cultivation in the region in which it is suitable. It enables the farmers to choose cultivars for cultivation in a region. In other words, release of a cultivar is in the nature of a recommendation to the farmers for its adoption. Therefore, notification of a variety is linked to the release of a variety, though the release process itself does not have statutory cover.

Notification of Cultivars

After official release (at State as well as Central levels), the cultivars are notified under the Seeds Act so that the quality of seeds can be regulated. The main purpose of notification is to bring the seeds of a particular crop/variety under the purview of Seed Law Enforcement, mainly to empower the Seed Inspectors to verify the quality of its seeds by sampling and analysis. The notification is made by the Central Government on the recommendation of the Central Seed Sub-Committee.

The proposals for notification of a State released variety are forwarded in the prescribed format by the State government after its release in that State to the Central Seed Sub-Committee for consideration for notification. It is made clear that state can only release a variety but cannot notify the same as notification is a statutory process vested only in the central government which exercises this power through Ministry of Agriculture and Farmers' Welfare. Notification means publishing certain information in the Gazette of India which is a public journal and an authorized legal document of government of India published weekly by the Department of Publication, Ministry of Housing and Urban Affairs. As a public journal, the Gazette prints official notices from the Government of India.

In notification, an S. O. E. (Statutory Order, Extra-ordinary) is allotted to the group of varieties which are covered under that order. The typical notification order is as follows:

NOTIFICATION

New Delhi, the 19th September 2013

S.O. 2817 (E)- In exercise of powers conferred by Section 5 of the Seed Act-1966 (54 of 1966), the Central government after consultation with the Central Seed Committee is of opinion that it is necessary and expedient to regulate the quality of the seed of the varieties specified in column (3) of the Table below of the kinds specified in the corresponding entries in column (2) of the said Table, hereby declares that the said varieties of the seeds shall be the notified varieties to be sold for purpose of agriculture for the whole of India for the purpose of the said Act with effect from the date of publication of this notification in the Official Gazette, Namely-

Table

S. No.	Kind	Variety	State/Area
65	Bottle Gourd Hybrid	Santosh (Santosh-20, KBGH-20)	Delhi, Uttar Pradesh, Bihar and Jharkhand

Official Release of the Cultivars at Central and State Levels

The practice of official release of cultivars started in October, 1964 with the formation of the Central Variety Release Committee (CVRC) at the Central government level and State Variety Release Committee at State Government level. The Central Variety Release Committee functioned up to 1969 when its functions were taken over by the Central Seed Committee (CSC) established under the Seeds Act, 1966. The Central Seed Committee constituted a Central Sub-Committee on Crop Standards Notification & Release of Varieties for Agricultural Crops and a similar separate committee for Horticultural Crops to discharge the functions of release/notification, provisional notification and de-notification of cultivars at Central level, while State Seed Sub-Committees (SSSC) were asked to discharge similar functions for release at State level. Initially, there was only one central variety release committee for all the crops covering field and horticultural crops. However, later on this committee was bifurcated into two and two independent committees were notified: These are as follows:

1. **Central Sub-Committee on Crop Standards, Notification and Release of Varieties for Agricultural Crops:** This committee is chaired by Deputy Director General (Crop Science) of Indian Council of Agricultural Research. The members are drawn from Directors/Crop Coordinators of ICAR, Directors Agriculture from States, Directors Seed Certification Agencies, Managing Directors of State Seed Corporations, Representatives from Private Seed Companies, etc. The crops under the purview of this committee are common cereals, pulses, oilseeds, millets, forage, cash crops and fibre crops.
2. **Central Sub-Committee on Crop Standards, Notification and Release of Varieties for Horticultural Crops:** This committee is chaired by Deputy Director General (Horticulture) of Indian Council of Agricultural Research and members are drawn from different organizations as indicated in the central seed sub-committee on agricultural crops. The crops included for release and/or notification by this committee are vegetable crops, tubers, fruit crops, plantation crops, and ornamental plants, medicinal and aromatic plants i.e. the crops which are included in Horticulture.

The Deputy Commissioner (Quality Control) under Government of India Ministry of Agriculture and Farmers Welfare (Department of Agriculture, Cooperation and Farmers Welfare), F1211, Shastri Bhawan, New Delhi is the Member Secretary of the committee to receive the release and notification proposals from the sponsoring agencies, to prepare the agenda of the meeting, to convene the meeting and to prepare and circulate the minutes of the meetings. The venue of the meeting is normally at the office of the ICAR-DDG (CS/Horticulture). However, the meetings can be organized at any other suitable place as desired by the Chairman of the committee.

It is necessary that all the proposals must be complete in respect of enclosures viz., acknowledgement certificate regarding submission of seed sample with NBPGR, detailed morphological characters of varieties, AICRP data on disease and pests reaction, source of material and pedigree, package of practices of the cultivar, DNA fingerprinting profile, photographs on glossy sheet etc. Now it has been decided that the proposals which are not verified by the Director/Project Coordinator will not be considered by the committee. Hence submission of one copy of the proposal to the Director/ Project Coordinator is mandatory for considering the release and notification proposals of all the state and central varieties by the Central Sub-Committee on Crop Standards, Notification and Release of Varieties for Agricultural/ Horticultural Crops.

Currently 40 hard copies of the release/notification proposal along with necessary enclosures are submitted to the Dy. Commissioner (QC) for further action at his secretariat. However, now efforts are on for development of portal for online submission of the varietal release and notification proposals to save time as well saving of stationery and postage. It has also been decided that once the online submission portal is made operational, all the ICAR Institutes/ State Agricultural Universities/other organizations shall submit their proposals online and only five hard copies shall be sent to the Member-Secretary of the sub-committee with one copy to concerned Project Director/Coordinator.

Difference between Release and Notification

The release is not a statutory function. Its main purpose is to make known the details of the newly evolved cultivar to the public and also the areas for which it is found suitable for cultivation. The notification is a statutory function performed under the Seeds Act so that the provisions of the Seeds Act could be applied to regulate the quality of seeds during sale.

Advantages of Notification

Under the Seeds Act, certified seeds can be produced only of notified varieties and thus, notification precedes the certification. Therefore, notification is compulsory for production of certified seeds. Unless the variety is notified, the seed cannot be certified. Seed Law Enforcement agency notified under the Seeds Act can draw and test samples of seeds of notified varieties. Therefore, for regulation of the quality of seeds, notification is a precondition. The farmers benefit from the notification of varieties because they get seeds of assured quality of notified varieties. As specified under the Seeds Act, seeds of notified varieties can be sold after proper labeling and packing indicating the minimum specified standards. Once the variety is notified, in general, it becomes government of India asset. The morphological characters of notified varieties are documented by the Central Seed Committee so as to curtail the bio-piracy. The notification of the varieties will help to trace out its genesis. Subsidies on seeds are considered based on the notification status. Seed planning/programmes are undertaken on the basis of statistical data. Notification facilitates the seed supply of designated varieties and hybrids of different crops by private sector seed companies to the state governments through the process of competitive bidding, etc.

Procedure for Release of Cultivars

The varieties which are identified by the all India coordinated research project of the concerned crop or group of the crops are eligible for release and notification by Central Sub-committee on Crop Standards, Notification and Release of Varieties for Agricultural Crops or Horticultural Crops as the case may be. However, the cultivars which fail to be identified by the coordinated crop group meetings but perform well only in one State are eligible for release by the particular state variety release committee. These varieties which have been evolved either by State Agricultural Universities/ICAR Institutes/Private Sector Seed Companies are treated as 'State Varieties'. Such varieties have to be considered by the State Seed Sub-committee of that particular State for release. The sponsoring authority intended to release the variety should furnish the relevant information in the prescribed pro-forma given as below and submit to the co-convener of the State Seed Sub-committee for consideration and release of cultivar. The varieties released by a particular state have to be notified by the Central Sub-Committee on Crop Standards, Notification and Release of Varieties for Agricultural/Horticultural Crops subsequently. The pro-forma for submission of proposals to central sub-committee or state seed sub-committee is given below. However, these need to be checked regularly and updated with the concerned crop coordinator/project director to use the pro-forma as applicable at a particular time.

Pro-forma for Submission of Proposal of Release of Crop Variety to Central Sub-Committee on Crop Standards, Notification and Release of Varieties (Central Varieties):

The general pro-forma along with the check-list is as follows:

Index

S. No.	Item	Page no.
1.	Proforma	
2.	Summary yield data of coordinated varietal trials	
3.	Adaptability to agronomic variables	
4.	Reaction to major diseases	
5.	Reaction to insect-pests	
6.	Data on quality characteristics	
7.	Data on other important characters	
8.	Copy of recommendation of workshop	
9.	Copy of allotment letter of IC/National Identity Number by NBPGR	

Summary of Proposal (in Bullets Only)

Proforma for Submission of Proposal for Release of Crop Variety to Central Sub-Committee on Crop Standards Notification and Release of Varieties as Applicable to Agricultural Crops:

1	Name of the crop and species	:	
2	a) Name of the variety under which tested in AICRP trials	:	
	b) Proposed name of the variety	:	
3	Sponsored by (institute)	:	
4	a) Institution or agency responsible for developing variety (with full address)	:	
	b) Name of the person who helped in the development of the variety Developers Collaborators	:	
5	a) Parentage (with details of its pedigree including source from which variety/ Inbred/ A, B and R lines of hybrid have been developed)	:	
	b) Source of material in case of introduction	:	
	c) DNA profile of variety/hybrid/ inbred/A, B, R line of hybrid vis-à-vis check variety/ line		
	d) Breeding method used	:	
	e) Breeding objective	:	
6	State the varieties which most closely resemble the proposed variety in general characters	:	
7	Recommended productions ecology (rain-fed/irrigated; high/low fertility; season)	:	
8	Specific area of its adaptation (zones and states for which variety is proposed) and recommended productions ecology	:	
9	Description of hybrid/variety	:	
	a) Plant height	:	
	b) Distinguishing morphological characters	:	
	c) Maturity (range in number of days) (from seedling/ transplanting to flowering, seed to seed)	:	
	d) Maturity group (early, medium and late wherever such classification exists)	:	

	e) Reaction to major diseases under field and controlled conditions (reaction to physiological strains/ races/ pathotypes/ bio-types to be indicated wherever possible)	:			
	f) Reaction to major pests (under field and controlled condition including store pests)	:			
	g) Agronomic features (e.g. resistance to lodging, shattering, fertilizer responsiveness, suitability to early or late sown conditions, seed rate etc.)	:			
	h) Quality of produce	:			
	Grain quality	:			
	Fodder quality	:			
	i) Reaction to stresses	:			
10	Description of the parents of the hybrid	:	A line/Inbred 1	B line/Inbred 2	R line
	a) Plant height (cm)	:			
	b) Distinguishing morphological characters	:			
	c) Days to flowering	:			
	d) Days to maturity (range in number of days – from seed to seed)	:			
	e) Is there any problem of synchronization? If yes, method to overcome it	:			
	f) Reaction to major diseases (under field and controlled conditions, reaction to physiological strains/ races/bio-types/ pathotypes to be indicated wherever possible)	:			
	g) Reaction to major pests (under field and controlled conditions including store pests)	:			
	h) Agronomic features (e.g. resistance to lodging, shattering, fertilizer responsiveness, suitability to early or late sown conditions, seed rate etc.)	:			
	i) Reaction to stresses	:			
11	a) Yield data in coordinated trials (breeding, agronomy, pathology, entomology, quality etc) regional/inter regional district trials year wise (levels of fertilizer application, density of plant population and superiority over local control/standard variety to be indicated (to be attached)	:			
	b) Yield data from national, demonstration/large scale demonstrations (to be attached)	:			

12	a) Agency responsible for maintaining breeder seed	:	
	b) Quantity of breeder seed in stock (kg) Variety A line B line R line Hybrid	:	
13	Specific recommendations, if any, for seed production (e.g. staggered sowing, plating ratio of parental lines of hybrids in foundation and certified seed production, probable area of seed production)	:	
14	Vivid presentation (field view, close-up of single plant and seed/economic parts)		
15	a) Whether recommended by any workshop, seminar, conference, state seed committee etc.	:	
	b) If so, its recommendations with specific justifications for the release of proposed variety	:	
16	c) Specific area of its adaptation	:	
17	Acknowledgement of submission of seed sample of variety/hybrid/inbred/ A, B and R lines of hybrid from NBPGR and IC numbers		
18	Package of practices along with attainable yield levels		(a) Sustainability of variety for the area; (b) Selection of field/land preparation; (c) Seed treatment; (d) Sowing time; (e) Seed rate/sowing method-line sowing with row to row and plant to plant distance; (f) Fertilizer doses; (g) Weed control; (h) Disease and pest control; (i) Irrigation; (j) Harvesting; (k) Quality characteristics of the variety, if any.
19	Information on acceptability of the variety by farmers/ consumers/industry	:	
20	Any other pertinent information	:	

Signature of all contributors Signature of Head of Institution

Checklist for proforma for submission of proposal for release of crop - variety to central sub-committee on crop standards notification and release of varieties:

Details/Documents	Attached	
Parentage with details of its pedigree including source from which variety/Inbred/A, B and R lines of hybrid have been developed.	Yes	No
Source of material in case of introduction (IC/EC numbers provided by NBPGR).	Yes	No
Flow chart of details of development of variety/ parental lines of hybrids.	Yes	No
Molecular/DNA profile of variety/hybrid/A, B, R line of hybrid vis-a-vis check variety/line (details of unique amplicons that distinguish markers along with photographs.	Yes	No
Detailed description of hybrid/variety.	Yes	No
Detailed description of the parental lines of hybrid.	Yes	No
Yield data and other data on diseases, insect-pest, quality etc. from coordinated trials.	Yes	No
Yield data from national, demonstration/large scale demonstrations.	Yes	No
Specific recommendations, if any, for seed production (e.g. staggered sowing, plating ratio of parental lines of hybrids in foundation and certified seed production, probable area of seed production etc.).	Yes	No
Vivid presentation (field view, close-up of single plant and seed) with the help of photographs of the variety).	Yes	No
Recommendation of workshop, conference.	Yes	No
Acknowledgement of submission of seed sample of variety/hybrid/ A, B and R lines of hybrid submitted to NBPGR.	Yes	No
Package of practices.	Yes	No
Proforma signed by all co-authors and Head of Organization.	Yes	No
Any other pertinent information.	Yes	No

Signature of Head of Institution

Similar proforma is to be followed by the state seed sub-committee both for release and notification. The specific heads of the proforma may vary as per specific nature of the crops for example in case of horticultural crops. Those details can be obtained from concerned crop coordinator(s).

Standard Operating Procedure (SOP) for Release and Notification of Crop Varieties

1. Enter the variety/hybrid in the all India coordinated research project trial of the concerned crop by participating in the annual all India coordinated

research project group meeting. In general, one can nominate 1-4 varieties/hybrids in each crop varietal or hybrid or disease resistance trials. This is not a fixed and mandatory number and keeps on changing as per situation. While nominating the product, one has to submit station trial data for two years to demonstrate that the proposed variety or hybrid has some superiority over the existing check(s). Also ensure that the names of the proposed entries are included in the technical programme formulated during the workshop.

2. In general, there is a three tier testing in the coordinated trials of most of crops where the first year trial is designated as Initial Evaluation Trial (IET), the second year trial as Advance Varietal Trial I (AVT I) and the third year trial as Advance Varietal Trial II (AVT II). Supply the seed as per details in the technical programme to the crop Coordinator well before the dead line.
3. Private sector seed companies must be prepared to deposit testing fee @ Rs. 60,000/entry/year to the Project Coordinator as and when demanded to ensure that the testing data is made available to the sponsor. Non-compliance with this requirement may lead to withdrawal of the entry from testing or if tested holding back the trial data. The testing fee keeps on changing and is normally revised upward.
4. Any variety/hybrid which has been successfully evaluated for three years in the coordinated trials, becomes eligible for identification by the annual coordinated crop workshop based on three years data and once the product has been identified, it becomes eligible for release and notification by Central Seed Sub-Committee on Crop Standards, Notification and Release for Agricultural Crops/Horticultural Crops as the case may be under the Chairmanship of Deputy Director General (Crop Sciences/ Horticulture, ICAR, New Delhi).
5. Ensure to have a copy of the minutes/proceedings of the identification committee meeting showing that the particular variety/hybrid has been identified.
6. Submit required quantity of seed in the prescribed manner to the National Bureau of Plant Genetic Resources (NBPGR), New Delhi to obtain proper acknowledgement with National Identity number mentioning that the required seed has been deposited with the NBPGR for long term conservation. While submitting the seed to NBPGR, ensure the following compliances:
   - Deposit at least 5000 seeds in cross pollinated crops and at least 3000 seeds in self pollinated crops.

- Seed should be freshly harvested and not more than 90 days old.
- Seed should be clean, sound, healthy, and fully mature and harvested from healthy plants.
- Seed should be dried under shade immediately after harvest.
- Potential viability of the seed should be more than 85% in most crops except in cotton and vegetables where 70-80% germination is acceptable.
- Seed should not be treated with chemicals.
- Seed should be packed in a good quality paper/muslin cloth/plastic packet with proper identity and preferably be repacked in a card board box to minimize damage and moisture absorption during transit.

7. After obtaining national identity number and acknowledgement from NBPGR, proceed to prepare the release proposal as per prescribed proforma for submission to Deputy Commissioner (QC), DAC, Shastri Bhavan, New Delhi (40 copies) along with following enclosures with each copy of proposal:
   - Relevant Minutes/relevant page of proceedings of the AICVIP-annual workshop showing that the variety/hybrid has been identified for release in the annual crop workshop. Preferably, this should be endorsed by Project Coordinator/Project Director.
   - Acknowledgement of receipt of seed by NBPGR with national identity number.
   - Glassy photograph of the variety/hybrid in A4 size.
   - Package of practices of the cultivar.
8. While preparing the release proposal, following points must be taken into consideration:
   - Source and genesis of parentage should be elaborated in detail.
   - The yield data should be presented year-wise where locations should be included under each year.
   - Each yield table (IET, AVT-I and AVT-II) should have marginal means, CD and CV, frequency of occurrence, and percentage increase or decrease over the checks.
   - There should be a summary yield table combining all the years and locations, again with CD, CV, mean, weighted mean, frequency of occurrence, percentage increase or decrease over the checks.
   - Include adequate data from coordinated trials on reaction to diseases and insect pests, and abiotic stresses.

- Include data on days to first harvesting and characters related to fruit size, shape, colour, weight, etc. as applicable to vegetable crop under reference.

9. Send 40 copies of the proposal complete in all respect with all relevant enclosures to Deputy Commissioner (Quality Control), Department of Agriculture and Cooperation (DAC) with one copy to Crop Project Coordinator-All India Coordinated Research Project.
10. Follow the progress regularly through appropriate correspondence with concerned authorities and visit the relevant website frequently (www.seednet.gov.in).
11. The sponsoring breeder may be called to make a power point presentation about the proposed cultivar in the central seed sub-committee meeting.
12. Once the cultivar is recommended for release and notification by the central seed sub-committee, one should look for minutes of the meeting on www.seednet.gov.in.
13. Take a print out of the minutes as a documentary proof of the recommendation for release and notification of the cultivar.
14. After release, there is Gazette Notification under certain Statutory Order (SO) number and date and that notification also appears in the www.seednet.gov.in and this happens a few months later after the release and usually before the next meeting on release and notification.
15. Take a print-out of this notification and file for reference and official use.
16. The release proposal of varieties which are not identified for release by the concerned crop workshop but have done exceptionally well in a particular state can be routed through Director-Agriculture/Horticulture of that particular state. In this situation, the cultivar is further tested in the state at regional research stations and the complete proposal having station trial data, coordinated trial data and state trial data is put-up to Director Agriculture/Horticulture for consideration and release in that particular state. Once the proposal is approved for release by the state variety release committee, then the entire release proposal (40 copies) along with recommendation of the state variety release committee meeting is again sent to Deputy Commissioner (QC), DAC, New Delhi with copies to concerned Crop Project-Coordinator for notification. During this submission, the requirements as mentioned under SN 7 and 8 must be met and there should be strong follow-up action as mentioned in case of release and notification of cultivars by central seed sub-committee.

A few Possible Causes of Rejection of Notification Proposals of the State Variety Release Committees

- No DNA fingerprinting profile.
- No glossy colour photo in A4 size.
- High disease severity.
- No package of practices.
- Not sufficient data on nutritional contents if the varieties are claimed to be bio-fortified.
- Lack of AICRIP data particularly on reaction to diseases and insect-pests.
- Varietal ownership dispute.
- Non-submission of copy of minutes of state seed sub-committee.
- Long years gap (about more than 4 years) between the date of release by the state seed sub-committee and the submission of notification proposal to the central seed sub-committee.

A flow chart diagram on crop release and notification is presented in Fig, 27.1.

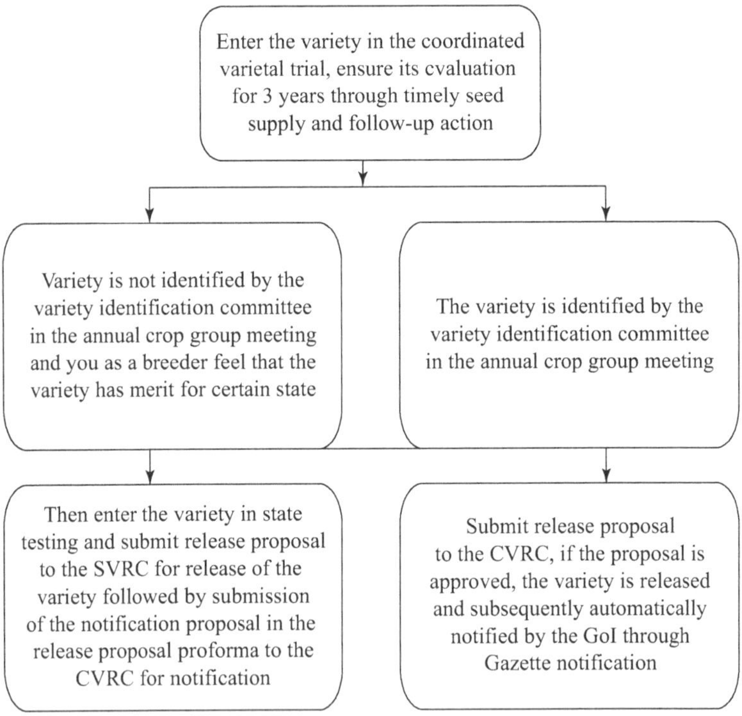

Fig. 27.1. Diagrammatic representation of variety release and notification.

Part-II
GENETICS

28

Genetics and Its History

Genetics and Gene

Genetics is defined as the branch of biology dealing with heredity and variation. The hereditary units which are transmitted from one generation to the next are called genes. Genes have been defined in various ways. These are units of inheritance located in linear order on chromosomes. The genes are considered as hereditary determinants of a specific biological function located in a fixed place on the chromosome. Chemically, gene is a segment of DNA (deoxyribonucleic acid) encoding one polypeptide and defined operationally by the cis-trans or complementation test. Gene is the functional unit of inheritance (a sequence of DNA) which is located on a certain position (locus, plural loci) on a chromosome. The locus is sometimes used interchangeably for gene. DNA, in conjunction with a protein matrix, forms nucleoprotein and becomes organized into structures with distinctive staining properties which are known as chromosomes. The chromosomes are located in the nucleus of the cell. Since genes are located on the chromosomes, in many ways behavior of genes is parallel to the behavior of the chromosomes. Two or more alternative forms of a gene are called as alleles. All genes on a chromosome are said to be linked and they all belong to the same linkage group.

Historical Developments in Genetics

Preformation and Epigenesis

In 1694 an observer endowed with too lively an imagination and a faulty microscope claimed that he could see inside human spermatozoa a miniature figure of a man- a "homunculus" Swammerdam and especially the Swish naturalist Bonnet (1720–1793) seized upon this spurious observation and used it as a basis to explain all heredity and even all creation. They thought that men's body were already preformed in the spermatozoon implying that the development of fetus and of the child required only growth to turn this tiny homunculus into a full-sized man resulting into continuity of generations. The ovists supported this proposition but insisted that the homunculus was located in the egg rather than in the spermatozoon. Later on this proposition was considered to be a figment of the imagination and K. E. von Baer (1792–1876) overthrew this doctrine of pre-formation and replaced it with the theory of epigenesist which states that the process of development consists of more than growth. Epigenesis postulates that sex cells are mere drops of structure-less liquid containing nothing whatsoever resembling the body which is produced later on. There are no pre-existing rudiments of the body. Wolff believed that the organs arise due to the action of mysterious vital forces. Von Baer was the first to see the egg cell of dog and the first to produce a detailed description of the embryonic development of a chick and he described the series of changes from fertilization to the formation of fetus and ultimately an adult body. Now is fully established that the early ideas of pre-formation and epigenesist were merely imagination and naive and far from the reality which we know today.

Darwin's Hypothesis of Pangenesis

Biologists of the nineteenth century, including Charles Darwin noted that characters observed in the individuals of one generation are often passed on to the individuals of the next generation. Since it was already established that neither eggs nor sperms contain anything like homunculus, Darwin tried to explain the operation of heredity (inheritance) by means of an interesting speculation and since his speculation had no solid factual basis, he called it "provisional hypothesis of pangenesis". Darwin's conjecture was that each cell of the body produces its own rudiments/diminutive copies called gemmules which in animals are shed into blood stream. The blood transports these gemmules into the testes and ovaries where gemmules of the different body parts are gathered together to form the gametes. When these give rise to a new organism, the different kinds of gemmules give rise to the same kinds of organs, tissues and cells which produced them in the parents. Heredity then was supposed to be due to the transmission of gemmules representing different organs, parts, and components of body.

This hypothesis in its original form was invalidated by Galton (1822-1911) who transfused blood between white and black rabbits expecting to produce a mixture of gemmules which would make the offspring of parents transfused spotted or dappled. However, this could not be observed. Nevertheless, variants of the hypothesis of pangenesis were adopted by believers in the inheritance of so called acquired characters by Lysenko in Russia.

Inheritance of Acquired Characters

Lamarck, the French Biologist (1774-1829) who was the father of the first theory of biological evolution, considered the inheritance of acquired characters to be the most important mechanism of evolutionary change and the type of evolutionary thought in which this view is accepted is known as Lamarckism. According to Lamarck, variations are induced in organisms in response to urgent need and striving on the part of individuals and through use or disuse of organs. Lamarck considered these changes as heritable. According to Lamarck, all variations were acquired and all were heritable. Darwin accepted the inheritance of acquired characters as an important though subsidiary factor in evolution and considered natural selection to be the directing agent of evolutionary process. It should be understood that all these explanations were spurious.

Weismans's Germplam Theory of Modern Genetics

August Weismann (1834-1914) challenged the popular belief on inheritance of acquired characters. He carried out controlled experiments to test the validity of the inheritance of acquired characters and developed germplasm theory which represented a forward step in understanding of heredity. He described the reproductive tissues-as the germ cells and the cells from which they arise as the germ plasm in contrast to the other tissues of the body as somatoplasm. The proponents of pangenesis believed the reproductive cells to be compounded of representative particles secreted by the somatoplasm. Weismann showed that on the contrary the germ plasm perpetuates itself and incidentally engenders the formation of the rest of the body (soma). In modern days, it amounts to the fact that the genes reproduce themselves and the rest of the body is a by-product of their self-reproduction. This explains as to why environment induced changes are not transferred to the germ plasm as would have happened if pangenesis occurred. Acquired characters cannot therefore be inherited. Weismann's famous experiments worth mentioning here is that he cut off tails of mice for 22 generations but observed that the progeny always had tails of normal length. This again reaffirmed non-validity of inheritance of acquired characters.

By the end of nineteenth century a number of important discoveries came to light which set the stage for a more precise material characterization of the source of heredity. Many details of cell structures and cell division were already known through the researches of a large number of biologists like Schleiden (1804-1881), Schwann (1810-1882), Nageli (1817-1891), Virchow (1821-1902), Flemming (1843-1915) and Buetschli (1848-1920). The cell nucleus, named and described by Robert Brown (1773-1858) in 1833, was shown by O. Hertwig (1849-1922) to be directly involved in sea urchin fertilization through union of the sperm and egg nuclei. Similarly, Strasburger (1844-1912) had shown such union to occur for plants and had invented the terms nucleoplasm and cytoplasm to refer to the protoplasmic material in the nucleus and its surrounding cell body, respectively. Dark staining nuclear threads named as chromatin were found to be dividing longitudinally during cell division as described first by Schneider (about 1873) and later by Flemming. These were shown also to pass on in equal portions to the two daughter cells by van Beneden (1845-1910). The total number of these individual threads or chromosomes remained constant in all cells of an organism except during gamete formation. In gametes, this number was reduced but later on restored during fertilization to form first embryonic cell. Thus, a link could be established between the cells and the chromosomes between generations.

By the early twentieth century, most of the morphological features of the cell had been seen under the light microscope. Mitosis and meiosis were understood. In somatic mitosis, the number of chromosomes remained identical in the daughter cells to that of the parental cell. In meiosis, gametes were found having exactly half the parental number. When the opposite gametes combined, the full chromosome number was restored. It was one step further to hypothesize that the constancy of chromosomes in a species and their precise partitioning in inheritance was a reflection of the similar constancy and inheritance pattern of more easily observable biological characters. Next step was full understanding of inheritance of traits when a series of fundamental discoveries by the Austrian monk, Gregor Johann Mendel (1822-1884) was brought to notice in 1900 by three botanists, namely, De Vries in Holland, Correns in Germany and Tschermak in Austria independently. Mendel and his law of segregation and principles of independent assortment are dealt with separately.

Lysenkoism- Bizarre Explanation Similar to the Inheritance of Acquired Characters

T. D. Lysenko, a young plant breeder from Ukraine argued that plant development and agricultural productivity could be improved by manipulating the environment. This could not be verified even when he proposed this. However, he got a powerful supporter, Joseph Stalin who was desperate to improve the Soviet Union's struggling agricultural programmes and saw an opportunity in Lysenko's ideas in the context of Marxism. The idea that environmentally induced changes could produce specific directed changes was quite interesting and appealing to Stalin. This idea was in a sense compatible with the concept of Marxism. It implied that proper social conditions would induce permanent and heritable changes in the human behavior. Because of this, Lysenko became a key member of Stalin's team and was appointed as Director of Institute of Genetics of the USSR Academy of Science in 1940. Because of support from Stalin, Lysenko suppressed all the genetic research that was based on Mendelism. Soviet scientists who were against Lysenkoism and advocated Mendelism were identified and imprisoned where many of them died. Lysenko remained a prominent figure in Soviet genetics and plant breeding until 1964. Soviet crop plant varietal improvement in agriculture based on the theory of acquired characteristics was in shambles. Soviet genetic research was not accepted internationally and very rightly. Nikita Khrushchev replaced Stalin in the mid-1950s and he also continued to support Lysenko's policies. However, Khrushchev was not as enthusiastic in the Lysenko's policy as Stalin. By the time Khrushchev was replaced, the bankruptcy of Lysenkoism was evident to all and Lysenko was removed from his post and relegated to obscurity. Incidentally, 1964 was the year when domination of Lysenko in Soviet agriculture and genetics ended, coincided with the eleventh anniversary of the discovery of the structure of DNA by Watson, Crick and Wilkins.

29

Cell, Chromosome Mitosis and Meiosis

Cell

Cells (small, membrane-bounded structures filled with a variety of chemicals in an aqueous solution that interact to give life) are the basic unit of life. Cells are capable of acquisition and use of energy, have the capacity to reproduce and respond to the environment and are able to carry out a variety of chemical reactions. They maintain a relatively constant internal environment despite changes in the external environment. There are two types of cells, namely, prokaryotes and eukaryotes.

The prokaryotic (pro = before, karyote = nucleus) cells are the simple form of cell types which include the various types of cells of bacteria and their relatives, all belonging to kingdom Monera. They possess a cell wall constructed of peptidoglycan, a substance unique to prokaryotes. Beneath the cell wall, there is a plasma membrane that encloses a cytoplasm containing DNA, RNA, proteins and various small molecules. The cytoplasm does not appear to have any distinct internal organization except the DNA sequestrated in a poorly demarcated region of the cytoplasm called a nucleoid. The nucleoid is not surrounded by any membrane. Simple structure of prokaryotic cell is shown in Fig. 29.1.

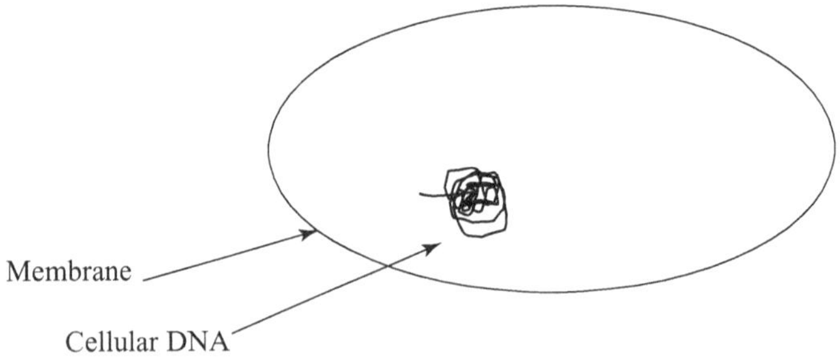

Fig.29.1. Diagrammatic sketch of a prokaryotic cell.

Eukaryotic (eu = true, karyote = nucleus) cell is generally much larger than the prokaryotic cell and far more complex in terms of it structure and functions. The single most unique and distinctive feature of a eukaryotic cell is membrane-bound true nucleus that houses the genetic material. Between the nucleus and the plasma membrane, there is cytoplasm which is an aqueous matrix containing various membranous and non-membranous organelles that perform a variety of complex metabolic functions. Sometimes, a cell wall lies just exterior to the plasma-membrane. Unlike prokaryotic cell, the eukaryotic cell wall is made of cellulose and other molecules but not peptidoglycan. Vast majority of living species including animals, plants, fungi and protists are made of eukaryotic cells. The eukaryotic cell constituents and their functions are summarized as follows:

1. **Plasma Membrane:** The cell is surrounded by plasma membrane which is constructed of a fluid phospholipid bilayer penetrated by a variety of proteins. It forms a barrier between the extracellular and intracellular matrices and regulates the flow of molecules into and out of cell. Membrane bound proteins are glycoproteins. Those proteins and glycoproteins that receive signals from the external world and communicate with the cell's interior are called receptors. A variety of organelles are distributed throughout the cytoplasm. Some of these organelles are surrounded by membranes and some remain free-floating without membranes.
2. **Nucleus:** Membrane-bound largest structure inside the cell is nucleus. Nucleus contains chromosomes which are made of DNA, RNA and various proteins.
3. **Nucleolus:** Further inside nucleus and associated with the chromosome, there are one or more nucleoli (singular = nucleolus) structures that function in the synthesis of RNA. The chromosome and nucleoli are contained in a nuclear matrix called the nucleoplasm. The nuclear membrane separating the cytoplasm from the nucleoplasm has numerous pores in it through which the molecules can travel both the ways. The RNA synthesized in the nucleus is transported to the cytoplasm through the nuclear pores.
4. **Endoplasmic Reticulum:** The vast network of tubular membranes inside the cytoplasm is known as endoplasmic reticulum (ER). Some of the ER is rough in appearance because ribosomes (cytoplasmic structures made of RNA and protein involved in protein synthesis) are bound to it. Ribosomes are assembled in the nucleolus. Rough ER is the site of protein synthesis. Proteins synthesized in the ER are transported across the ER membrane. The ER which is not complexed with the ribosomes is called smooth ER. Smooth ER functions in the synthesis of certain types of hormones and other cellular reactions.

5. **Golgi Complex:** It is an organized pile of flattened membranous sacs. It receives, stores and often modifies proteins synthesized on the rough endoplasmic reticulum. For example, the insulin protein that regulates blood sugar levels is synthesized as large precursor protein called as pro-insulin on the rough ER and is then modified to its functional form, the insulin in the golgi complex. Inherited defect in modification of pro-insulin to insulin in the Golgi complex leads to diabetes.
6. **Peroxisomes:** These are membrane-bounded vesicles containing a different set of enzymes that use molecular oxygen to oxidize organic molecules, especially fatty acids. Some of the genetically encoded enzymes in peroxisomes generate dangerously reactive hydrogen peroxide (H_2O_2) during chemical reactions and these are degraded by other enzymes. Hydrogen peroxide is a powerful oxidizing agent.
7. **Vacuoles:** These are membrane-bound, liquid filled vesicles occupying about 90% of the total cell volume and are found in plant cells. The plant vacuole may contain several dissolved molecules like salts, sugars, toxic waste and pigments that are responsible for the colour of the flowers and leaves. Vacuoles often function to maintain high internal cell water pressure in plant cells which aids in the physical support of the plant tissues.
8. **Mitochondria:** Mitochondrion has a double membrane. Proteins embedded in the inner membrane function in the reactions related to the energy transfer. The inner membrane has a specific structure and organization to carry out the reactions in a precise and orderly fashion. The inner membrane is folded into cristae that project into a matrix thus increasing the surface area of these reactions. Aging is associated with degenerative changes in the mitochondrial DNA. Mitochondria are considered as the "power plants" of the cells. They convert energy stored in energy-rich macromolecules (lipids and carbohydrates) into adenosine tri-phosphate (ATP) which is used to power virtually all the activities of the cell. The most energetically active cells have the largest number of mitochondria.
9. **Chloroplasts:** These are found only in the plant cells and are the sites of photosynthesis. Like mitochondria, chloroplasts have a double membrane system.

Diagrammatic representation of a typical eukaryotic cell is given in Fig. 29.2.

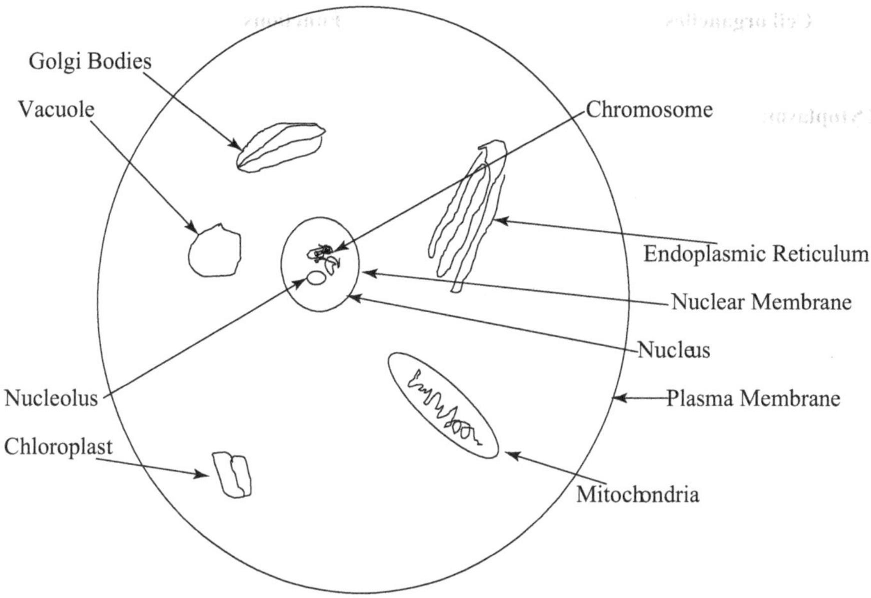

Fig. 29.2. Representative diagram of a plant cell.

In a summarized form, the functions of cellular organelles are given in Table 29.1 (Stansfield, 1969; Acquaah, 2007).

Table 29.1. Functions of cellular organelles.

Cell organelles	Functions
Cell or plasma membrane	Differentially permeable membrane through which extra-cellular substances may be selectively sampled and cell products may be liberated. It delimits the cell from its immediate external environments. The surface may contain specific receptor molecules and may elicit an immediate response.
Cell wall (plants only)	Thick cellulose wall surrounding the membrane giving strength and rigidity to the cell.
Nucleus:	Regulates growth and reproduction of the cell. It contains DNA and proteins that are condensed in strands called chromosomes (called chromatin when uncoiled).
Chromosomes	Bearers of hereditary instructions, regulation of cellular processes (seen clearly during nuclear division).
Nucleolus	May synthesize ribosome, disappears during cellular replication.
Nucleoplasm (nuclear sap)	Contains materials for building DNA and messenger molecules which act as intermediates between nucleus and cytoplasm.

Cell organelles	Functions
Nuclear membrane	Provides selective continuity between nuclear and cytoplasmic materials.
Cytoplasm:	Part of the cell excluding the nucleus and enclosed by the plasma membrane. It is made up of a colloidal material called cytosol and contains machinery for carrying out the instructions sent from the nucleus.
Endoplasmic reticulum	Greatly expanded surface area for biochemical reactions which normally occur at or across membrane surfaces. This membranous structure is of two kinds. Smooth without ribosomes and rough which has ribosomes.
Ribosomes	Organelles that contain RNA and are sites of protein synthesis.
Centrioles	Form poles for the cell division process, capable of replication, usually not seen in the plant cells.
Mitochondria	Organelles that are the site of respiration and contain DNA. These are involved in energy production (Kreb's cycle, electron transport chain, beta oxidation of fatty acids, etc.).
Plastids (plants only)	Structures for storage of starch, pigments and other cellular products. Photosynthesis occurs in chloroplasts.
Golgi bodies or apparatus	Production of cellular secretions, sometimes called dictyosomes in plants.
Vacuoles	Storage depots for excess water, waste products, soluble pigments, etc. They help to regulate water pressure in the cell and maintain cell rigidity.

Chromosomes

The genetic material is organized into structures called chromosomes (chromo = colour and soma = body). In prokaryotes, the chromosome is simply a double-stranded DNA. The eukaryotic chromosome is more complex. Each eukaryotic chromosome is a single DNA molecule packaged in a protein matrix. Chromosomal DNA is compacted by coiling and super-coiling. If all DNA molecules of a human cell were uncoiled and lined up end to end, they would measure more than a meter in length. A condensed chromosome as observed in cell division may be as short as 1/4 micron (1 micron = 0.001 mm) in fungi and birds or as long as 30 microns in *Trillium* plants. As a rule, most metaphase chromosomes fall within a range represented by about 3.5 microns in *Drosophilla*, 5 microns in man and 8-10 microns in corn. As seen under light microscope, the chromosome has certain distinct morphological features. When mitotic chromosomes are stretched out, the knoblike chromomere regions along its length give each non-homologous chromosome a morphologically distinct appearance. Some areas of chromosomes stain very darkly and these are called as heterochromatin. The other areas staining lightly are called as euchromatin. The replicated chromosome is made up of two sister chromatids joined together by a centromere. The kinetochore, a protein structure at the centromere functions in the chromosome movement during cell division. The position of the centromere varies from the middle to the end of the chromosomes. The ends of the chromosomes are called telomeres. The diagrammatic structure of a typical mitotic chromosome is shown in Fig. 29.3.

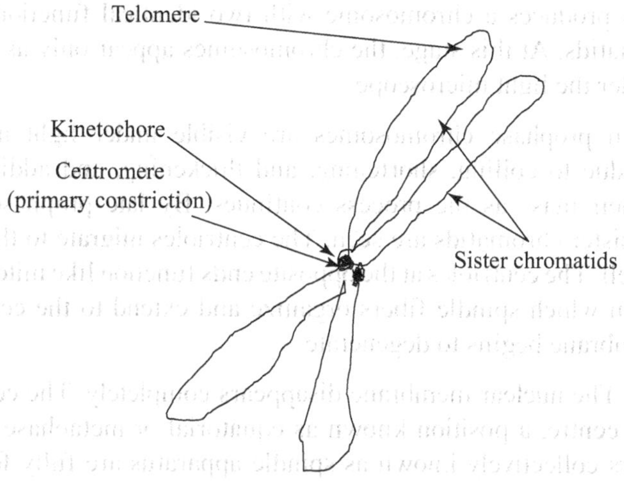

Fig. 29.3. The structure of a mitotic chromosome.

A chromosome with median centromere is known as metacentric and has almost equal length arms. A submetacentric chromosome has centromere between the central position of the chromosome and the end of the chromosome. A telocentric chromosome has centromere at or very near one end of the chromosome. Each chromosome of the genome (with the exception of sex chromosome) is numbered consecutively according to length, beginning with the longest chromosome first.

At the molecular level, several key DNA sequence elements are required to produce a stable chromosome. Various DNA sequences signal replication region, the starting points for DNA replication.

Mitosis

Mitosis, a mode of nuclear and cytoplasmic division, provides for the production of two daughter nuclei that contain identical chromosome sets and are genetically identical to each other and to the parent cell from which they arose. The function of mitosis is first to construct an exact copy of each chromosome and then to distribute through division of the original (mother) cell, an identical set of chromosomes to each of the two daughter cells. The division is divided into five phases, namely, interphse, prophase, metaphase, anaphase and telophase. Cytokinesis, the division of the cytoplasm, follows telophase. A brief description of each of these stages is as follows:

Interphase: Inter phase or inter-mitotic period is the period between division cycles. When the cell is ready to undergo mitosis, each DNA molecule which is an integral part of the chromosome, replicates or makes an exact copy of itself. This process produces a chromosome with two identical functional strands called chromatids. At this stage, the chromosomes appear only as chromatin granules under the light microscope.

Prophase: In prophase chromosomes are visible under light microscope presumably due to coiling, shortening, and thickening, and adding protein matrix to their mass as the process continues. By late prophase, the two identical or sister chromatids are seen. The centrioles migrate to the opposite ends of the cell. The centrioles at the opposite ends function like mitotic centres or poles from which spindle fibers organize and extend to the centromeres. Nuclear membrane begins to degenerate.

Metaphase: The nuclear membrane disappears completely. The centromeres move to the centre, a position known as equatorial or metaphase plate. The spindle fibers collectively known as spindle apparatus are fully formed and these are attached to the centromers.

Anaphase: The shortest of the mitotic phases, anaphase, occurs when centromere splits into two allowing sister chromatids to separate and move towards opposite poles led by the centromeres attached with the spindle fibers. Now each separated sister chromatid may be treated as a new chromosome. The appearance of this event is that of an active repulsion between sister centromeres, each dragging its chromatid along with it. The arms of each chromosome drag behind their centromeres, giving them characteristic shape depending upon the position of centromere on the chromosome. Metacentric chromosomes appear as V-shaped, sub-metacentric chromosomes as J-shaped and telocentric chromosomes as rod-shaped.

Telophase: An identical set of chromosomes assemble at each pole of the cell. The chromosomes begin to uncoil and return to an inter-phase condition. There is degeneration of spindle followed by formation of nuclear membrane and division of cytoplasm through a process called as cytokinesis which gives rise to cell plate in plant cells and indented cell furrow in animal cells. Nucleoli are reformed. The two products of mitosis are called as daughter cells which contain the same type and number of chromosomes as the mother cell and possess exactly the same genetic constitution.

The diagrammatic representation of mitosis is shown in Fig. 29.4.

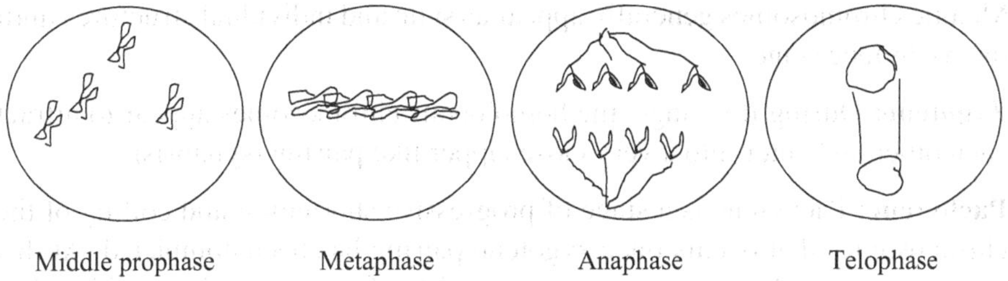

Fig. 29.4. Mitosis.

Meiosis

Meiosis is the process by which the chromosomes are separated in the sex cells and their number reduced from diploid to the haploid condition. In this process, each individual chromosome usually has a pairing mate called as homologue so that parental constitution of four chromosomes (diploid number) would consist of two homologous pairs, i.e. AA' and BB'. Through meiosis, each parental gamete becomes haploid containing two individual chromosomes, one from each pair (A or A' together with B or B'). As a result of fertilization, a diploid zygote is formed with two homologous pairs. In this process, each parent contributes one of a homologous pair of chromosomes to the offspring. Meiosis actually involves two divisions. The first meiotic division (meiosis I) is a reductional division producing two haploid cells from a single diploid cell. The second meiotic division (meiosis II) is an equational division which separates the sister chromatids of the haploid cells.

Prophase I

The prophase of meiosis I differs from the prophase of mitotic division in that homologous chromosomes come to lie side by side in a pairing process called synapsis. Each pair of synapsed homologues is called a bivalent. Since it consists of four chromatids, a bivalent is also called a tetrad. During synapsis, non-sister chromatids may break and reunite with one another in a process called crossing over. The point of exchange appears in the microscope as an overlapping region called a chiasma (plural, chiasmata).

From all indications, this initial pairing occurs during the fairly lengthy first meiotic prophase which has been sub-divided into five stages as given below:

Leptotene: Chromosomes first appear as long slender threads with many beadlike structures (chromomeres) along their length. Although considerable biochemical evidence indicates the replication of chromosome material during this period, it is usually difficult to observe any morphological duplication.

Meiotic chromosomes generally appear as sigle and individual structures until the pachytene stage.

Zygotene: During this stage, the homologous chromosomes appear to attract each other and enter into a very close zipper like pairing (synapsis).

Pachytene: Pachytene is a stage of progressive shortening and coiling of the chromosomes that occurs once zygotene pairing has been completed. At this stage, each homologous chromosome consists of two sister chromatids. Thus each bivalent (pair of two homologous chromosomes) consists of a total of four chromatids and therefore this structure is also called as tetrad. Series of exchanges of genetic material between homologous chromatids (non-sister chromatids) signifying crossing over occur during this stage.

Diplotene: At diplotene stage, each chromosome appears to be moving away from its homologous chromosome partner except the regions where crossing over appears to have taken place between homologous chromatids. These crossed areas chiasmata are X-shaped attachments between the chromosomes and appear to be the only remaining force holding each bivalent together until metaphase I.

Diakinesis: Coiling and condensation/contraction of the chromosomes continues and they become thick and heavy staining. The bivalents migrate close to nuclear membrane and appear evenly distributed. The nucleolus disappears or detaches from the associated chromosome. During later part of this stage, nuclear membrane dissolves and bivalents attach themselves by their centromere to rapidly formed spindle.

Metaphase I

In metaphase I, the chromosomes reach their most condensed state and appear relatively smooth in outline. There is movement of chiasmata towards end of each chromosome (terminalization) leaving only the single terminal attachment between formerly paired arms of homologous chromosomes. These chiasmata prevent the separation of bivalents (seen as tetrad) which now lie at the equatorial plate of the spindle attached by their respective centromeres toward opposite poles.

Anaphase I

Each centromere now splits into two and there is separation of one homologous chromosome from its partner chromosome and both the sister chromatids (dyad in contrast to tetrad) are dragged toward the opposite poles. Chromatids are now bound together at only one point, the centromere. Since two chromatids compose each dyad, their appearance depends on the position of the centromere. If the centromere is metacentric or acrocentric, a double V-shape is seen and if the centromere is telocentric, single V-shape is seen.

It should be noted that it is purely a matter of chance as to which of the parental homologous chromosomes are separated to each daughter cell. More the pairs of chromosomes, greater is the chance that a gamete will contain material from both the parents and lesser the chance that it will have all the chromosomes from one parent. Let us presume that there is an organism with two homologous chromosome pairs, AA' and BB'. The genotype of this individual can be written as AA'BB'. This individual can form four types of haploid gametes AB, AB', A'B and A'B' in equal proportions of which two are parental types. On the other hand, if any individual like Drosophila has four pairs of chromosomes, then on an average, only one out of eight gametes will carry all the chromosomes from a single parent. In general, for 'n' pairs of homologous chromosomes, the proportion of gametes containing parental combination is given by the following formula:

$$1/2^{n-1}$$

In humans, with 23 homologous chromosome pairs, the proportion of meiotic daughter cells to have all the chromosomes from one parent is only one out of 4194304 (= 2^{22}). Rest of the gametes will have a mixed proportion of chromosomes from both the parents. Thus, one can visualize that the random distribution of parental chromosomes and the crossing over between them leads to the most important source of variability among the gametes of sexually reproducing organisms.

Telophase I and Interphase: In most cases, once dyads reach to spindle poles, a nuclear membrane is formed around them and chromosomes pass into a short interphase before Meiosis II begins. Cytokinesis may occur during this stage in some cases like in corn or it may be postponed to until formation of four daughter cells at the end of meiosis II.

Meiosis II: Chromosomes enter into the prophase of second meiotic division as dyad or two sister chromatids connected together with a single functional centromere. As soon as this centromere splits into two, each chromatid (= monad) separates from its sister and moves to the opposite pole in the second anaphase. The second telophase and cytokinesis follow rapidly giving rise to four haploid cells originating from each initial diploid cell that entered meiosis.

In a cell consisting of four chromosomes or two pairs of homologues, the meiotic events leading to reduction in chromosome can be summarized as follows:

First Meiotic Division

Metaphase I: There is formation of two bivalents or tetrads consisting of four chromosomes or eight chromatids on metaphase spindle.

Anaphase I: Two dyads or two chromosomes or four chromatids pass on to each pole.

Second Meiotic Division

Metaphase II: There are two dyads on the spindle in each daughter cell.

Anaphase II: Two monads (two chromatids) pass on to each pole.

A diagrammatic illustration of meiosis is shown in Fig. 29.5.

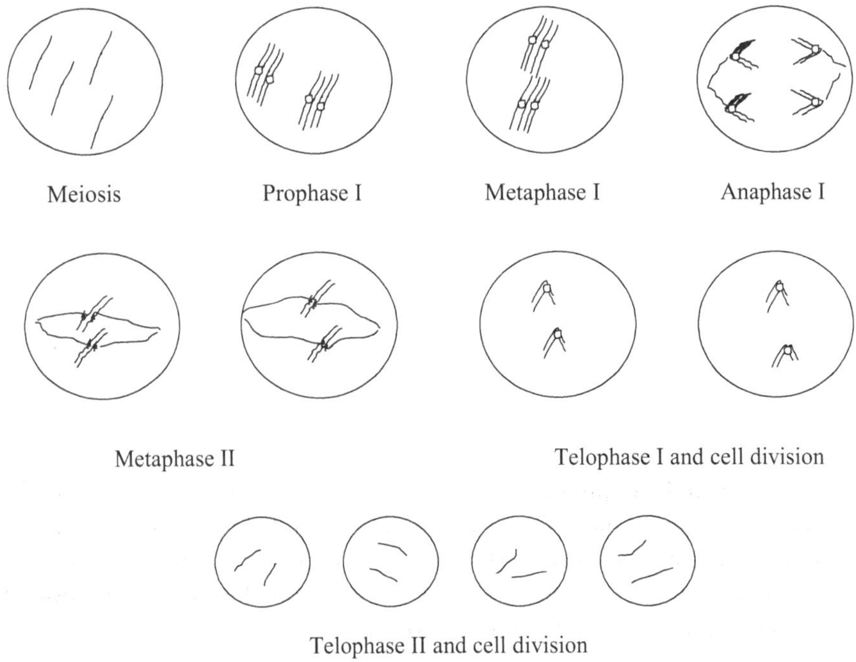

Fig. 29.5. Diagrammatic representation of meiosis.

Comparative characteristics of mitosis and meiosis are summarized in Table 29.2.

Table 29.2. Differences between mitosis and meiosis.

S. No.	Mitosis	Meiosis
1.	An equational division which separates sister chromatids.	The first stage is a reductional division which separates homologous chromosomes at first anaphase, sister chromatids separate in an equational division at second anaphase.
2.	One division per cycle i.e. one cytoplasmic division per equational chromosomal division.	Two divisions per cycle i.e. two cytoplasmic divisions per cycle, one following reductional chromosomal division and one following equational chromosomal division.
3.	Chromosomes fail to synapse, no chiasmata form, genetic exchange between homologous chromosomes does not occur.	Chromosomes synapse and form chiasmata, genetic exchange occurs between homologues.
4.	Two products (daughter cells) produced per cycle.	Four cellular products (gametes or spores) produced per cycle.
5.	Genetic contents of mitotic products are identical.	Genetic contents of meiotic products different, centromeres may be replicas of either maternal or paternal centromeres in varying combinations.
6.	Chromosome number of the daughter cells is same as that of other cell.	Chromosome number of meiotic products is half of that of the mother cell.
7.	Mitotic products are capable of undergoing additional mitotic divisions.	Meiotic products cannot undergo another meiotic division, although they may undergo mitotic division.
8	Normally occurs in most all somatic cells.	Occurs only in the specialized cells of the germ line.
9.	Begins at the zygote stage and continues through the life of the organism.	Occurs only after a higher organism has begun to mature, occurs in the zygote of many algae and fungi.

30

Chi-Square, Probability and Binomial

Chi-Square

The Chi-square test (χ^2) proposed by Karl Pearson in 1899 is used to measure goodness of fit of observed and theoretical or expected data and is a continuous distribution.

Formula for Chi-Square

1. General Formula

The general formula for Chi-square is as follows:

$$\chi^2 = \sum_{i=1}^{n} \frac{(oi - ei)^2}{ei}$$

Where o = number observed in any group of the experimental distribution.

e = expected value or theoretically calculated number for the same group.

It is obvious that Chi-square becomes smaller as the differences between the observed and expected values become smaller. Further, all the differences in frequency (o-e) are squared, whether positive or negative and hence Chi-square is always a positive quantity. The number of degrees of freedom is one less than the number of groups when the only restriction imposed is that the total frequencies of the observed and the theoretical distribution shall be equal.

The illustration on calculation of Chi-square is demonstrated as follows in a cross that involves two independently inherited factor pairs where a 9:3:3:1 is expected. Taking example from independent segregation for yellow vs green seed and round vs wrinkled seed in garden pea let us presume that the following values were observed in the F_2 from a cross:

Yellow-Round Seed = 38
Yellow-Wrinkled Seed = 11
Green-Round Seed = 13
Green-Wrinkled Seed = 2

The Chi-square from the above given data can be computed as follows:

Classes	Theoretical ratio	Observed values (O)	Expected values (E)	(O-E)	$(O-E)^2$	$(O-E)^2/E$
Yellow-Round	9	38	36	2	4	0.11
Yellow-Wrinkled	3	11	12	-1	1	0.08
Green-Round	3	13	12	1	1	0.08
Green-Wrinkled	1	2	4	-2	4	0.00
Total	16	64	64	0		0.27

Degrees of freedom = 4–1 = 3, Table value of Chi-square at 3 DF at 0.05 $P = 7.82$

Conclusions: Since calculated value of Chi-square is less than the table value of chi-square at the given degrees of freedom, the differences/deviations from the expected values are considered as non-significant and the data afford a good fit to the 9:3:3:1 ratio.

2. Special Formula of Chi-Square Calculation for Two Classes

The Chi-square may be calculated directly using the observed values if the data pertains only to two classes. Let us presume that '*A*' is the number of one class and '*a*' is the number in another class and '*n*' is the total of '*A*' and '*a*'. The formula used under such situations is as follows:

Ratio of '*A*' and '*a*'	Chi-square value
1:1	$\frac{(A-a)^2}{n}$
3:1	$\frac{(A-3a)^2}{3n}$
9:7	$\frac{(7A-9a)^2}{63n}$
j:k	$\frac{(kA-ja)^2}{jkn}$

The computation could be illustrated with F_2 data which appear to fit a 3:1 ratio where $A = 2903$, $a = 936$ and $n = 3839$.

$$\chi^2 = \frac{(A-3a)^2}{3n} = \frac{(2903-2808)^2}{3x3839} = \frac{(95)^2}{11517} = \frac{9025}{11517} = 0.784$$

The calculated value of χ^2 is less than the table value of χ^2 at 1 degree of freedom at 0.05 P and hence the data afford a good fit to the 3:1 ratio.

Limitations of Chi-Square

It has two limitations.

1. It has to be used on numerical data itself and never on ratios and percentages derived from the data.
2. It cannot be properly used for the experiments wherein the expected frequency within any phenotypic class is less than 5.

Correction for Small Samples

The formula from which the Chi-square table is derived is based upon a continuous distribution (normal curve). However, the kind of genetic problems which we encounter involve separate or discrete phenotypic classes. A correction should therefore be applied in calculation of Chi-square to correct for this lack of continuity. The "Yates' Correction for continuity is applied as follows:

$$\chi^2\,(\text{corrected}) = \sum_{i=1}^{n} \frac{(!oi - ei! - 0.5)^2}{ei}$$

This correction makes little difference in the calculated Chi-square of most problems but may become of importance when the calculated Chi-square is near the critical value. This correction should be routinely applied whenever there is only one degree of freedom or when the sample size is so small that each expected value is between 5 and 10. In most case, Chi-square is applied as such without using this correction.

Probability

Mathematically probability of an event is the ratio of the number of ways the event may occur to the total number of ways it either may occur or fail to occur on the assumption that all such ways are equally likely. For example, probability of drawing any individual card from an ordinary deck is 1/52, while that of drawing any card of a given suit is 13/52 or 1/4.

Experimental results seldom conform to the expected ratios in many genetic investigations. Genetic probabilities derive from the operation of chance events during meiosis when the gametes are produced and random union of gametes gives rise to zygotes. Samples from small population of individuals often deviate from the expected ratios rather widely. The results approach the expected ratios with increasing number of individuals in the population.

Let us understand this with an example. Let us assume that a testcross of round seeded pea (Rr x rr) produces 3 round seeded and 2 wrinkled seeded plants. Theoretically we expect half of the total number to be round and half to be wrinkled seeded i.e. each class should be two and half and that is not possible so the observed result conforms to the theoretical expectation as closely as possible biologically.

In another experiment, numerous testcrosses of a black guinea pig produced a total of 10 offspring, 8 of which were black and 2 were white. Theoretical expectation was 5 black and 5 white but the results obtained in case of 10 offspring (8 black and 2 white) should not be any more surprising than the results which one can get after tossing a coin 10 times and observing 8 heads and 2 tails. The fact that at least one white offspring appeared is sufficient to classify black parent as genetically heterozygous (Bb).

There are rules to combine the probabilities depending upon whether the events are independent or mutually exclusive. Two are more events are said to be independent if occurrence of any one of them does not affect the probabilities of occurrence of any of the other events. When two independent events occur with the probabilities p and q respectively, then the probability of their joint occurrence is given by the multiplication of probabilities of individual events i.e. pq. If the word "and" is used or implied in phrasing of problem solutions, a multiplication of independent probabilities is usually required. For example if a coin is tossed 2 times, then the probability of getting head both the times will be given by the multiplication of the probability of getting head in both the events. Probability of getting head or tail in one toss is 1/2 and hence after two tosses, probability of getting head in both the tosses is given by $1/2 \times 1/2 = 1/4$.

Mutually exclusive events are those events in which occurrence of any one of them excludes the occurrence of the other events. In such cases, the word "or" is generally used or implied in phrasing of the problem solutions and the probability of joint occurrence of more than one event in such cases is given by the addition of the probabilities of the individual mutually exclusive events. That is whenever alternative possibilities exist for the satisfaction of the condition of a problem the individual probabilities are combined by addition. This is demonstrated numerically as follows:

In two tosses of a coin, there are two ways to get a head and a tail.

	First toss		**Second toss**	**Probability**
First alternative	Head (p)	and	Tail (q)	pq
Second alternative	Tail (q)	and	Head (p)	$qp = pq$
		Combined probability		$2\,pq$

Since $p = q = 1/2$, hence the combined probability is 2 (1/2) (1/2) = 1/2

The rules of combining the probabilities based on either the multiplication or addition is the basis of binomial distribution which is briefly described.

Binomial

In $(p + q)^n$, the p and q represent the probabilities of alternative independent events and the power 'n' to which the binomial is raised, tells the number of events. The sum of the factors in the binomial must add up to unity.

Thus, $p + q = 1$

It can be recalled from the rules on combining the probabilities that when two independent events are occurring with the probabilities p and q, then the probability of their joint occurrence is given by pq, i.e. the combined probability is the product of the independent events. When alternative probabilities exist, the probabilities are combined by addition.

In two tosses of a coin with p = head = 1/2 and q = tail = 1/2, there are four probabilities as illustrated below:

First toss		Second toss	Probability
Head (p)	and	Head (p) =	P^2
Head (p)	and	Tail (q) =	pq
Tail (q)	and	Head (p) =	pq
Tail (q)	and	Tail (q) =	q^2
		Total	1.0

The above situation may be expressed as follows:

$$p^2 \quad + \quad 2\,pq \quad + \quad q^2 = 1.0$$

(Two heads) (One head and one tail) (Two tails)

Expanding the binomial $(p + q)^2 = p^2 + 2pq + q^2$ gives the same expression as given in above table.

Likewise, when a coin is tossed three times, the probabilities for any combination of heads and/or tails can be found from $(p + q)^3 = p^3 + 3p^2q + 3pq^2 + q^3$.

Let the probability of head = 1/2 and that of tail = 1/2. This situation can be summarized as follows:

Number of heads	Number of tails	Term	Probability
3	0	p^3	$(1/2)^3 = 1/8$
2	1	$3p^2q$	$3(1/2)^2 (1/2) = 3/8$
1	2	$3pq^2$	$3(1/2) (1/2)^2 = 3/8$
0	3	q^3	$(1/2)^3 = 1/8$

The expansion of $(p + q)^3$ is obtained by multiplying $(p^2 + 2pq + q^2)$ by $(p + q)$. This process can be extended for higher powers but the process will become increasingly tedious and laborious. A short method for expanding $(p + q)$ to any power n can be performed by following the following rules:

(i) The coefficient of the first term is one.

(ii) The power of the first factor (*p*) is n.

(iii) The power of the second factor (*q*) is zero (note: any factor to the zero power is unity).

(iv) In each term, multiply the coefficient by power of *p* and divide by the number of that term in the expansion. The result is the coefficient of the second term.

(v) Thereafter, the power of p will decrease by one and the power of q will increase by one in each term of the expansion.

(vi) The fully expanded binomial will have $(n + 1)$ terms.

The coefficients for each term in a binomial expansion raised to the n power when n goes from 0 to 10 and total number of combinations for each expansion are given in Table 30.1.

Table 30.1. Coefficients for each term in a binomial expansion raised to the n power when n goes from 0 to 10 and total number of combinations for each expansion.

N	Binomial coefficients	Total number of combinations
0	1	1
1	1, 1	2
2	1, 2, 1	4
3	1, 3, 3, 1	8
4	1, 4, 6, 4, 1	16
5	1, 5, 10, 10, 5, 1	32
6	1, 6, 15, 20, 15, 6, 1	64
7	1, 7, 21, 35, 35, 21, 7, 1	128
8	1, 8, 28, 56, 70, 56, 28, 8, 1	256
9	1, 9, 36, 84, 126, 126, 84, 36, 9, 1	512
10	1, 10, 45, 120, 210, 252, 210, 120, 45, 10, 1	1024

Let us assume that the above table relates to the probability of offspring from a cross of Aa x aa which will give rise to Aa and aa in equal frequencies of $p = q = 1/2$. In sib-ships of four, the binomial distribution a shown in Table 30.1 shows that one would obtain an average of one sib-ship out of 16 with all four offspring having the genotype Aa whose probability would be 1/16. Similarly, in a sib-ship of four individuals, the probability of having offspring as Aa, Aa, Aa and aa will be 4/16 and so on. This probability can be derived using the rules of combining the probabilities also as given below.

1. Aa, Aa, Aa, aa $= 1/2 \times 1/2 \times 1/2 \times 1/2 = 1/16$
2. Aa, Aa, aa, Aa $= 1/2 \times 1/2 \times 1/2 \times 1/2 = 1/16$

3. Aa, aa, Aa, Aa = 1/2 × 1/2 × 1/2 × 1/2 = 1/16
4. aa, Aa, Aa, Aa = 1/2 × 1/2 × 1/2 × 1/2 = 1/16

Total = 4/16

In general, wherever p is the probability that a particular genotype will appear and q or $1 - p$ is the probability of an alternative form or event, so that $p + q = 1$, the probability for each combination in which a succession of such events may occur is described by the binomial distribution. A formula which is used to calculate the probability for any special combination of genotypes or phenotypes disregarding their order of appearance is given as follows:

$$\frac{n!}{w!x!}p^{w}q^{x} = \frac{n!}{(n-x)!x!}p^{n-x}q^{x} = \frac{n!}{(n-k)!k!}p^{n-k}q^{k}$$

Where ***n*** refers to the total number of children in the family (sib-ship size), w to the number of children with a specific genetic or phenotypic constitution having probability p, e.g., p = probability of Aa offspring, and x to the number of children in this sib-ship bearing a genotype or phenotype having probability q, e.g. q = probability of aa offspring. The ! symbol means 'factorial' or multiplication of the number by all integers down to 1. For example, 5! Would mean 5 × 4 × 3 × 2 × 1 = 120. It should be noted that the zero factorial (0!) equals 1 and any number raised to the zero power [$(1/2)^0$] also equals 1.

Using this formula, let us calculate the probability of all the four offspring with genotype Aa produced from mating of Aa × aa. The probability that any offspring from Aa × aa will be Aa is 1/2. The binomial formula applicable to this case will be as follows:

$$p = \frac{4!}{4!0!}p\left(\frac{1}{2}\right)^{4}\left(\frac{1}{2}\right)^{0} = \frac{4\times3\times2\times1}{4\times3\times2\times1(1)}\times\frac{1}{16} = \frac{1}{16} = 0.0625$$

Number of combinations (c) of n things taken at a time is given by:

$$n_k^C = \frac{n!}{(n-k)!k!}$$

where ***n***! (n factorial means n (***n*** – 1) (***n*** – 2).......1 and (0! =1).

Just to illustrate this formula, let us find out number of ways to obtain two heads in three tosses of a coin which

$$= \frac{3!}{(3-2)!2!} = \frac{3\times2\times1}{(1)\times2\times1} = 3$$

The three combinations having two heads could be HHT, HTH, THH.

A few more applications of the above formulas can be illustrated in the following examples.

Let us find out the middle term of expansion $(p + q)^{10}$.

The middle term of expansion $(p + q)^{10}$ is the sixth term as there are $(n + 1 = 11)$ terms in the expansion. The power of q starts at zero in the first term and increases by one in each successive term so that the sixth term would have q^5 and so $k = 5$. Then the sixth term is as follows:

$$= \frac{n!}{(n-k)!k!} p^{n-k} q^k = \frac{10 \times 9 \times 8 \times 7 \times 6 \times 5 \times 4 \times 3 \times 2 \times 1}{(10-5)!5!} p^{10-5} q^5$$

$$= \frac{10 \times 9 \times 8 \times 7 \times 6 \times 5 \times 4 \times 3 \times 2 \times 1}{5 \times 4 \times 3 \times 2 \times 1 \times 5 \times 4 \times 3 \times 2 \times 1} p^5 q^5$$

$$= \frac{3628800}{14400} = 252\, p^5 q^5$$

Let us find out the probability of obtaining 6 heads and 3 tails in 9 tosses of coin using the applicable formula:

$$= \frac{n!}{(n-k)!k!} p^{n-k} q^k = \frac{9!}{6!3!} p^6 q^3 = \frac{9 \times 8 \times 7 \times 6 \times 5 \times 4 \times 3 \times 2 \times 1}{6 \times 5 \times 4 \times 3 \times 2 \times 1 \times 3 \times 2 \times 1} p^6 q^3$$

$$= \frac{362880}{4320} p^6 q^3 = 84 p^6 q^3$$

If probability of getting a head (p) is equal to that of getting a tail (q) = 1/2, then $84p^6q^3$

$$= 84\left(\frac{1}{2}\right)^6 \left(\frac{1}{2}\right)^3 = \frac{84}{512} = \frac{21}{128}$$

31

Mendel and Mendelian Principles (Segregation and Independent Assortment)

Historically the first and still the most conclusive evidence for the existence of genes comes from the phenomenon of segregation of traits observed in the offspring of hybrids between individuals or strains that differ in some recognizable respect. The principle of segregation was formulated by Gregor Johann Mendel in 1866 under such peculiar circumstances that the scientific world failed to recognize it until after a lapse of 34 years. In the first place Gregor Mendel was not primarily a biologist but a Monk in the Augustinian monastery at Bruenn, Austria (now Brno, Moravia in Czech Republic). During his childhood, Mendel worked as a gardener and studied beekeeping. He became part time monk because it enabled him to obtain an education without having to pay for it himself. From 1840 to 1843 he studied Practical and Theoretical Philosophy and Physics at the Philosophical Institute of the University of Olomouc. Upon recommendation of his teacher, Mendel entered the Augustinian St. Thomas Abbey in Brno and began his training as a Priest. Born Johann Mendel, he took the name Gregor upon entering religious life. Mendel worked also as a substitute High School teacher. In 1850, he failed the oral part of his examination and could not become a certified High School Teacher. In 1851, he was sent to the University of Vienna to study further so that he could get formal education. Mendel returned to his Abbey in 1853 as a Physics teacher. In 1856, he took the examination to become a certified teacher but again failed the oral part.

In 1857, he began to collect the varieties of the garden pea which seedsmen offered for sale and to study the differences among them. After seven years of experimental work on the garden pea in the monastery gardens, he presented the results of his experiment (Experiments on Plant Hybridization), together with generalizations we now know as "Mendel's Laws" at two meetings of the Natural History Society of Bruenn, Austria (now Brno, Moravia in Czech Republic) in the spring of 1865. The results and theory were printed in the Annual Proceedings of the Society which were distributed to libraries in Europe and America in 1866. It is safe to say that no one who heard Mendel's paper and no one who read it in the ninth century appreciated its significance until 1900 when the law of segregation was rediscovered almost simultaneously by three different investigators who had obtained results like Mendel. These three scientists, namely, De-Vries in Holland, Correns in Germany and Tschermak in Austria found Mendel's forgotten paper and proclaimed its importance. Immediately his conclusions began to be confirmed and extended by experiments carried out in various parts of the world on many kinds of animals and plants. In his later years, Mendel turned to experiments with other plants and bees and to meteorological observations but gradually he became more and more concerned with the administration of the monastery of which he had become Abbot in 1868. Mendel "the Founder of Genetics or the Father of Genetics" died on 6 January, 1884 at the age of 61 in Brno, unrecognized and unsung long before his scientific work came into its own. Mendel's museum in Brno, Czech Republic has been an institution of Masaryk University founded in 1919, now named as Mendel University since 1994 after Gregor Mendel, the Botanist and the "Father of Genetics" who has been active in the city of Brno during his life time. Photos of Mendel, Mendel Museum and Mendel University are shown in Figs. 31.1, 31.2 and 31.3 respectively.

Figs. 31.1, 31.2, 31.3 (Mendel-Father of Genetics, 20 July 1822-6 January 1884), Mendel Museum and Mendel University, respectively at Brno, Czech Republic.

Mendel's Method

Mendel avoided the complexities that had troubled the earlier researchers and simplified the problem as far as possible. Mendel confined his attention to a single character at a time, flower colour, for example in garden pea unlike earlier workers who made observations on plants and animals as a whole studying all the characters in which the hybrids differed from the parents and from each other. When behaviour of each single trait studied by Mendel was established, he then studied two traits together such as flower colour and plant height. Furthr, he counted the number of each type of progeny from crosses of garden pea thus reducing the phenomenon of inheritance to a measurable quantitative basis.

Mendel's procedure was to cross two plants differing in a pair of contrasted characters and to observe the appearance of the first hybrid or "F_1" generation. He then crossed hybrid plants or allowed them to effect self-fertilization and raised a second generation or F_2. He made counts of plants possessing each of contrasting traits in the F_2 generation, such as different flower colours or vine lenth in which the parents differed. Most important of all, from a study of such comparative simple data, Mendel formulated working hypothesis or theoretical explanations of the events he observed. The validity of these hypothetical explanations was tested by further experiments first by Mendel himself and later by other experiments using other plants and animals.

Mendel's theoretical explanations of his results are now firmly established as Mendel's laws of inheritance. These include:

1. Principles of segregation and independent assortment (major principles)
2. Dominance and recessiveness of traits in the hybrids (less fundamental generalization)

Seven pairs of characters studied by Mendel in garden pea were as follows:

1. Seed shape: round vs wrinkled
2. Seed cotyledons colour: yellow vs green
3. Seed coat colour: white vs grey equivalent to white flowers vs violet flowers
4. Pod surface: full vs constricted
5. Pod colour: yellow vs green
6. Pod position: axial vs terminal
7. Stem length: tall vs dwarf

Mendel obtained the results as given in Table 31.1.

Table 31.1. Results of Mendel's monohybrid crosses.

Parental crosses	F1 progeny	F2 progeny	Ratio
Tall x dwarf plants	Tall	787 tall, 227 dwarf	2.84:1
Round seeds x wrinkled seeds	Round seeds	5474 round, 1850 wrinkled seeds	2.96:1
Yellow seeds x green seeds	Yellow seeds	6022 yellow, 2001 green	3.01:1
Violet flowers x white flowers	Violet flowers	705 violet, 224 white	3.15:1
Inflated pods x constricted pods	Inflated pods	882 inflated, 299 constricted	2.95:1
Green pods x yellow pods	Green pods	428 green, 152 yellow	2.82:1
Axial flowers x terminal flowers	Axial flowers	651 axial, 207 terminal	3.14:1

Uniformly, Mendel's crosses between two different varieties for each character always produced an F_1 that was of only one parental type. However, when these F_1 plants were allowed to reproduce by self-fertilization, both the parental types appeared in the F_2 in very close to 3:1 ratio which can also be written as 3/4:1/4 or 0.75:0.25 or 75% : 25%. The point worth noting is that the trait which had disappeared or hidden in the F_1, reappeared in the F_2 but only in a frequency one quarter of the total number.

Thus each trait that Mendel studied seems to be controlled by a heritable factor that existed in two forms, the one dominant and the other recessive. These factors are now called as genes, a world coined by the Danish plant breeder, W. Johannsen in 1909. The dominat and recessive forms are called as alleles-from the Greek word meaning "of one another".

The regular numerical relationships that Mendel observed in these crosses led him to another important conclusion that genes come in pairs. Mendel proposed that each of the parental strains that he used in his experiments, carried two identical copies of a gene-in modern terminology, they are diploid and homozygous. However, during production of gametes, Mendel proposed that these two copies are reduced to one, that the gametes that emerge from meiosis carry a single copy of a gene-in modern terminology, they are haploid. Mendel recognized that the diploid gene number would be restored when sperm and egg unite to form a zygote. Furthermore, Mendel understood that if the pollen and egg came from genetically different plants, as they did in his crosses, the hybrid zygote would inherit two different alleles, one from the mother and one from the father. Such an offspring is said to be heterozygous. Mendel realized that the different alleles that are present in the heterozygotc must co-exist even though one is dominant and the other recessive and each of these alleles would have an equal chance of entering a gamete when the heterozygote reproduces. Furthermore, Mendel realized that random fertilization with a mixed population of gametes-half carrying the dominant allele and the half carrying the recessive allele-would produce some zygotes in which both alleles were recessive. Thus, he could explain the reappearance of the recessive characteristic in the progeny of the hybrid plants.

Mendel used symbols to represent the hereditary factors that he postulated. With symbols, he could describe the hereditary phenomenon clearly and concisely and he could explain the results of the crosses mathematically. He could even make predictions about outcome of the future crosses and validated the results. The symbols used by Mendel and for genes (or more precisely for their alleles) and they are manipulated according to the rules of inheritance that Mendel discovered. These manipulations are the essence of formal genetic analysis.

As an illustration, let us consider symbolic representation through letters of the cross between red and white flowered peas to visualize how the genes are transmitted and distributed. Conventionally dominant genes are represented by capital letters and their recessive alternatives by corresponding small letters. In the cross of red and white peas, let C stand for the gene for the red flowers and c for the alternative, or the allelic, form of the gene, which gives white flowers. C and c are allelic genes or alleles. Since an individual arises from the union of two gametes, it receives a gene for flower colour from both the parents. The true breeding red flowered parent may therefore be represented by CC and its gametes as C. In contrast, the true breeding white parent may therefore be represented by cc and its gametes as c. When the two contrasting parents are crossed, an egg 'C' is fertilized by a male gamete 'c' or vice-versa and resulting hybrid zygote will have both 'C' and 'c' and its genetic formula will be Cc. Now the allelic constitution of each strain is said to be its genotype and the appearance of each strain, that is the red and white flower characteristic is said to be its phenotype. The hybrid progeny are referred to as the first filial generation or F_1, from a Latin word meaning "offspring". During meiosis, the F_1 plant produces two kinds of gametes, C and c in equal proportions. Neither allele is changed by having co-existed with the other in a heterozygous genotype (Cc), rather they separate or segregate from each other during gamete formation. This process of allele segregation is perhaps the most important discovery that Mendel made. Upon self-fertilization, two kinds of gametes produced by heterozygotes can unite in all possible combinations producing four kinds of zygotes, namely, CC, Cc, cC and cc in equal proportion of 1:1:1:1. However, Cc and cC will be similar and can be written as Cc and thus the genotypic ratio in the F_2 will be 1CC:2Cc:1cc. However, because of dominance, CC and Cc will have same phenotype (red flowers) and thus the red and white flowered plants in the F_2 will be in a ratio of 3:1 (Fig. 31.4).

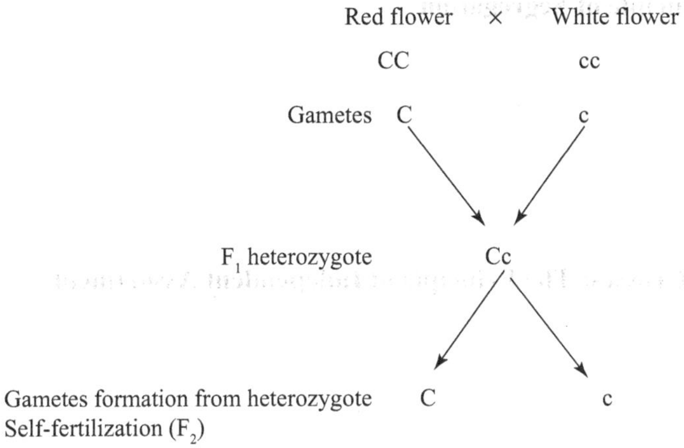

	C	c
C	CC Red flower	Cc Red flower
c	Cc Red flower	cc White flower

F2	Phenotype	Genotype	Genotypic ratio	Phenotypic ratio
	Red flower	CC =1 Cc = 2	1 2	3
	White flower	cc = 1	1	1

Fig. 31.4. Symbolic representation of the cross between red and white flowered peas and summary of phenotypic and genotypic results.

Mendel took this analysis one step further. F_2 plants were self-fertilized to produce F_3 generation. All the white flowered F_2 plants (cc) produced only white flowered plants confirming that they were homozygous for the c allele. The red flowered F_2 plants were of two types. Approximately one third of them produced only red flowered offspring, whereas the remaining two thirds produced a mixture of red flowered and white flowered offspring. Mendel concluded correctly that the one third were true breeding homozygotes (CC) and the two thirds that were segregating into red and white flowered plants were heterozygotes (Cc) and this matches with the predicted genotypic ratio as shown in Fig. 31.4.

Mendel's results and analysis of this example and similar results as observed in other monohybrid crosses can be summarized by stating two key principles that Mendel discovered. These are as follows:

1. The Principle of Dominance

In a heterozygote, one allele may conceal the presence of another. This principle is a statement about genetic function. Some alleles evidently control the phenotype even when they are present in a single copy.

2. The Principle of Segregation

In a heterozygote, two different alleles segregate (separate) from each other during the formation of gametes. This principle is a statement about genetic transmission. An allele is transmitted faithfully to the next generation even if it was present with a different allele in a heterozygote. The biological basis of this phenomenon is the pairing and subsequent separation of homologous chromosomes during meiosis.

Dihybrid Crosses: The Principle of Independent Assortment

Mendel also carried out experiments with plants that differed in two traits. He made crosses between plants having yellow, round seeds with those having green, wrinkled seeds. The idea behind this experiment was to see if the two seed traits, colour and texture, were inherited independently. The F_1 seeds were all yellow and round showing that the alleles for these two characteristics were dominant. Mendel grew F_2 plants and counted F_2 seeds by phenotype and observed the results as follows:

Yellow, round = 315
Green, round = 108
Yellow, wrinkled = 101
Green, wrinkled = 32

This data fitted to an approximate ratio of 9:3:3:1 and Mendel offered a simple explanation that each trait was controlled by a different gene segregating two alleles, and the two genes were inherited independently. The symbolic representation of this cross is shown in Fig. 31.5.

Parents	Yellow, round	×	Green, wrinkled
	GGWW		ggww
Gametes	GW		gw
Heterozygote F_1		GgWw	
Gametes		GW, Gw, gW, gw	
Self-fertilization			

	GW	Gw	gW	gw
GW	GGWW Yellow, Round	GGWw Yellow, Round	GgWW Yellow, Round	GgWw Yellow, Round
Gw	GGWw Yellow, Round.	GGww Yellow, Wrinkled	GgWw Yellow, Round	Ggww Yellow, Wrinkled
gW	GgWW Yellow, Round	GgWw Yellow, Round	ggWW Green, Round	ggWw Green, Round
gw	GgWw Yellow, Round	Ggww Yellow, Wrinkled	ggWw Green, Round	ggww Green, Wrinkled

F2

Phenotypes	Genotypes	Genotypic ratio	Phenotypic ratio
Yellow, round	GGWW GGWw GgWW GgWw	1 2 2 4	9
Yellow, wrinkled	GGww Ggww	1 2	3
Green, round	ggWW ggWw	1 2	3
Green, wrinkled	ggww	1	1

Fig. 31.5. Symbolic representation of the results of a cross between yellow-round and green-wrinkled seeded peas.

The haploid gametes produced by a diploid plant contain one copy of each gene and thus the gametes from GGWW will be symbolized by GW. Likewise, gametes from ggww will be gw type. The reunion of these two will give rise to an F_1 representd by GgWw and is a double heterozygote. Its yellow-round phenotype indicates that G and W alleles are dominant. The principle of segregation predicts that the F_1 hybrid will produce four different gametic genotypes as GW, Gw, gW and gw. In case each gene segregates its alleles independently, then each gamete will be equally present and will be 25 percent of the total. On this basis, self-fertilization in the F_1 will produce 16 equally frequent zygotic genotypes through all possible combinations among four types of female and four types of male gametes. The genotypic and phenotypic frequencies of such zygotes is given in Fig. 31.5. The absolute frequencies of each of the four phenotypes are as follows:

Yellow, round = 9/16
Yellow, wrinkled = 3/16
Green, round = 3/16
Green, wrinkled = 1/16

This analysis is based on two assumptions, namely, that each gene segregates its alleles and these segreations are independent of each other. The second assumption implies that there is no linkage between the segregating events of the two genes which would mean that a gamete which receives W through the segregation of the texture gene (round vs wrinkled seed) is just as likely to receive G as it is to receive g through the segregation of the colour gene (yellow vs green cotyledon). Through Chi-square test, it can be demonstrated that there is no significant difference between the observed values and the expected values for phenotypic ratios in the F_2. Thus, it can be concluded that the segregation of the seed colour and seed texture genes are independent of each other.

3. The Principle of Independent Assortment

Mendel conducted similar experiments with other combination of traits and in each case observed that the genes segregated independently. The results of these experiments led mendel to a third principle, that is the "Principle of Independent Assortment" which is defined as follows:

The alleles of different genes segregate or as we sometimes say assort independently of each other. However, it must be understood that only those alleles assort independently which are located on different chromosomes or on the same chromosome but are not linked that is both the genes located on the same chromosomes show maximum recombination of 50 percent.

Application of Mendelian Principles

If the genetic basis of a trait is known to fall in line with principles of segregation, three general procedures can be applied to calculate zygotic genotypes or phenotypes as given below:

The Punnett Square or Checkerboard Method: If the two parents differ for one gene or more gene pairs and the genes segregate independently, it is possible to write down all the gametes and combine them systematically in all possible combinations to generate all possible zygotic genotypes. Principle of dominance can be used to determine the associated phenotypes. This procedure, called the Punnett Square Method after the British geneticist R. C. Punnett is straightforward procedure of predicting the phenotypic and genotypic ratios/outcome of the crosses as already shown in Figs. 31.4 and 31.5.

The Forked Line or Branching Method: In this case, instead of enumerating progeny in square or rectangle, they are tallied in a diagramme of banching lines as illustrated in Fig. 31.6.

Cross Dd Gg Ww × Dd Gg Ww

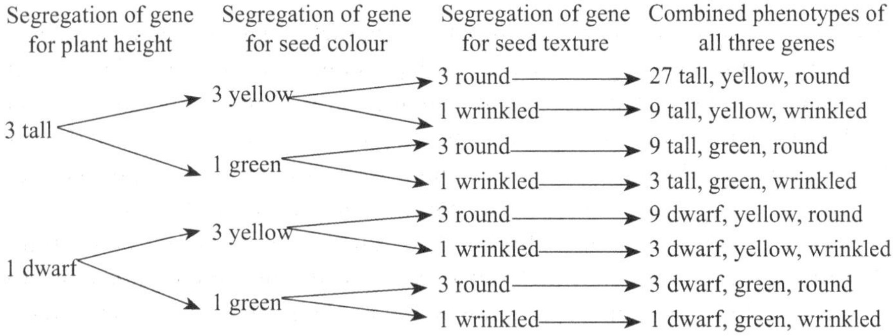

Fig. 31.6. The forked line method of predicting the outcome of an intercross involving three independently assorting genes in peas.

The Probability Method: The probability method based on principle of probability using multiplicative and additive rules of the probability as elaborated in chapter 27 of this book is quicker method than the Punnett Square and forked line methods to predict outcome of crosses in terms of frequencies of certain phenotypes and the genotypes. It is worthwhile to recollect that when the events are independent, then the probability of the joint occurrence of the events is given by the multiplication of the probabilities of the individual independent events and when the events are mutually exclusive, then the probability of the joint occurrence of such events is given by the addition of the probabilities of the such individual events. Mendelian segregation is like a coin toss for production of gametes. For example, a monogenic heteroygote (Aa) produces gametes where half of the gametes contains allele A and the other half contains allele a. If two such heterozygotes are crossed, the zygotic genotypes are produced by random combination of both the gametes. In a cross of Aa x Aa, the chance that a zygote will be AA will be given simply by the probability of each of the uniting gametes containing A. The probability of gamete containing A is 1/2 and when two A's will combine, only then 'AA' will be produced. Therefore, the probability of AA will be 1/2 x 1/2 = 1/4 since the two gametes are produced independently. The probability for 'aa' homozygote is also 1/2 x 1/2 = 1/4. However, the probability for 'Aa' heterozygote is 1/2 because there are two ways of creating Aa. In one method, 'A' may come from egg and 'a' from the sperm and the vice versa. The probability of the first event is 1/2 x 1/2 = 1/4 and that of the second event is also 1/2 x 1/2 = 1/4 and their joint occurrence will be given by applying the additive principle of probability and hence it will be 1/4 + 1/4 = 1/2. Thus, the intercrossing of Aa x Aa will give rise to the following probability distribution of the genotypes.

AA = 1/4

Aa = 1/2

aa = 1/4

From this, it is also concluded that 1/4 + 1/2 = 3/4 of the progeny will be representing the dominant phenotype and the 1/4 progeny will have the recessive phenotype.

The application of rules of probability to deal with crosses involving heterozygosity for more number of genes can be demonstrated with some more examples:

1. Let there be a cross between plants heterozygous for four different genes (AaBbCcDc), each assorting independently. What fraction of progeny will be homozygous for all four recessive alleles?

To answer this, let us take one gene at a time. For the first gene, the fraction of offspring that will be recessive homozygous (aa) will be 1/4. Similar would be the case for second, third and fourth genes. Therefore, based on principle of independent assortment and applying the rule of multiplication of probability, the offspring that will be homozygous recessive at all the four loci will be (1/4) x (1/4) x (1/4) x (1/4) = 1/256. Certainly, this is a better method than the long process of a Punnett Square with 256 entries.

Let us set up another question. What fraction of offspring will be homozygous for all four genes? If we recall a single gene case (1/4 AA : 2/4 Aa :1/4 aa), then 1/4 + 1/4 (AA + aa) = 1/2 is the fraction of offspring homozygous for both 'A' and 'a' alleles. Same would be the situation for other three genes. Thus, the fraction of offspring homozygous for all the four genes will be (1/2) x (1/2) x (1/2) x (1/2) = 1/16.

2. Suppose there is a cross AaBb x AaBb. What will be the fraction of the progeny that will show recessive phenotype for at least one gene? This can be solved as follows:

Cross AaBb × AaBb

↓

Segregation of A gene

Segregation for B gene	A - (3/4)	aa (1/4)
B - (3/4)	A-B- = (3/4) × (1/4) = 9/16	aaB- = (1/4) × (3/4) = 3/16
bb (1/4)	A-bb = (3/4) × (1/4) = 3/16	aabb = (1/4) × (1/4) = 1/16

Progeny:

Genotype	Frequency	Phenotype	Frequency
A-B-	9/16	Dominant for both genes	9/16
aaB-	3/16	Recessive for at least one gene	7/16
A-bb	3/16		
aabb	1/16		

(-) dash stands either for A or a.

In this case, each gene segregates dominant and recessive phenotypes with the probabilities 3/4 and 1/4, respectively. Because segregations occur independently, the frequencies of the combined phenotypes are obtained by multiplying the probabilities of individual and independent events that is the marginal probabilities. The events responsible for presence of at least one recessive gene are mutually exclusive and hence the frequency of progeny showing the recessive phenotype for at least one of the genes is obtained by adding the frequencies in the relevant cells and this comes out to be 3/16 + 3/16 + 1/16 = 7/16.

The Relation Between the Number of Genes Involved in a Cross and the Number of Phenotypic and Genotypic Classes in F2

This relationship is presented in Table 31.2.

Table 31.2. The relationship between the number of genes involved in a cross and the number of phenotypic and genotypic classes in F2.

Number of genes involved in the cross	Number of visibly different F2 classes of individuals if dominance is complete	Number of different kinds of gametes formed by the F1 hybrid	Number of genotypically different combinations	Number of possible combinations of F1 gametes
1	2	2	3	4
2	4	4	9	16
3	8	8	27	64
4	16	16	81	256
5	32	32	243	1024
n	2^n	2^n	3^n	4^n

From Table 31.2, it is obvious that when the number of genes (n) involved in a cross reaches tens or hundreds, the total number of genotypes, 3^n, and even the number of homozygous genotypes, 2^n, which may arise in the progeny of a heterozygote become enormous. In fact, these numbers are likely to be greater than the existing number of individuals in a species. It follows that the process of sexual reproduction with its twin consequences of gene assortment and gene recombination, creates enormous amount of genetic variability.

Modified Di-hybrid Ratios

The classical phenotypic ratio resulting from the mating of dihybrid genotypes is 9:3:3:1 which appears in the situations where alleles at both loci display complete dominance and recessive relationship. This classical di-hybrid ratio may get modified if one or both loci have co-dominant alleles or lethal alleles. A summary of these modified phenotypic ratios in adult progeny is as shown below.

Allelic relationship in di-hybrid parents		Expected adult phenotypic ratio
First locus	**Second locus**	
Dominant-recessive	Co-dominants	3:6:3:1:2:1
Co-dominants	Co-dominants	1:2:1:2:4:2:1:2:1
Dominant-recessive	Recessive lethal	3:1:6:2
Co-dominants	Recessive lethal	1:2:1:2:4:2
Recessive lethal	Recessive lethal	4:2:2:1

Distinguishing Between heterozygous and Homozygous Individuals

In a segregating population where genotypes TT and Tt produce same phenotypes because of dominance, then the exact genotypes of the parents are known by two procedures as given below:

Test Cross

Initially used by Mendel, it amounts to crossing of plants with same phenotype but possibly with different genotypes (TT and Tt) with the recessive parent. These crosses produce following progeny.

TT (tall) × tt (dwarf) and Tt (tall) × tt (dwarf)

↓ Tt –tall (100%)

↙↘ Tt -tall tt -dwarf (1:1 ratio)

Thus, if the parent has the genotype (TT), all the test-cross progeny will be non-segregating and tall types. However, in case the parent is heterozygous (Tt), then the progeny will segregate in a ratio of 1:1 for tall and dwarf phenotypes.

Progeny Test

Unlike a test-cross, a progeny test does not include a cross with a special parent but selfing of F_2. Each F_2 plant is harvested and separately bagged and subsequently planted in individual plant-progeny rows in the F_3. In F_3 stage, plants that are homozygous dominant will produce progenies that are uniformly tall (in the present case) whereas plants that are heterozygous will produce a segregating progeny row.

Plant breeders use the progeny test for a number of purposes. In breeding methodologies, in which selection is based on phenotype, a progeny test will allow a breeder to select superior plants from among a genetically mixed population. Following an environmental stress, biotic or abiotic, a breeder may use a progeny test to identify superior individuals and further ascertain if the phenotypic variation is due to genetic effects or just caused by environmental factors.

Incomplete Dominance and Co-Dominance

Mendel worked with traits that exhibited complete dominance. Post-Mendelian studies revealed that frequently, the masking of one trait by another is partial (incomplete dominance or partial dominance). A cross between a red-flowered (RR) and white-flowered (rr) snapdragon produces pink-flowered plants (Rr). The genotypic ratio remains 1:2:1 but a lack of complete dominance also makes the phenotypic ratio 1:2:1 instead of 3:1 as expected under complete dominance.

There is another situation in which no dominance occurs in which both the alleles of a heterozygote are expressed in equal degrees. The two alleles code for two equally functional and detectable gene products. Commonly observed and useful examples for plant breeding technology are alloenzymes, the production of different forms of the same enzyme by different alleles at the same locus. Allozymes catalyze the same reaction. This pattern of inheritance comes under co-dominant inheritance. Some molecular markers exhibit co-dominance. Incomplete dominance produces a blended phenotype but co-dominance produces distinct and separate phenotypes.

Multiple Alleles of the Same Gene

The concept of multiple alleles can be studied only in a population. Any individual diploid organism can have at most two homologous gene-loci that can be occupied by different alleles of the same gene. However, in a population, member of a species can have many alternative forms of the same gene. A diploid by definition can have only two alleles at each locus (e.g. C1C1, C7C10, C4C6). However, mutations may cause additional alleles to be created in a population. Multiple alleles of allozymes are known to occur. The mode of inheritance by which individuals have access to three or more alleles in the population is called multiple allelism and this set of alleles is called an allelic series. A common example of multiple-allelism is ABO blood group system in humans. An allelic series of importance in plant breeding is the S-alleles that condition self-incompatibility (inability of a flower to be fertilized by its own pollen). Self-incompatibility is a constraint to sexual biology and can be used as a tool in plant breeding to produce hybrid seed on commercial scale in certain brassicas.

Mendelian Principles in Human Genetics

The application of Mendelian principles to human genetics began soon after rediscovery of these principles in 1900. Since planned crosses cannot be done in human being, the progress was obviously slow. Genetic analysis of human heredity depends largely on family records which are often incomplete.

Further, human beings do not produce many progeny making it further difficult to fit Mendelian ratios. Mistaken paternity is another problem while dealing with human genetics. Time is also a factor because some genetic conditions do not manifest themselves until a person reaches middle age. Despite these difficulties but in face of strong drive to understand human heredity, thousands of human genes are known. A few important genes as understood in human beings are listed in Table 31.3.

Table 31.3 Inherited conditions in human beings.

Dominant Traits	Recessive Traits
Achondroplasia (dwarfism) Brachydactyly (short fingers) Congenital stationary night blindness Ehler-Danlos syndrome (a connective tissue disorder) Hutington's disease (a neurological disorder) Marfan syndrome (tall, gangly stature) Neurofibromatosis (tumor like growths on the body) Phenylthiocarbamide (PTC) tasting Widow's peak Wooly hair	Albinism (lack of pigment) Alkaptonuria (a disorder of amino acid metabolism) Ataxia telangiectasia (a neurological disorder) Cystic fibrosis (a respiratory disorder) Duchenne muscular dystrophy Galactosemia (a disorder of carbohydrate metabolism) Phenylketonuria (a disorder of amino acid metabolism) Sickle cell anaemia (a hemoglobin disorder) Tay-Sachs disease (a lipid storage disorder)

Pedigrees

Pedigrees are diagrammmes that show the relationships among the members of a family. A conventional pedigree is shown in Fig. 31.7.

Traits caused by dominant alleles are the easiest to identify as every individual who carries the dominant allele manifests the trait making it possible to trace the transmission of the dominant allele through pedigree analysis. Every affected individual is expected to have at least one affected parent unless, of course, the dominant allele has just appeared in the family due to new mutation. However, the frequency of new mutant is very low, almost one in one million. Thus, the chance of spontaneous appearance of a dominant condition is an extremely rare event. Recessive traits are not easy to identify because they may occur in individuals whose parents are not affected. Sometimes several generations of pedigree data are needed to track the transmission of a recessive allele. Nevertheless, a large number of recessive traits (more than 4000) have been observed in human beings. Rare recessive traits are more likely to appear in the progeny when spouses are related to each other. This happens because relatives share alleles by virtue of their common ancestry. Siblings share one half their alleles, half-siblings one fourth their alleles and first cousins, one eighth their alleles. Thus, when such relatives marry, they have a greater chance of producing a child who is homozygous for a particular recessive allele than what is possible after marriage between unrelated parents.

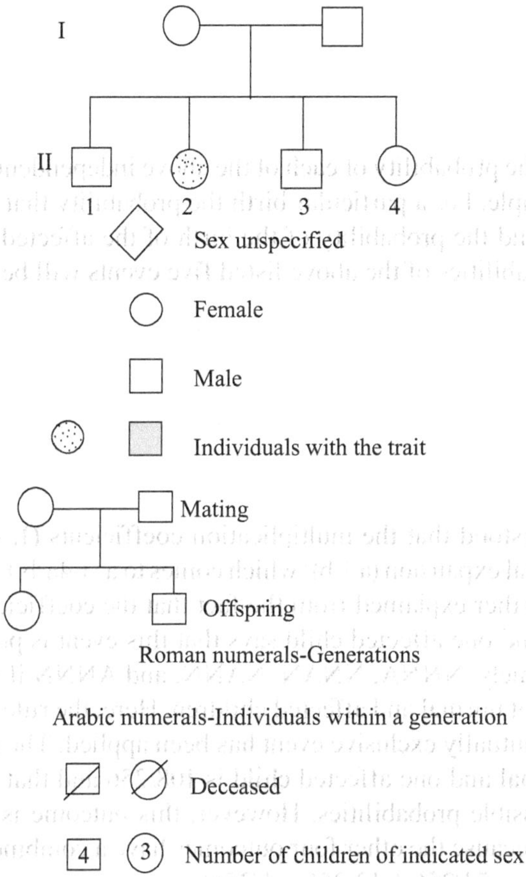

Fig. 31.7. Pedigree convention.

Mendelian Segregation in Human Families

In human being, the number of children produced from a marriage is small (approximately 2-6) and Mendelian segregation ratios cannot be applied successfully in such a small number of individuals. Consequently, phenotypic ratios in human families often deviate significantly from expected Mendelian ratios. This can be demonstrated from a typical example where the parents are heterozygous for a recessive allele that in homozygous condition causes cystic fibrosis. Let us say that there are four children from this parentage. Can we expect exact fit to 3:1 (3 normal: 1 cystic fibrosis) ratio in this case.

The answer is no, although the possibility of this event cannot be ruled out either. There are five distinct possibilities in this case.

1. Four normal, none affected children.
2. Three normal, one affected child.
3. Two normal, two affected children.
4. One normal, three affected children.
5. None normal, all four affected children.

We can calculate the probability of each of the above independent events based on Mendel's principle. For a particular birth the probability that the child will be normal is 3/4 and the probability of the birth of the affected child will be 1/4. Thus the probabilities of the above listed five events will be as follows:

S. No.	Normal child	Affected child	Probability of the event
1.	4	0	$1 \times (3/4) \times (3/4) \times (3/4) \times (3/4) = 81/256$
2.	3	1	$4 \times (3/4) \times (3/4) \times (3/4) \times (1/4) = 108/256$
3.	2	2	$6 \times (3/4) \times (3/4) \times (1/4) \times (1/4) = 54/256$
4.	1	3	$4 \times (3/4) \times (1/4) \times (1/4) \times (1/4) = 12/256$
5.	0	4	$1 \times (1/4) \times (1/4) \times (1/4) \times (1/4) = 1/256$

It should be understood that the multiplication coefficients (1, 4, 6, 4, 1) are taken from binomial expansion $(a + b)^4$ which comes to $a^4 + 4a^3b + 6a^2b^2 + 4ab^3 + b^4$. This can be further explained from the fact that the coefficient 4 for three normal children and one affected child says that this event is possible in four possible ways, namely, NNNA, NNAN, NANN, and ANNN if we ignore the sequence of birth of normal and affected children. Here, the rule of addition of probabilities for mutually exclusive event has been applied. The probability of having three normal and one affected child is 108/256 and that is the highest among all the possible probabilities. However, this outcome is not expected majority of time because the other four outcomes have a combined probability of 148/256 (81/256 + 54/256 + 12/256 + 1/256).

ABO Blood Groups in Human Being

This is a multi-allelic system discovered by Bernstein in 1925 and consists of three alleles of a single gene, I^A, I^B and I^O which form four different phenotypic groups as given below:

Blood group A: $I^A I^A$ or $I^A I^O$

Blood group B: $I^B I^B$ or $I^B I^O$

Blood group AB: $I^A I^B$

Blood group O: $I^O I^O$

The blood serum of man himself manufactures the antibodies that react with the blood cell antigens of other individuals. The transfusion of blood between individuals of different ABO blood group types may result in an agglutination reaction especially if large quantities of different blood type are being transfused. Either the antibodies of the recipient will agglutinate the antigens of the donor or the antibodies of the donor will agglutinate the antigens of the recipient. Various possibilities on blood transfusion are depicted in Fig. 31.8 (Strickberger, 1968). The A serum carries antibodies for B and AB cells and the B serum has antibodies for A and AB, AB serum has no antibodies for any of the cells, whereas the O serum carries antibodies for all types of cells except its own. On this basis, people of O blood type have been described as "universal donors" (absence of A and B antigens) and people of AB blood type as "universal recipients" (absence of A and B antibodies).

Donor (antigens) Red blood cells from		**Recipient (antibodies) Reaction of serum from group**			
Blood group	**Genotype**	**A**	**B**	**AB**	**O**
A	$I^A I^A$(AA) or $I^A I^O$ (AO)				
B	$I^B I^B$ (BB) or $I^B I^O$ (BO)				
AB	$I^A I^B$ (AB)				
O	$I^O I^O$ (OO)				

Fig. 31.8. Action of serum from persons of various blood groups (recipients) on the different kinds of red blood cells (donors).

32

Genetic Interaction

Phenotype is a result of gene products brought to expression in a given environment. Genes specify the structure of proteins. All known enzymes are proteins. Enzymes perform catalytic function, causing the splitting or union of various molecules. All of the chemical reactions which occur in the cell constitute the subject of intermediary metabolism. These reactions occur as stepwise conversions of one substance into another, each group being mediated by a specific enzyme. All of the steps which transform a precursor substance to its end product constitute a biosynthetic pathway. Several genes are usually required to specify the enzymes involved in even the simplest pathways. Before gene-enzyme relationship could be postulated, soon after rediscovery of Mendelian principles of inheritance, it was discovered that genes were not merely separate elements producing distinct individual effects, but they could interact with each other to produce completely novel phenotypes. The simplest example of gene interaction could be the fact that allele A could be characterized by phenotype A, and the allele B by the phenotype B and an organism having both alleles A and B at the same time might show phenotype C. The occurrence of such interactions means that the phenotype we observe is not present in the genes themselves but arises from a complicated developmental process resulting from a biosynthetic pathway as illustrated below (Stansfield, 1969):

$$\begin{array}{ccccccc} & g_1^+ & & g_2^+ & & g_3^+ & \\ & \downarrow & & \downarrow & & \downarrow & \\ \text{P (precursor)} & e_1 & \text{A} & e_2 & \text{B} & e_3 & \text{C (end product)} \end{array}$$

In this diagrammatic illustration, each metabolite (A, B, C) is produced by the catalytic action of different enzymes, each specified by wild type alleles g_1^+, g_2^+, and g_3^+. Obviously, genetic interaction occurs whenever two or more genes specify enzymes which catalyze steps in a common pathway. If substance C is essential for a the production of a normal phenotype and recessive mutant alleles gl, g2, and g3 produce defective enzymes, then a mutant phenotype will be produced from a genotype homozygous recessive at any of the three loci (gl, g2, and g3). If g3 is mutant, the conversion of substance B to C will not happen and there will be excessive accumulation of substance B. On the other hand, if g2 is mutant, then there will be accumulation of substance A. These mutants are said to produce 'metabolic blocks'. If the genotype is homozygous recessive for g_2 allele, then the pathway ends with substance A. Thus the genotype g_2g_2 has the capacity to mask the phenotypic expression of alleles at the g_3 locus. Originally a gene or locus which suppressed or masked the action of a gene at another locus was termed epistatic. The gene or locus suppressed was termed as hypostatic. Later on it was found that both loci could be mutually epistatic to one another. Now the term epistasis has come to be synonymous with almost any type of gene interaction. Dominance involves intra-allelic gene suppression, or the masking effect of one allele on the other allele at the same locus. Epistasis involves inter-allelic gene suppression or the masking effect which one gene locus has upon the expression of another gene. The classical phenotypic ratio of 9:3:3:1 observed in the progeny of di-hybrid parents gets modified into ratios which are various combinations of the 9:3:3:1 ratio. Strickberger (1968) has summarized various possibilities in terms of different phenotypic ratios arising from a cross of AaBb x AaBb where each gene pair is assorting independently (Table 32.1) including examples related to gene interaction.

Table 32.1. Phenotypic ratios in crosses between individuals heterozygous for two gene pairs , each gene pair assorting independently.

AaBb × AaBb

↓

gametes

		AB	**Ab**	**aB**	**ab**	
gametes	**AB**	AABB (1)	AABb (2)	AaBB (3)	AaBb (4)	genotypes
	Ab	AABb (5)	AAbb (6)	AaBb (7)	Aabb (8)	
	aB	AaBB (9)	AaBb (10)	aaBB (11)	aaBb (12)	
	ab	AaBb (13)	Aabb (14)	aaBb (15)	aabb (16)	

Each gene pair affecting a different character	**Offspring expected in F2**		**Numbered genotypes in the above checker board**
	Ratio	**Phenotype**	
1. Complete dominance at both gene pairs	9/16	yellow, round	1, 2, 3, 4, 5, 7, 9, 10, 13
Example: Mendel's peas	3/16	yellow, wrinkled	6, 8, 14
Gene pair A: (seed colour), yellow dominant over green	3/16	green, round	11, 12, 15
Gene pair B (seed shape), round dominant over wrinkled	1/16	green, wrinkled	16

Each gene pair affecting a different character	Offspring expected in F2		Numbered genotypes in the above checker board
	Ratio	**Phenotype**	
2. Complete dominance at one gene pair, partial or over-dominance at the other Example: cattle Gene pair A: polled (hornless dominant over horned Gene pair B: red hair colour partially dominant over white producing "roan" heterozygotes	6/16 3/16 3/16 2/16 1/16 1/16	polled, roan polled, red polled, white horned, roan horned, red horned, white	2, 4, 5, 7, 10, 13 1, 3, 9 6, 8, 14 12, 15 11 16
3. Partial or co-dominance at both gene pairs Example: human blood group Gene pair A: (ABO), A and B co-dominant Gene pair B: (MN), M and N co-dominant	1/16 2/16 1/16 2/16 4/16 2/16 1/16 2/16 1/16	AAMM AAMN AANN ABMM ABMN ABNN BBMM BBMN BBNN	1 2, 5 6 3, 9 4, 7, 10, 13 8, 14 11 12, 15 16
Each gene pair affecting the same character (gene interaction)			
4. Complete dominance at both gene pairs, new phenotypes resulting from interaction between dominants and also from interaction between recessives Example: comb shape in poultry Gene pair A: rose comb dominant over non-rose Gene pair B: pea comb dominant over non-pea Interaction: Dominants for rose and pea produce walnut comb and homozygous recessives for rose and pea produce single comb	9/16 3/16 3/16 1/16	walnut rose pea single	1, 2, 3, 4, 5, 7, 9, 10, 13 6, 8, 14 11, 12, 15 16
5. Complete dominance at both gene pairs, but one gene, when recessive, is epistatic to the other			

Each gene pair affecting a different character	Offspring expected in F2		Numbered genotypes in the above checker board
	Ratio	Phenotype	
Example: mouse coat colour Gene pair A: colour dominant over albino Gene pair B: agouti colour dominant over black Interaction: homozygous albino is epistatic to agouti and black This 9:3:4 phenotypic ratio is typically called as recessive epistasis. If the recessive genotype at one locus (e.g. aa) suppresses the expression of alleles at the B-locus, the A-locus is said to exhibit recessive epistasis over the B-locus. Only if the dominant allele is present at the A-locus, can the alleles of hypostatic B-locus be expressed. The genotypes A-B- and A-bb produce two additional phenotypes. The 9:3:3:1 ratio becomes 9:3:4 ratio.	9/16 3/16 4/16	agouti black albino	1, 2, 3, 4, 5, 7, 9, 10, 13 6, 8, 14 11, 12, 15, 16
6. Complete dominance at both gene pairs, but either recessive homozygote is epistatic to the effect of other gene. This is also known as duplicate recessive gene interaction producing a 9 : 7 phenotypic ratio. In this case, identical phenotypes are produced by both homozygous recessive genotypes. The genotypes, aaB-, A-bb and aabb produce one phenotype. Both the dominant alleles when present together produce the same phenotypic effect giving rise to 9:7 phenotypic ratio			

Each gene pair affecting a different character	Offspring expected in F2		Numbered genotypes in the above checker board
	Ratio	Phenotype	
in the F_2. This amounts to complementary interaction of two independent pairs of alleles Example: sweet pea flower colour	9/16	purple	1, 2, 3, 4, 5, 7, 9, 10, 13
Gene pair A: purple dominant over white Gene pair B: colour dominant to colourless (white)	7/16	white	6, 8, 11, 12, 14, 15, 16
7. Complete dominance at both gene pairs, but one gene when dominant, epistatic to the other Example: fruit colour in summer squash	12/16	white	1, 2, 3, 4, 5, 6, 7, 8, 9, 10, 13, 14
Gene pair A: white dominant to colour	3/16	yellow	11, 12, 15
Gene pair B: yellow dominant to green This kind of gene interaction is typically called as dominant epistasis where the dominant allele at one locus, for example the A allele, produces a certain phenotype regardless of the allelic condition at the other locus. Then the A locus is said to be epistatic to the B locus. Furthermore, since dominant allele A is able to express itself in the presence of either B or b, this is a case of dominant epistasis. Only when the genotype of individual is homozygous recessive at the epistatic locus (aa) can the alleles of hypostatic locus (B or b) be expressed. Thus, the genotypes A-B- and A-bb produce the same	1/16	green	16

Each gene pair affecting a different character	Offspring expected in F2		Numbered genotypes in the above checker board
	Ratio	Phenotype	
phenotype whereas aaB- and aabb produce two different phenotypes and the classical 9:3:3:1 ratio becomes modified into a 12:3:1 ratio.			
8. Complete dominance at both gene pairs, but either gene, when dominant, epistatic to the other, also known as duplicate dominant genes Example: seed capsule of shepherd's purse (*Bursa*) Gene pair A: triangular shape dominant to ovoid Gene pair B: triangular shape dominant to ovoid Interaction: a dominant allele at either gene pair hides the effect of ovoid	15/16 1/16	triangular ovoid	1-15 16
9. Complete dominance at both gene pairs, but one gene, when dominant, epistatic to the second, and the second gene, when homozygous recessive, epistatic to the first, also known as dominant and recessive interaction, also known as typical epistasis in earlier days Example: feather colour in fowl Gene pair A: colour inhibition is dominant to colour appearance Gene pair B: colour is dominant to white Interaction: dominant colour inhibition prevents colour even when colour is present, colour gene, when homozygous recessive, prevents colour even when dominant inhibitor is absent	13/16 3/16	white colour	1, 2, 3, 4, 5, 6, 7, 8, 9, 10, 13, 14, 16 11, 12, 15

Each gene pair affecting a different character	Offspring expected in F2		Numbered genotypes in the above checker board
	Ratio	Phenotype	
10. Complete dominance at both gene pairs, interaction between both dominants to give new phenotypes, also known as duplicate genes with cumulative effect Example: fruit shape in summer squash Gene pair A: sphere shape dominant over long Gene pair B: sphere shape dominant over long Interaction: dominants at A and B, when present together, form disc shape fruit	9/16 6/16 1/16	disc sphere long	1, 2, 3, 4, 5, 7, 9, 10, 13 6, 8, 11, 12, 14, 15 16
11. Complete dominance at one gene pair, partial dominance at the other gene pair, first gene, when homozygous recessive, is epistatic to the second gene Example: hair direction in guinea pig Gene pair A: rough hair dominant over smooth Gene pair B: modifies rough to smooth (A-BB mostly smooth, A-Bb partly rough, A-bb full rough	7/16 6/16 3/16	smooth partly rough full rough	1, 3, 9, 11, 12, 15, 16 2, 4, 5, 7, 10, 13 6, 8, 14
12. Complete dominance at one gene pair, partial dominance at the other gene pair, either gene when homozygous recessive, is epistatic to the other gene, when both genes are homozygous recessive, the second gene is epistatic to the first gene Example: body colour in the flour bettle	6/16	sooty	2, 4, 5, 7, 10, 13

Each gene pair affecting a different character	Offspring expected in F2		Numbered genotypes in the above checker board
	Ratio	Phenotype	
Gene pair A: normal (red) is completely dominant over jet (a darker colour with tinge of red) Gene pair B: normal (red) is partially dominant over black, the black heterozygote appears sooty in colour Interaction: when jet is homozygous recessive (aa), the effect of BB or Bb does not appear, the double recessive (aabb) is black	3/16 3/16 4/16	red jet black	1, 3, 9 11, 12, 15 6, 8, 14, 16
13. Complete dominance at one gene pair, partial dominance at the other gene pair, the partially dominant heterozygote has the same phenotype as the recessive homozygote at the other gene pair and an additive effect when they are both combined Example: body colour in the flour beetle Gene pair A: normal (red) is completely dominant over sooty Gene pair B: normal (red) is partially dominant over black, the black heterozygote is sooty in colour and when combined with the sooty homozygote produces a dark sooty phenotype	7/16 4/16 3/16 2/16	sooty black red dark sooty	2, 4, 5, 7, 10, 11, 13 6, 8, 14, 16 1, 3, 9 12, 15
14. Partial dominance at both gene pairs, additive effects for each partially dominant gene Example: flower colour in beans (Mendel) Gene pair A: purple flower colour partially dominant to white, additive effect on colour for each A gene (i.e. value of 3)	1/16 AABB 2/16 AABb 2/16 AaBB 1/16 AAbb 4/16 AaBb 1/16 aaBB 2/16 Aabb 2/16 aaBb	purple shade 10 purple shade 8 purple shade 7 purple shade 6 purple shade 5 purple shade 4 purple shade 3 purple shade 2	1 2, 5 3, 9 6 4, 7, 10, 13 11 8, 14 12, 15

Each gene pair affecting a different character	Offspring expected in F2		Numbered genotypes in the above checker board
	Ratio	Phenotype	
Gene pair B: purple flower colour partially dominant to white, additive effect on colour for each B gene (i.e. value of 2)	1/16 aabb	purple shade 0 (white)	16

Modifiers

In the early days of historical developments in genetics, phenomenon of gene action and interaction were little known. To many geneticist, action of a particular gene appeared to produce a particular phenotype and that was considered to be independent of action of all other genes. The discoveries of examples where effect of one gene changed or modified the effect of another gene were always of special interest. As more examples of gene interaction came to light, several terms came to be used to describe some of these interactions. Some of the earlier terms used like complementary genes and epistasis are still in use because they help us to understand the phenotypic behavior of particular genes. However, now it is well accepted that the phenotypes as they are observed are the result of more complex processes than were first imagined. Epistasis, for example, is not to be taken literally as one gene hiding or masking another gene but may be consequence of increased or decreased enzyme activity, change in the acidity that affect the appearance of colour or there could be various other complex events.

With this reservation in mind, it can be safely mentioned that there could be a large group of genes which come under general heading of modifiers but whose particular form of action are not always known. Broadly defined, modifiers are genes which change the phenotypic expression of other genes in a quantitative fashion. Many of the genes considered in Table 32.1 are of this kind. For example, the dilution and sooty factors involved in guinea pig coat colour modify the effect of other gene through dilution or enhancement. Similarly, the inhibitor gene that prevents the appearance of feather colour in chickens can be considered as modifier gene of the normal fowl colour genes.

Lethality

Perhaps the most serious effect of a gene could be the death of the organism carrying that gene. In 1905, a French geneticist (Cuenot) reported the inheritance of a mouse-body colour gene 'Yellow' that did not fit to the expected Mendelian segregation pattern. Yellow body colour was found to behave like a dominant gene but it could not be bred as a pure yellow strain because crosses of Yellow x Yellow always produced the normal agouti and as well as yellow offspring . The yellow offspring were found to be heterozygous for the Yellow gene through back-crossing and no Yellow homozygous individuals could be found. A cross of Yellow x Yellow always produced offspring in the ratio 2/3 yellow : 1/3 agouti instead of normally expected 3/4 yellow : 1/4 agouti as shown below in Fig. 32.1.

Yellow x Yellow
y/+ y/+

↓

		y
+	++ dies	+y yellow
y	+y yellow	yy agouti

Phenotypic ratio = 2 yellow : 1 agouti

Fig. 32.1. Segregation of yellow gene (y) in mice in a cross of y/+ x y/+.

Soon after discovery of lethal gene in yellow mice, Baur reported a similar lethal gene in plants. In snapdragon, *Antirrhinum majus*, a yellow leaved dominant genotype, *aurea*, when crossed *aurea* x *aurea*, always produced normal green plants and *aurea* heterozygotes. *Aurea* homozygotes lacked the ability to make chlorophyll and these plants died either before germination or as seedlings.

Since then, several other lethal genes have been reported which have a dominant phenotypic effect in the heterozygote but are lethal in homozygous conditions. Recessive lethal genes have also been reported.

33

Genetics of Sex

In most of higher organisms including animals, the aspect of sexes has been normally confined to males and females which exist separately or even in the same individual. An individual possessing both male and female reproductive organs is usually referred to as hermaphrodite. In plants, where staminate (male) and pistillate (female) flowers occur on the same plant, the term used is monoecious. The most common plants under monoecious form are cucurbits. However, majority of the flowering plants have male and female parts within the same flower and this situation is referred to as hermaphrodite. There are relatively less numbers of plants having male and female flowers located on different plants and this is termed as dioecious condition. Among the common cultivated crops known to be dioecious are asparagus, date palm, hemp, hops and spinach.

Mechanisms of Sex Determination

Most mechanisms of sex determination are genetically determined and may be classified into one of the following types:

1. Sex Chromosome Mechanism

(i) Heterogametic Males: In man and probably in all other mammals, the presence of Y chromosome may determine a tendency toward maleness. This system is broadly working in Drosophila also. Normal males are chromosomally XY and the females are XX. Since male produces two kinds of gametes as far as the sex chromosomes are concerned, the male is said to be heterogametic sex. On the other hand, the females producing only one kind of gametes considering the sex chromosomes are said to be of homogametic sex. This mode of sex determination is commonly referred to as XY type. This is demonstrated as follows:

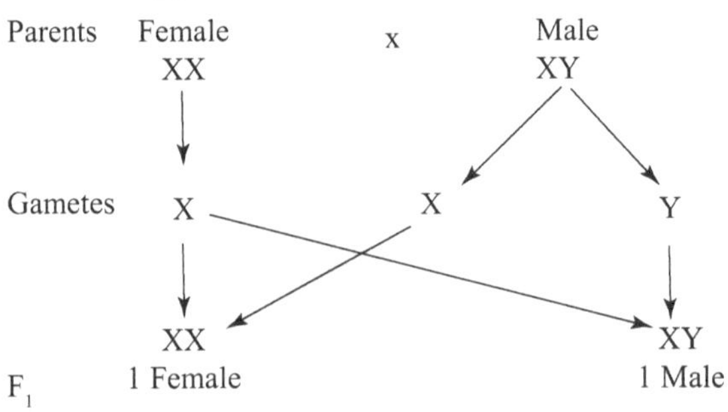

In some insects of orders Hemiptera (true bugs) and Orthoptera (grasshoppers), males are heterogametic but produce X-bearing sperms or gametes without a sex chromosome. In males of these species, the X chromosome has no homologous pairing chromosome because no Y chromosome is present in the males. Thus, males exhibit odd chromosome numbers in their chromosome complement. The one-X and two-X conditions determine maleness and femaleness, respectively. This mode of sex determination commonly referred to as XO type is illustrated as follows:

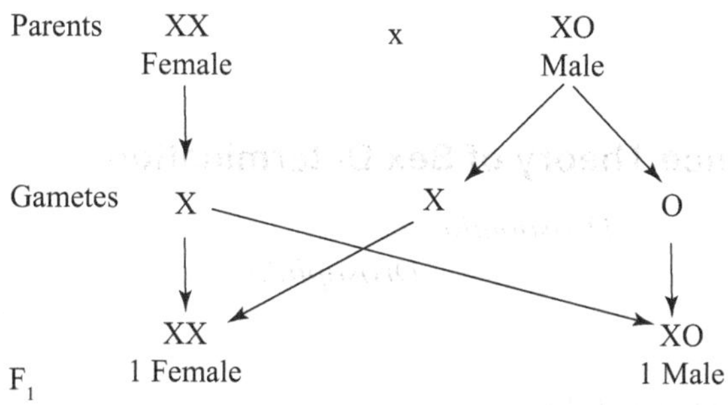

(ii) Heterogametic Female: This method of sex determination is found in comparatively large number of animals including butterflies, moths, silk worms and in some birds and fish. The one-X and two-X conditions determine femaleness and maleness, respectively. Females of some species have a chromosome similar to that of Y in man. In these cases, chromosomes are sometimes labeled as Z and W instead of X and Y, respectively to highlight the fact that the female (ZW) is the heterogametic sex and the male (ZZ) is the homogametic sex. The females of domestic chicken have no homologue to the single X chromosome as in the case of XO mechanism of sex determination described previously. In this case the designated symbols are ZZ for males and ZO for the females. A 1: 1 sex ratio is expected in both the cases. Both the methods of sex determination are illustrated as follows:

ZO Method of Sex Determination:

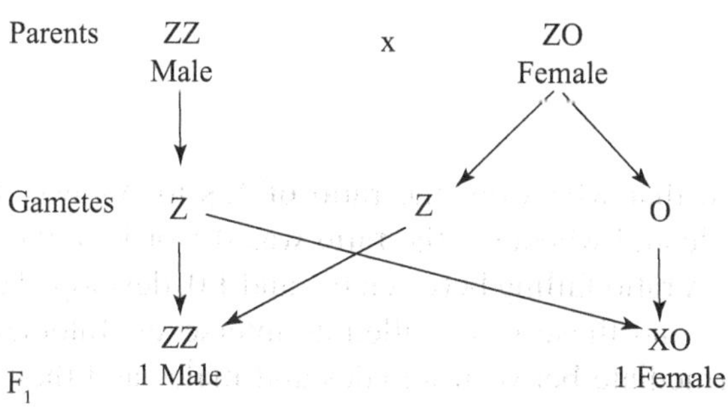

ZW Method of Sex Determination:

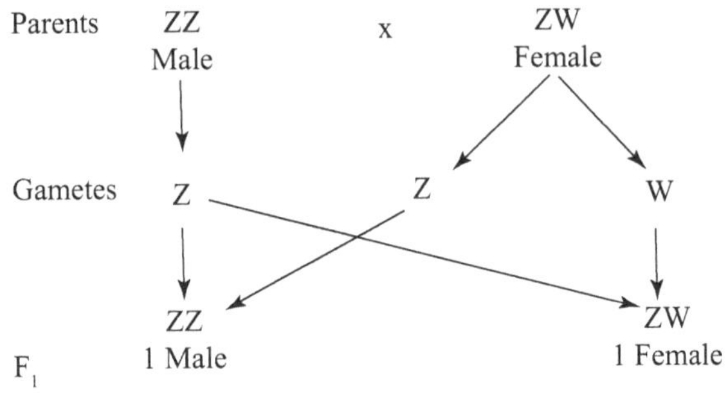

Genetic Balance Theory of Sex Determination

The Y chromosome in *Drosophila* unlike that in humans plays no role in sex determination. Instead, the sex of *Drosophila* is decided by the ratio of X chromosomes to autosomes. This mechanism was first demonstrated by Bridges in 1921 through an analysis of flies with unusual chromosome constitutions. Normal diploid *Drosophila* has a pair of sex chromosomes (XX or XY) and three pairs of autosomes usually denoted by AA where each A represents one haploid set of autosomes. Through various genetic manipulations, Bridges produced flies with abnormal number of chromosomes and sex types as given in Table 33.1.

Table 33.1. Ratio of X chromosomes to autosomes and corresponding sexual types in Drosophila.

X-chromosomes	Sets of autosomes	X:A ratio	Sex form
1X	2A	0.5	Male
2X	2A	1.0	Female
3X	2A	1.5	Meta female or super female
4X	3A	1.33	Meta female or super female
4X	4A	1.0	Tetraploid female
3X	3A	1.0	Triploid female
3X	4A	0.75	Intersex
2X	3A	0.67	Intersex
2X	4A	0.5	Tetraploid male
1X	3A	0.33	Meta male or super male

Bridges observed that whenever, the ratio of X's to A's was 1.0 or greater, the fly was female and whenever the ratio was 0.5 or less, the fly was male. Flies with an X : A ratio falling between 0.5 and 1.0, developed characteristics of both the sexes and those were called as intersexes. Intersexes are sterile individuals intermediate between females and males and they should not be

confused with gynandromorphs which are individuals having some parts of the body pure female and others male. Super females and super males are also sterile individuals that show some relatively minor differences from ordinary females and males. Bridges gave the following interpretation to his findings:

The intersexes and 3n (triploid) types lead to the conclusion that sex in *Drosophila melanogaster* is determined by the autosomes as well as by the X chromosomes, the ratio of autosomes to X's being the significant relation. The old formula of 2X = female is at once seen to be inadequate, for here there are individuals with two X chromosomes and yet, they are not females. They are shifted out of female class by the presence of an extra set of autosomes, and thereby the autosomes are proved to play a positive role in deciding the sex. Since the intersexes differ from female by the assumption of certain male characters, this effect of the autosomes is due to an internal preponderance of male tendency genes. The sex relation could be reformulated as follows: Both sexes are due to the simultaneous action of two opposed sets of genes, one set tending to produce the characters called female and the other to produce the characters called male. These two sets of genes are not equally effective. The male tendency genes in the autosomes are more numerous or more effective than those in the X chromosomes, while the net effect of genes in the X chromosome is tendency to the production of female characters. When in a diploid zygote the relative effectiveness of the female-tendency genes is lowered by the absence of one X, the male-tendency genes outweigh the female and the result is a normal male. When the two sets of genes are acting in a ratio between these two extremes, as is the case in the ratio of 2X : 3 sets of autosomes (2X : 3A), the result is sex intermediate- the intersex.

Now it is known that an X-linked gene called Sex-lethal (*Sxl*) is a key player in the ***Drosophila*** sex determination system (Fig. 33.1, Snustad ***et al.***, 1997). An elaborate network of other X-linked genes, working with factors already present in the ***Drosophila*** eggs, sets the level of ***Sxl*** activity in the zygote. If the X:A ratio is greater than or equal to one, the ***Sxl*** gene is activated and the zygote develops as female. If the ratio is less than or equal to 0.5, the ***Sxl*** gene is inactivated and the zygote develops as a male. A ratio between 0.5 and 1.0 causes a mix of signals and the zygote develops both male and female characteristics. Inappropriate expression of ***Sxl***-off in female or on in male leads to embryonic death and hence the gene has been named as Sex-lethal. Molecular details of sex determination in ***Drosophila*** and the nature of lethality associated with inappropriate expression of the ***Sxl*** are now fully understood.

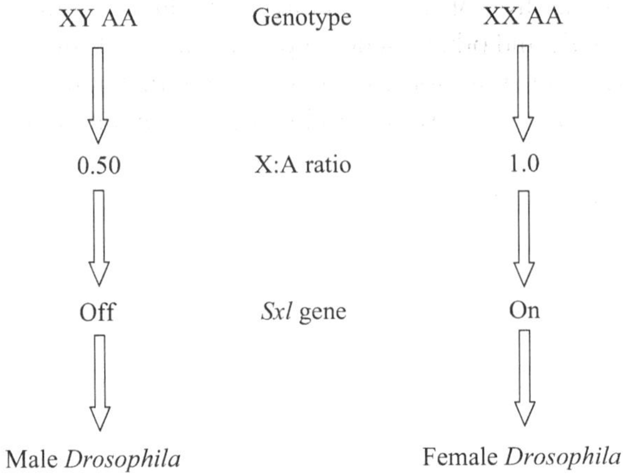

Fig. 33.1. The Sxl sex determination switch in Drosophila.

Sex Determination in Human Beings

The fact that XX individuals in human beings are females and XY as the males suggests that sex might be determined by the number of X chromosomes or by the presence or absence of Y chromosome. Now it is understood that the second hypothesis is correct. In humans and other placental mammals, maleness is due to a dominant effect of the Y chromosome evidenced by the fact that XO animals develop as females and XXY animals develop as males. The dominant effect of Y chromosome is expressed early in development when it directs the primordial gonads to develop into testes. Once the testes have formed, they secrete testosterones, a hormone that stimulates the development of male secondary sexual characteristics. Further, it has been known that the testes determining factor (TDF) is present on Y chromosome in human beings. This factor appears to correspond to a gene called *SRY* (for sex-determining region Y) which is located just outside the pseudoautosomal region in the chromosome's short arm. The discovery of *SRY* was made possible by the identification of unusual individuals whose sex was inconsistent with their chromosome constitution where XX males and XY females were identified. Some of the XX males were found to carry a small piece of the Y chromosome inserted into one of the X chromosomes. This piece evidently carried a gene responsible for maleness. On the other hand, some of the XY females were found to have an incomplete Y chromosome. The part of the Y chromosome that was missing in XY females corresponded with the piece that was present in XX males. Thus, the absence of this critical piece of Y chromosome in XY females prevented them from developing testes. These complementary lines of evidence showed that a particular segment of Y chromosome was needed for maleness. Molecular approaches later on identified the *SRY* gene in this male determining segment in the Y chromosome. Additional research has shown that an *SRY* gene is present on the Y chromosome of mouse. As in human beings, this gene was also found to specify maleness in the mouse. Both human and mouse *SRY* genes encode a protein that can bind to DNA resulting into regulation of the activity of the other genes that are directly involved in the process of testes formation.

Sex-Linked Inheritance

Any gene located on the X chromosome (mammals, ***Drosophila***, and others) or on the analogous Z chromosome (birds, etc with ZO or ZW mechanism of sex determination) is said to be sex-linked. The first sex-linked gene found in Drosophila was the recessive white eye mutation. Reciprocal crosses involving autosomal traits yield comparable results but this does not hold true with sex-linked traits as shown below:

Parents	Xw/ Xw White eye-female	x	X+/Y Wild type (red eye)-male
Gametes	Xw		X+, Y
F_1	Xw/X+ Red eye-female		Xw/Y White eye-male

In this case when white eyed females are crossed with the red eyed males, all the male offspring have white eyes like their mother and all the female offspring have red eyes like their father. This criss-cross method of inheritance is called as sex-linked inheritance and the particular genes located on sex chromosomes (for example X chromosome) are called as sex-linked genes. This peculiar type of inheritance is due to the fact that the Y chromosome carries no allele's homologues to those at the white eye locus on the X chromosome. In fact in most organisms, the Y chromosome is virtually devoid of known genes excluding those related to maleness in certain cases, for human beings. Thus, males carry only one allele for sex-linked traits. This onc allcle condition is termed as hemizygous in contrast to the homozygous or heterozygous conditions in the females. If the F_1's of the above case are allowed to mate among themselves, the F2 generation will be as follows:

F_1 X+/Xw x Xw/Xy

↓

	Xw	Xy
X+	X+/Xw red, female	X+/Xy red, male
Xw	Xw/Xw white, female	Xw/Xy white, male

F_2 = 1 red: 1 white phenotypic ratio in both males and female.

In diploid organisms with sex determining mechanism based on XY system (XX = female and XY = male), a trait governed by sex-linked recessive gene usually exhibits itself in the following manner:

- It is usually found more frequently in male than in the female.
- It fails to appear in the female unless it also appeared in the paternal parent.
- It seldom appears in both father and son, then only if the maternal parent is heterozygous.

On the other hand, a trait governed by sex-linked dominant gene usually manifests itself by:

- Being found more frequently in the female than the male.
- Being found in all the female offspring of a male which shows the trait.
- Failing to be transmitted to any son from a mother which did not exhibit the trait herself.

Sex-Linked Genes in Human Beings

Development of chromosome theory depended on the discovery of white eye mutation in *Drosophila*. Morgan demonstrated it to be an X-linked gene i.e. sex-linked gene as explained above. It is also true that such mutants are easiest to detect because they show up immediately in hemizygous males. In contrast, autosomal recessive mutations show up only after two mutant alleles have been brought together in a homozygote which is much more unlikely to happen. In human beings too, recessive X-linked traits are more easily identified than the recessive autosomal traits. A male only needs to inherit one recessive allele to show an X-linked trait whereas a female needs to inherit two, one each from her parents. Thus preponderance of people who show X-linked traits are males. Two typical examples of sex-linked traits in human beings are as follows:

Hemophilia-Blood Clotting Disorder: In human beings, hemophilia is one of the best known examples of an X-linked trait. People with this disorder are unable to produce a factor needed for blood clotting. Thus, people with such disorder continue to bleed after cuts and wounds and may die if the bleeding is not stopped by therapeutic treatment. Nearly all the affected individuals in the population are males. Very few females with this disorder have been reported. The most famous case of hemophilia occurred in the Russian imperial family at the beginning of the twentieth century. Czar Nicholas and Czarina Alexandra had four daughters and one son. The son, Alexis, suffered from hemophilia. The X-linked mutation responsible for Alexis's disease was transmitted to him by his mother who was a heterozygous carrier. Czarina Alexandra was a grand-daughter of Queen Victoria of Great Britain, who was also a carrier. Pedigree records show that Victoria transmitted the mutant allele to three of her nine children, namely, Alice, who was Alexandra's mother, Beatrice, who had two sons with the disease and Leopold, who had the disease himself. The allele that Victoria carried evidently arose as a new mutation in the germ cells of one of her parents or in those of a more distant maternal ancestor. Some authors have even argued that this mutation changed the course of history by contributing to downfall of the Romanov dynasty in Russia and subsequent rise of Soviet state.

Colour Blindness-an X-linked Vision Disorder: In human being colour perception is mediated by light absorbing proteins in the specialized cone cells of the retina in the eye. Three such proteins have been identified, one to absorb blue light, second to absorb red light and the third one to absorb green light. Colour blindness results due to abnormality in any of these three receptor proteins. The classical case of colour blindness involves red-green colour blindness where red and green are seen as the same colour. This abnormality in colour recognition follows an X-linked pattern of inheritance. About 5-10 percent of human males are red-green colour blind. However, among females, this frequency is much lower, close to less than 1 percent suggesting that the mutant allele for colour blindness is recessive. Molecular studies have shown that there are two distinct genes for colour perception on the X chromosome. One gene encodes the receptor for green light and the other gene encodes the receptor for red light. A third gene for colour perception, the one encoding the receptor for blue lights is located on an autosome.

Fragile X Syndrome and Mental Retardation: In human beings, many cases of mental retardation appear to follow an X-linked inheritance. Most of these are associated with a cytological anomaly that is detectable in cells that have been cultured in the absence of certain nucleotides, the building blocks of DNA. This anomaly, a constriction near the tip of the long arm of the X chromosome gives the impression that the tip is ready to detach from the rest of chromosome. The clinical features of the fragile X syndrome vary considerably resulting into poor diagnosis. Many patients show significant mental retardation and some show facial and behavioural abnormalities.

This can occur both in males and females. The incidence of this syndrome is about one in 2000.

Sex-Influenced Traits

The genes governing sex-influenced traits may reside on any of the autosomes or on the homologous portions of the sex chromosomes. The expression of dominance or recessiveness by the alleles of the sex-influenced loci is reversed in males and females due in large part to the differences in the internal environment provided by sex hormones. Thus, such examples are found mostly in higher animals with well developed endocrine system. A typical example of sex-influenced trait is given in Table 33.2.

Table 33.2. The gene for pattern baldness in man exhibiting dominance in men but recessiveness in women.

Genotypes	Phenotypes	
	Men	Women
b'b'	Bald	Bald
b'b	Bald	Non-bald
bb	Non-bald	Non-bald

Sex-Limited Traits

Some genes may express in one of the sexes because of differences in the internal hormonal environment or because of anatomical differences. For example, it is known that bulls have many genes for milk production which they may transmit to their daughters but the males/sons are unable to express this trait. The production of milk is therefore limited to variable expression in only the female sex. In another words, when the penetrance of a gene is zero in one sex, then the trait is termed as sex-limited trait. This is illustrated in Table 33.3.

Table 33.3. Expression of cock-feathering in chickens.

Genotypes	Phenotype	
	Males	Females
HH	Hen-feathering	Hen-feathering
Hh	Hen-feathering	Hen-feathering
hh	Cock-feathering	Hen-feathering

Sexual Phenomenon in Plants

Most of the plants are monoecious and are devoid of sex chromosomes. Ability of mitotically produced cells with exactly the same genetic endowment to produce tissues with different sexual functions in a perfect flower (male and female organs in the same flower) indicates clearly the bipotentiality of such plant cells. Well known examples of dioecism usually are under the control of single gene locus. However, there is at least one case of well documented chromosomal sexuality in plant. In the genus *Melandrium*, Y chromosome determines tendency to maleness just as it does in human beings. Pistillate plants are XX and the staminate plants are XY.

The ability of gametes produced by same individuals to unite and to produce viable and fertile offspring is common among many families of flowering plants. This amounts to self pollination. Self-fertilization is obligatory in some major agricultural crops like wheat, barley, rice, beans, peas, soybean, oats, tobacco, tomato and many other crops. In some species, self and cross-fertilization may occur to varying degrees. For example, cotton and sorghum commonly exhibit more than 10% cross-pollination and are classified as often cross-pollinated crops. Self-incompatibility enforces cross-fertilization in several otherwise self-pollinated crops.

Non-Disjunction as Proof of the Chromosome Theory

Morgan showed that the gene for white colour in *Drosophila* was located on the X chromosome through correlating the transmission of gene with that of X chromosome. However, it was C. B. Bridges (one of students of Morgan) who secured the proof of chromosome theory by showing that exceptions to the rules of inheritance could also be explained by chromosome behavior. Bridges performed one of Morgan's experiments on a large scale by crossing white-eyed females with red-eyed males and observing the F_1 progeny. All the F_1 flies were either red-eyed-females or white-eyed males as expected. However, Bridges found a few exceptional flies, i.e. white-eyed females and red-eyed males. He crossed these exceptional flies to understand as to how they could have arisen. All exceptional males proved to be sterile but the exceptional females were fertile. When crossed to normal red-eyed males, they produced many progeny including large numbers of white-eyed daughters and red-eyed sons. Thus exceptional F_1 females though rare in number, were able to produce many exceptional progeny.

Bridges explained these results by proposing that the exceptional F_1 flies were the result of abnormal X chromosome behaviour during meiosis in the females of the parent generation. Normally X chromosomes in these females should disjoin or separate from each other during meiosis. However, occasionally they might fail to separate producing an egg with two X chromosomes or an egg with no X chromosome at all. Fertilization of such eggs with normal sperm would produce zygotes with an abnormal number of sex chromosomes as shown in Fig. 33.2.

White-eyed female x Red-eyed male

XwXw Xw+Y

Normal eggs Nondisjunctional eggs

Xw XwXw Nullo-X

Xw+	Xw+Xw XX red-eyed female	Xw+XwXw XXX metafemale, usually dies	Xw+ XO exceptional red-eyed male
Y	XwY XY white-eyed male	XwXwY XXY exceptional white-eyed female	YO dies

Fig. 33.2. X-chromosome non-disjunction in Drosophila.

X chromosome non-disjunction is responsible for the exceptional progeny that appeared in Bridges' experiment. Non-disjunctional eggs that contain either two X chromosomes or no X chromosome unite with normal sperm that has either an X chromosome or a Y chromosome to produce four types of zygotes. The XXY zygotes develop into white-eyed females, the XO zygotes develop into red-eyed sterile males and the XXX and YO zygotes die.

Bridges' ability to explain the exceptional progeny that came from these crosses showed the power of the chromosome theory. Each of these exceptions was due to anomalous chromosome behaviour during meiosis. Bridges called the anomaly non-disjunction because it involved a failure of the chromosomes to disjoin during one of the meiotic divisions. This failure could result from faulty chromosome movement, imprecise or incomplete pairing or centromere malfunction.

34

Linkage, Crossing Over and Chromosome Mapping in Diploids

Linkage

When two or more genes are located on the same chromosome, they are said to be linked. Genes on different chromosomes are distributed to the gametes following the law of independent assortment. However, genes located on the same chromosome, tend to stay together during gamete formation. For example, results of test-crossing dihybrid individuals will produce different results depending upon whether the genes are linked or are located on the same chromosome. This is illustrated as follows:

Genes on Different Chromosomes Assort Independently Giving a 1:1:1:1 Testcross Ratio

Parents AaBb x aabb

Gametes AB, Ab, aB, ab ab

F_1 1/4 AaBb : 1/4 Aabb : 1/4 aaBb : 1/4 aabb

Genes Located on the Same Chromosome (Linked Genes) Do Not Assort Independently But Tend to Stay Together (For Notation, Genes to the Left of the Slash Line (/) Are Assumed to Be on One Chromosome and Those to the Right are Considered on Homologous Chromosome)

Example of linked gene:

Parents AB/ab x ab/ab

Gametes AB, ab x ab

F_1 1/2 AB/ab, 1/2 ab/ab

Thus, if the test-cross ratio for a dihybrid is 1 : 1 : 1 : 1, then law of independent assortment is operating and apparently there is no linkage between the two genes. On the other hand, significant deviation from a 1 : 1 : 1: 1 ratio in the testcross progeny of a dihybrid could be used as an evidence of linkage. If the two genes are tightly linked and there is no crossing over between the two genes, the testcross ratio will be 1: 1 for AB/ab and ab/ab. However, linked genes always do not remain together because of crossing over between non-sister chromatids resulting into exchange of varying length of chromosomes with one another during meiotic prophase. The homologous chromosomes pair with one another during 'synapsis' in meiosis and the point of exchange is called as "chiasmata". The net result of this process is production of recombinant gametes through crossing over.

Crossing Over

During meiosis I, each chromosome replicates and forms two identical sister chromatids. There is pairing between the two homologous chromosomes and crossing over occurs between the non-sister chromatids. The crossing over involves the breakage and reunion of two of the four strands at any given point on the chromosome. Fig. 34.1 shows the crossing over in the region of chromosome between the two genes A and B.

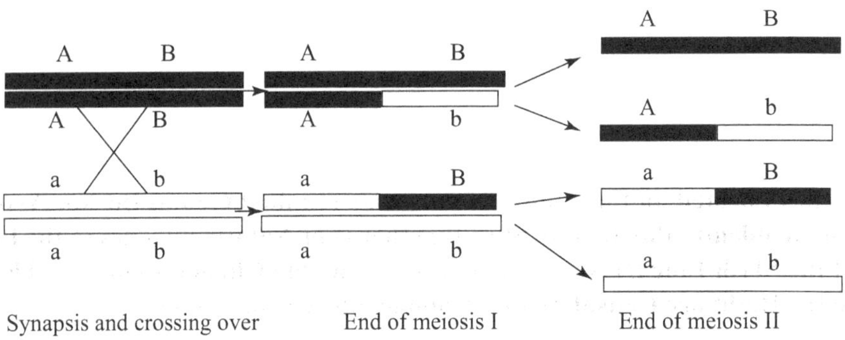

Fig. 34.1. Crossing over between two non-sister chromatids.

If the two dominant or wild type alleles are on one chromosome and the two recessivs or mutants on the other (AB/ab), the linkage relationship is called as coupling phase. When the dominant allele of one locus and the recessive allele of the other locus occupy the same chromosome (Ab/aB), the relationship is termed as repulsion phase. Parental and recombinant gametes will be of different types, depending upon coupling or repulsion phase linkage.

Coupling phase parents: AB/ab
Parental gametes: AB and ab

Recombinant gametes:	Ab and aB
Repulsion phase parents:	Ab/aB
Parental gametes:	Ab and aB
Recombinant gametes:	AB and ab

A pair of synapsed chromosomes (bivalent) consists of four chromatids called a tetrad as depicted in Fig. 34.1. Each tetrad usually experiences at least one chiasma along its length. Longer the distance between the two genes, greater is the probability of formation of chiasma. If the genes are closer then the probability of chiama formation is smaller. In other words, frequency of chiama formation between the two linked genes is proportional to the distance between the two genes. The percentage of crossover (recombinant) gametes formed by a given genotype is a direct reflection of the frequency with which a chiasma forms between the genes under consideration. Only when there is a crossover between the gene loci under consideration, only then recombination will be detected. When a chiasma forms between two genes, only half of the meiotic products will be crossover/recombinant types. Therefore, chiasma frequency is twice the frequency of crossover products. This relationship is shown as follows:

Chiasma% = 2 (crossover%) OR Crossover% = 1/2 (chiasma%)

For example, if a chiasma forms between the loci of genes A and B in 20% of the tetrads of an individual with the genotype AB/ab, then 10% of the gametes will be recombinant type (Ab, aB) and 90% will be of parental type (AB, ab).

Multiple Crossovers

When two strand double crossovers occur between two genes, the products detected through the progeny phenotypes are only the parental types as shown below:

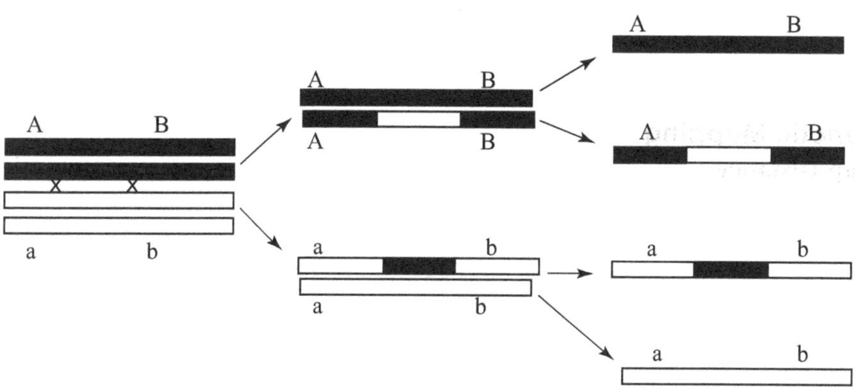

Synapsis and crossing over (X) End of meiosis I End of meiosis II

Thus, the parental genes combinations remained unchanged in the gametes despite two chiasmata (double crossovers) between the two genes. In order to detect these double crossovers, a third gene C between the two outside marker genes must be present and used as illustrated below:

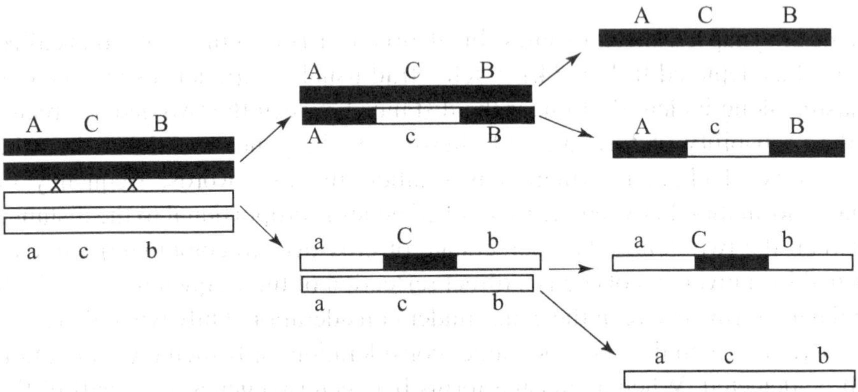

If there is a certain probability of occurrence of crossover between the genes A and C and that of another between the genes C and B, then the probability of double crossover between and A and B is given by the products of the two independent probabilities (one between A and C and another between C and B). For example, there is crossover in 20% of tetrads for gene A and C and the corresponding value for C and B is 10% in the genotype, ACB/acb, then 0.2 x 0.1 = 0.02 = 2% of the gametes are expected to be of double crossover types with the genotype, AcB/aCb.

Limits of Recombination

If the two genes are far apart in the chromosome such that the probability of chiasma formation between them is 100%, then 50% of gametes will be parental types (non-crossover) and 50% of the gametes will be recombinant types (crossover). Thus, recombination between two linked genes cannot exceed 50% even when multiple crossovers occur between the two genes.

Genetic Mapping

Map Distance

The genes are positioned in the chromosomes in the linear order analogous to beads on a string. The genetic mapping deals with determining the linear order of the genes on the chromosomes and determining the relative distance between the genes. The unit of distance which has the greatest utility in predicting the outcome of certain types of mating is an expression of the probability that crossing over will occur between the two genes under consideration. One unit of map distance (centimorgan) is equivalent to 1% crossing over.

A few examples are given to clarify the points. If the genotype Aa/Bb produces 10% of each of the crossover/recombinant gametes, AB and ab, then the distance between the two genes A and B will be 20 map units. Similarly, if the distance between the two genes A and B is 14 map units, then 14% of the gametes will be crossover/recombinant types i.e. Ab and aB each will be 7 percent.

Two-Point Testcross

The easiest way to detect crossover gametes in a dihybrid is through the testcross progeny. Suppose AB/ab, a dihybrid is testcrossed with the ab/ab, the recessive type parent, and the testcross progeny obtained are as follows:

1. Phenotypes dominant at both the loci (AB/ab or AB) = 36%
2. Phenotypes dominant at first and recessive at the second locus (Ab/ab or Ab) = 14%
3. Phenotypes recessive at first and dominant at second locus (aB/ab or aB) = 14%
4. Phenotypes recessive at both the loci (ab/ab or ab) = 36%

This example shows that the 14% + 14% of the progeny were crossover/recombinant types making a total of 28% and hence the distance between the two genes A and B is 28 map units.

Three-Point Testcross

Double crossovers generally do not occur between the two genes less than 5 map units apart. For the genes located beyond this distance, it is advisable to use a third marker between the two genes to detect any double crossovers. Suppose we cross a trihybrid individual of the genotype ABC/abc and find the progeny as follows:

ABC/abc:	34%	68% parental types
abc/abc:	34%	
Abc/abc:	8%	16% single crossover between A and B (region I)
aBC/abc:	8%	
ABc/abc:	6%	12% single crossover between B and C (region II)
abC/abc:	6%	
AbC/abc:	2%	4% double crossovers
aBc/abc:	2%	

To find the distance between A and B, all crossovers (single and double) that occurred in the region I must be added and this comes to 16 + 4 = 20% or 20 map units. Likewise the distance between B and C comes to 12 + 4 = 16% or 16 map units. Thus, the distance between A and C is 20 + 16 = 36 map units when the double crossovers are detected in contrast to 16 + 12 = 28 map units when the double crossovers remain undetected.

Thus, for any three linked genes, whose distances are known, the amount of detectable crossovers between the two outer genes A and C when the middle marker gene B is missing is given as follows:

= (Crossover percentage between A and B) plus (crossover percentage between B and C) minus (2 x double crossover percentage). In the present given case, the distance between A and B is 20 map units, the distance between B and C is 16 map units, then percentage of detectable crossover from the dihybrid testcross AC/ac × ac/ac = 0.20 + 0.16 – (2 × 0.04) = 0.28 (14% Ac/ac and 14% aC/ac).

Gene Order

The additivity of map distance allows us to place genes in specific linear order on the chromosome. Three linked genes may be in any one of the three different gene orders depending upon which gene is in the centre. Let there be gene distances as A-B = 12 map units, B-C = 7 map units and A-C = 5 map units. It is obvious that since the maximum distance is between the genes A and B, these two genes have to be the distal genes with only option of gene C being in the middle. Accordingly, only the following linear order of the genes satisfies the additivity of the map distances.

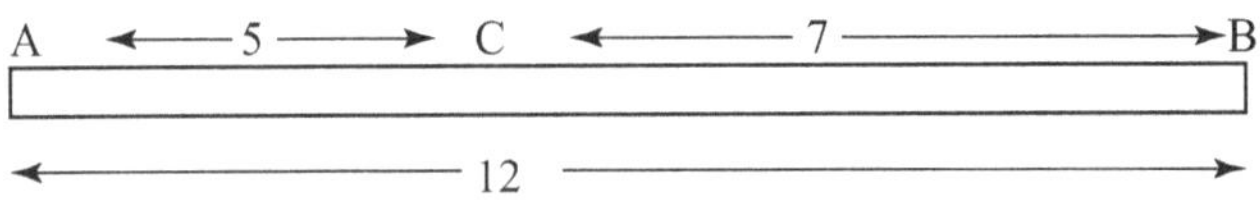

Linkage Relationship from a Three-Point Testcross Data

Let there be a three-point testcross data as given below:

Trihybrid parent x Testcross parent

Aa, Bb, Cc abc/abc

(Linkage relationship unknown)

F_1 Aabbcc = 36%, aaBbCc = 36%, aabbCc = 9%, AaBbcc = 9%

AabbCc = 4%, aaBbcc = 4%, AaBbCc = 1%, aabbcc = 1%

What are gene order and the map distance between the genes?

Before solving this kind of problem, one should always keep the following facts into account:

1. The most frequent classes are the parental types.
2. The least frequent classes are the double crossover types.
3. Any two alleles of a particular gene, if exchanged give rise to the parental combination, then that gene is in the centre.

The double crossover types are: AaBbCc, aabbcc (because of the lowest frequency).

The parental types are: Aabbcc, aaBbCc. If the A and a of the double crossover types are exchanged, we get the parental types disregarding the gene order and therefore, the gene A is in the centre. Further, the only contribution from the testcross parent to all the progeny is abc. Thus, the trihybrid parent must have A, b, c on one chromosome and a, B, C on the other chromosome. Therefore, the gene order will be as follows:

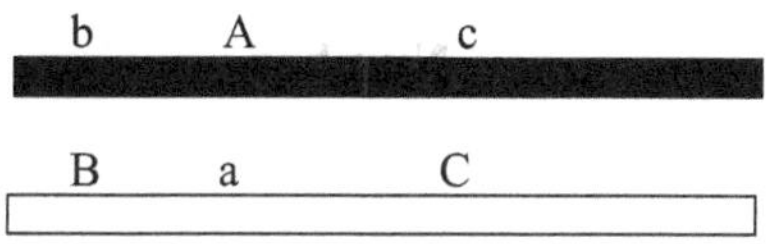

Now that we know the gene order and the parental linkage relationship, we can deduce the single crossovers. Let us designate B-A as region I and A-C as region II. Single crossover in region I and region II will produce the following results:

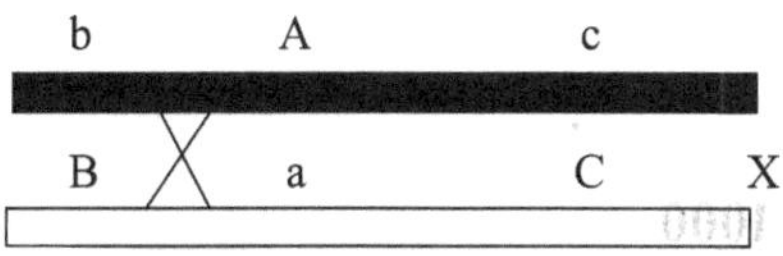

X bac/bac = baC/bac and BAc/bac

Region I

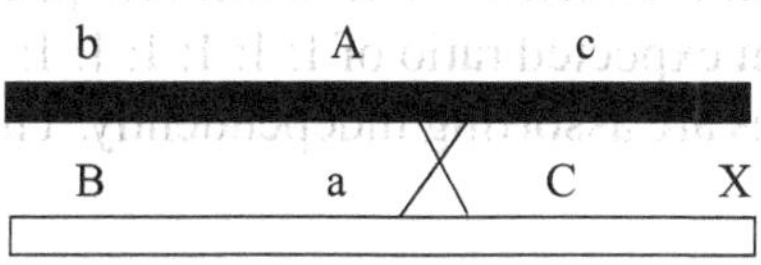

X bac/bac = bAC/bac and Bac//bac

Region II

Interference and Coincidence

In many higher organisms, formation of one chiasmata in a particular region of the chromosome, actually reduces the probability of another chiasmata formation in an immediately adjacent region. This reduction in chiasma formation could be due to physical inability of the chromatids to bend back upon themselves within certain minimum distances. The net result of this 'interference' is the observation of fewer double crossover types than would be expected according to map distances. The strength of interference varies in different segments of chromosome and is usually expressed in terms of 'coefficient of coincidence' which is equal to the ratio between observed and expected double crossovers as defined by Muller.

$$\text{Coefficient of coincidence} = \frac{\%\ \text{observed double crossovers}}{\%\ \text{expected double crossovers}}$$

Coincidence + Interference = 1.0

The entire process of three-point testcross data to answer various questions related to chromosome mapping is illustrated using an example from *Drosophila* as follows.

Female *Drosophila* heterozygous for three recessive mutations e (ebony body), st (scarlet eyes) and ss (spineless bristles) were test-crossed with the following progeny results:

Phenotype	**Number**
wild-type	67
ebony	8
ebony, scarlet	68
ebony, spineless	347
ebony, scarlet, spineless	78
scalet	368
scarlet, spineless	10
spineless	54
Total	**1000**

What Indicates That the Genes are Linked? The observed phenotypic values are significantly different from expected ratio of 1: 1: 1: 1: 1: 1: 1: 1 in a typical tri-hybrid testcross if the genes are assorting independently. Therefore, the genes are linked.

What Was the Genotype of Original Heterozygous Female and What is the Order of the Genes? This can be answered once we know the parental and double crossover types in the progeny. The most frequent class represented by ebony, spineless (347 progeny) and scarlet (368 progeny) can be considered as the parental types. In contrast, the least frequent class represented by ebony (8 progeny) and scarlet, spineless (10 progeny) will be considered as the double crossover types. These can be shown as follows:

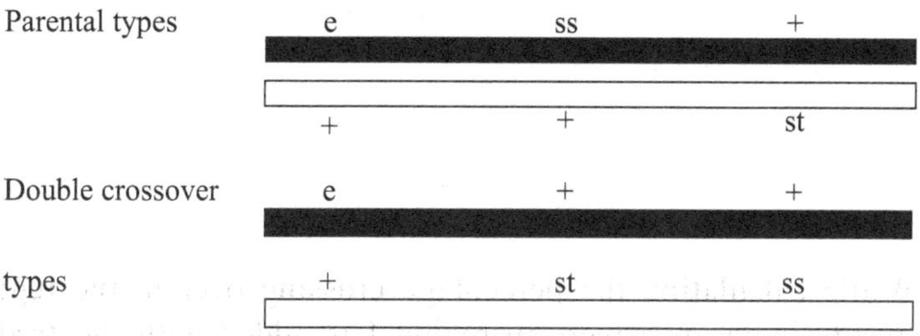

Now let us identify the alleles of a gene in double crossover types, which when exchanged will give rise to the parental gene combination disregarding the order of the genes. The gene which meets this requirement will be in the centre and the remaining two genes will be at the sides. This requirement is met by exchange of + and ss genes and hence the gene "ss" is in the centre. Thus the genotype of the original heterozygous female is st + +/+ ss e and the gene order is st-ss-e. Let us be clear that at this point of time, the genotype of the original heterozygous female could be written as + + st/e ss + also and the gene order as e-ss-st unless something more is kown. Crucially important issue is that the gene "ss" is in the centre.

What is Map Distance Between e and st? To answer this let us draw the chromosomal configuration of heterozygous female parent putting the gene "ss" in the centre.

The distance between the two distally located genes e and st can be derived by calculating the distance between the genes e and ss and between ss and st and that will be given by the sum of single crossovers in the region I plus sum of single crossovers in the region II plus double the double crossovers. This comes to as follows:

$$\frac{67+78+68+54+2\ (10+8)}{1000} = \frac{145+122+36}{1000} = \frac{303}{1000} = 30.3\% = 30.3 \text{ cM}$$

What is the Distance Between the Genes e and ss? This will be answered by totaling the progeny frequency resulting from single crossovers between the genes e and ss plus the double crossovers. This comes to 68 + 54 + 18 = 140/1000 = 14% = 14 cM.

What is the Coefficient of Coincidence? This is equal to the ratio of observed double crossovers and the expected double crossovers.

Observed double crossover percentage = 18/1000 = 1.8% = 0. 018

Expected double crossovers = Frequency of single crossovers in region I × frequency of single crossovers in region II = 145 + 18/1000 × 122 + 18/1000 = 0.163 × 0.140 = 0.02282.

Coincidence = 0.018/0.02282 = 0.789

(***Note:*** While calculating the percentage crossing over in the region I, the frequency of single crossovers in region I is added with the frequency of double crossovers and likewise the same thing is repeated for region II also).

Physical Maps

Physical maps are constructed based on nucleotides, the building blocks of DNA. Genetic distance on a linkage map expressed in centimorgans (cM) is not directly correlated with the physical distance expressed in terms of nucleotides.

35

Estimation of Recombination Value

When there is indication of linkage from F_2 or testcross data through significant deviation from 9: 3: 3: 1 and 1: 1: 1: 1 ratio, it is implied that independent assortment is not taking place and this is taken as an indication of linkage. The inverse of linkage strength is the recombination value and is often symbolized by '*p*' whose maximum value is 50% because out of four strands of a tetrad, two are always parental types and two are recombinant types when all the cells show crossing over.

Determination of Recombination Value from F_2 Data

Four methods are available and are commonly used. These are as follows:

Square Root Method

This is simpler method, free from complicated formula and tables. Further it is also applicable when gene interaction is involved. In coupling phase data, the parental and recombinant types of gametes are as given below assuming recombination value as '*p*'.

Types of gametes		Gametic frequencies
Parental types	→ AB	$(1-p)/2$
	→ ab	$(1-p)/2$
Recombinant types	→ Ab	$p/2$
	→ aB	$p/2$

The above illustration shows that the total recombination value is p and since there are two types of recombinant gametes, each is given a value of $p/2$. The remaining gametes (parental type gametes) are then equal to $1-p$ and again considering the fact that there are two types of parental gametes, each is given a value of half of $(1-p)$ which means $(1-p)/2$.

The frequency of double recessive homozygote, DRH (aabb) will be $\left(\frac{1-p}{2}\right)^2$ Therefore,

$$\left(\frac{1-p}{2}\right)^2 = DRH$$

$$\frac{1-p}{2} = \sqrt{DRH}$$

$$1-p = 2\sqrt{DRH}$$

$$-p = -1 + 2\sqrt{DRH}$$

$$p = 1 - 2\sqrt{DRH}$$

In repulsion phase, the parental and recombinant type gametes are exchanged and the recombination frequency (*p*) is equated as follows:

$$p = 2\sqrt{DRH}$$

In case, there is any confusion, value of *p* can be calculated using either of the formulae and the value of *p* below 0.5 is accepted as the maximum value of *p* is 0.5.

Emerson's Method

This was given by Emerson in 1916 in which a statistic ϕ is calculated and then this is converted to *p* as given below:

$$\phi = \frac{E-M}{N}$$

where

E = Total of end classes (a1 + a4)

M = Total of middle classes (a2 + a3) and a1, a2, a3, a4 are as follows:

F_2 phenotype	AB	Ab	aB	ab
Expected ratio	9	3	3	1
Observed values	a1	a2	a3	a4
	69	6	8	13

N = a1 + a2 + a3 + a4 = 96

$p = 1 - \sqrt{\phi}$ in coupling phase and

$p = \sqrt{\phi}$ in repulsion phase

$$\phi = \frac{69+13-(6-8)}{96} = 0.7083$$

$$p = 1 - \sqrt{0.7083} = 1 - 0.84 = 0.16 = 16\%$$

Product Ratio Method

This method originally proposed by Fischer and Balmukand has been found to be effective and widely used. The ratio of product of crossovers to that of parental types (product ratio) is a function of recombination.

For coupling phase data: $X = \frac{a2a3}{a1a4}$

For repulsion phase data $X = \frac{a1a4}{a2a3}$

Recombination value represented by the value of product ratio (X) is directly read from a product ratio table (Table 35.1) which has been taken from Stansfield (1969). The product ratio method utilizes all of the F_2 data available and not just the double recessive class as used in the square root method. The product ratio method should therefore yield more accurate estimates of recombination than the square root method.

Table 35.1. Recombination fraction estimated by the product ratio method.

Recombination fraction (p)	Ratio of product		Recombination fraction (p)	Ratio of product	
	Repulsion a1a4/a2a3	Coupling a2a3/a1a4		Repulsion a1a4/a2a3	Coupling a2a3/a1a4
0.00	0.000000	0.000000	0.26	0.1608	0.1467
0.01	0.000200	0.000136	0.27	0.1758	0.1616
0.02	0.000801	0.000552	0.28	0.1919	0.1777
0.03	0.001804	0.001262	0.29	0.2089	0.1948
0.04	0.003213	0.002283	0.30	0.2271	0.2132
0.05	0.005031	0.003629			
0.06	0.007265	0.005318	0.31	0.2465	0.2328
0.07	0.009921	0.007366	0.32	0.2672	0.2538
0.08	0.01301	0.009793	0.33	0.2892	0.2763
0.09	0.01653	0.0.01262	0.34	0.3127	0.3003
0.10	0.02051	0.01586	0.35	0.3377	0.3259
0.11	0.02495	0.01954	0.36	0.3643	0.3532
0.12	0.02986	0.02369	0.37	0.3927	0.3823
0.13	0.03527	0.02832	0.38	0.4230	0.4135
0.14	0.04118	0.03347	0.39	0.4553	0.4467
0.15	0.04763	0.03915	0.40	0.4898	0.4821
0.16	0.05462	0.04540	0.41	0.5266	0.5199
0.17	0.06218	0.05225	0.42	0.5660	0.5603

Recombina-tion fraction (p)	Ratio of product		Recombina-tion fraction (p)	Ratio of product	
	Repulsion a1a4/a2a3	**Coupling a2a3/a1a4**		**Repulsion a1a4/a2a3**	**Coupling a2a3/a1a4**
0.18	0.07033	0.05973	0.43	0.6081	0.6034
0.19	0.07911	0.06787	0.44	0.6531	0.6494
0.20	0.08854	0.07671	0.45	0.7013	0.6985
0.21	0.9865	0.8628	0.46	0.7529	0.7510
0.22	0.1095	0.9663	0.47	0.8082	0.8071
0.23	0.1211	0.1078	0.48	0.8676	0.8671
0.24	0.1334	0.1198	0.49	0.9314	0.9313
0.25	0.1467	0.1328	0.50	1.0000	1.0000

Source: F. R. Immer and M. T. Henderson. "Linkage studies in barley," Genetics, 28: 419-440, 1943.

This is illustrated numerically as follows (Stansfield, 1969):

Coupling phase data:

Parents:	RS/rs	X	rs/rs
F_1		RS/rs	
F_2	Phenotype		Number
	a1	R-S-	1221
	a2	R-ss	219
	a3	rrS-	246
	a4	rrss	243

$$\text{X for coupling data} = \frac{a2a3}{a1a4} = \frac{(219)(246)}{(1221)(243)} = \frac{53874}{296703} = 0.1816$$

From Table 35.1, it can be seen that 0.1816 lies between 0.1777 and 0.1948 which correspond to recombination values 0.28 and 0.29 respectively. Thus, without interpolation, the recombination value for this data is approximately 28%.

Maximum Likelihood Method

This method originally proposed by Haldane in 1919 is based on the following quadratic equation to get the value of ϕ:

$$n\phi^2 - \phi\{\text{a1} - 2(\text{a2} + \text{a3}) - \text{a4}\} - 2\text{a4} = 0 \text{ where n} = \text{a1} + \text{a2} + \text{a3} + \text{a4}$$

From the example given in Emerson's method, this equation becomes as follows:

$$96\phi^2 - \phi\ \{69 - 2(6 + 8) - 13\} - 26 = 0$$

$$96\phi^{2} - \phi\{69 - 12 - 16 - 13\} - 26 = 0$$

$$96\phi^{2} - 28\phi - 26 = 0$$

$$\phi = \frac{-(-28) \pm \sqrt{(-28)^2 - 4 \times 96 \times (-26)}}{2 \times 96} (ax^2 + bx + c = 0) \text{ and}$$

$$x = \frac{-b \pm \sqrt{b^2 - 4ac}}{2a}$$

$$= \frac{28 \pm \sqrt{784 + 9984}}{192} = \frac{28 \pm \sqrt{10768}}{192} = \frac{28 + 104}{192} = \frac{132}{192} = 0.6875$$

(negative value ignored).

$$p = \sqrt{\phi} \text{ or } 1 - \sqrt{\phi} = \sqrt{0.6875} \text{ or } 1 - \sqrt{0.6875} = 0.83 \text{ or } 0.17$$

Since 0.83 is more than 0.5, the maximum value of recombination, hence this is rejected and the recombination value accepted is 0.17 or 17%.

Determination of Recombination Value From Testcross Data

Let us have the following results from a dihybrid testcross:

Phenotype	A-B-	A-bb	aaB-	aabb
Observed values	a1	a2	a3	a4
	40	12	13	35 = 100

$$p \text{ in coupling phase } = \frac{a2 + a3}{a1 + a2 + a3 + a4} = \frac{12 + 13}{100} = \frac{25}{100} = 0.25 = 25\%$$

$$p \text{ in repulsion phase } = \frac{a1 + a4}{a1 + a2 + a3 + a4}$$

Predicting Results from a Dihybrid Cross Based on Map Distance Between the Two Genes

Let us say there are two genes A and B, 10 map units apart from each other in coupling phase. We wish to derive the phenotypic ratio in the F_2 based on this information. The procedure is simple and straightforward. The fact that both the genes are showing 10% recombination, indicates that the parental type gametes (AB and ab) constitute 90% and the recombinant gametes (Ab and aB) account for 5% each thus making a total of 10%. Thus, the gametic frequencies are as follows for deriving the phenotypic ratio in the F_2.

Parental gametes: AB = 0.45 and ab = 0.45

Recombinant gametes: Ab = 0.05 and aB = 0.05

The F_2 can be derived by use of the gametic checkerboard combining independent probabilities by multiplication as given below:

		Parental types		Crossover types	
		AB = 0.45	ab = 0.45	Ab = 0.05	aB = 0.05
Parental types	AB = 0.45	AB/AB 0.2025	AB/ab 0.2025	AB/Ab 0.0225	AB/aB 0.0225
	ab = 0.45	ab/AB 0.2025	ab/ab 0.2025	ab/Ab 0.0225	ab/aB 0.0225
Crossover types	Ab = 0.05	Ab/AB 0.0225	Ab/ab 0.0225	Ab/Ab 0.0025	Ab/aB 0.0025
	aB = 0.05	aB/AB 0.0225	aB/ab 0.0225	aB/Ab 0.0025	aB/aB 0.0025

Phenotypic summary: A-B- = 0.7025
Abb = 0.0475
aaB- = 0.0475
aabb = 0.2025
Total = 1.0

Predicting Results of a Trihybrid Testcross

Given a particular type of mating, the map distances involved and either the coincidence or interference for the given regions of the chromosome, one can predict the results in the offspring generation as illustrated by a solved example from Stansfield (1969).

Parents: AbC/aBc x abc/abc

a←10→b←20→c

Map:

Interference: 40%

From the above given example, one can determine the parental types, single crossover types in region I, single crossover types in region II and the double crossover types as described below:

1. Single crossover in region II (between b and c): Map distance between b and c – double crossover types and total to be divided into two as there will be two types of recombinants.
   = 20% – 1.2% = 18.8% and each crossover class represented by 9.4%.

2. Single crossover in region I (between a and b): Map distance between a and b – double crossover types and total to be divided into two as there will be two types of recombinants.

   = 10% -1.2% = 8.8% and each crossover class represented by 4.4%.

3. Frequency of expected double crossovers: This is obtained by multiplying the two decimal equivalents of the map distances and the coefficient of coincidence and this comes to:

   $0.1 \times 0.2 \times 0.6 = 0.012 = 1.2\%$ (coefficient of coincidence = 1 – interference = 1 – 0.4 = 0.6). This percentages has to be equally divided (0.6% each) between the two double crossover types.

4. Parental types: Total – (single cross over types in region I + single cross over types in region II + double crossover types) = 100 – (8.8 + 18.8 +1.2) = 100 – 28.8 = 71.2% and each parental type will account for 71.2/2 = 35.6%.

   Based on above information, the parental types, single crossover types in region I, single crossover types in region II and double crossover types are as follows:

Parental types:	AbC	35.6%
	aBc	35.6%
Single crossover types in region I:	ABc	4.4%
	abC	4.4%
Single crossover types in region II:	Abc	9.4%
	aBC	9.4%
Double crossover types:	ABC	0.6%
	abc	0.6%
Total:		100%

36

Translocations and Inversions

Translocations

Chromosomes occasionally undergo spontaneous rupture or can be induced to rupture in high frequency by ionizing radiation. The broken ends of such chromosomes behave as though they were 'sticky' and may rejoin into non-homologous combinations giving rise to translocations. The most common type of translocation is reciprocal translocation which involves the exchange of segments between two non-homologous chromosomes. Translocations change only the arrangement of genes in the chromosomes, not the quality or quantity of genes. For this reason, they are sometimes referred to as chromosomal rearrangements. Individuals carrying such rearrangements should be phenotypically entirely normal unless the relations of a gene or genes to adjacent genes affect the phenotypic expression (position effect). The first translocation was discovered by Bridges in 1923 in *D. melanogaster*.

The genetic techniques for detecting and studying translocations will be more easily understood when the cytological phenomena produced by translocations are known. Suppose two chromosomes having respectively the genes ABCDEF and GHIJKL exchange sections and give rise to translocation chromosomes ABCJKL and GHIDEF. The individual thus formed receives from one of the parents the normal and from the other parent the translocation chromosome. Such an individual is a translocation heterozygote (Fig. 36.1).

Fig. 36.1 shows that cross-shaped configuration is formed at the pachytene stage of meiosis. The row below cross-shaped configuration shows the twisted rings and two open-ring configurations formed by the chromosomes involved in the meiotic metaphase. The two lower rows are the chromosomal complements in the six types of the gametes formed. Only the two types on the left contain the normal sets of genes whereas the other four contain some genes in duplicate (duplication) and fail to carry some genes (deficiency).

Occurrence of crossing over in each of the four arms of the cross will result in formation of chiasmata in each arm. Instead of bivalent formation, there will be formation of quadrivalent, a group of four associated chromosomes where each member of the group is partially homologous to the two other chromosomes in the group. The quadrivalent will appear at diakinesis and at metaphase of the first meiotic division as a ring or circle of four chromosomes, which may be either twisted as figure of 8 as shown in Fig. 36.1 left or open as in the centre of the same figure. If chiasmata fail to be formed in one arm of the pachytene cross, the ring is transformed into an open chain of four chromosomes.

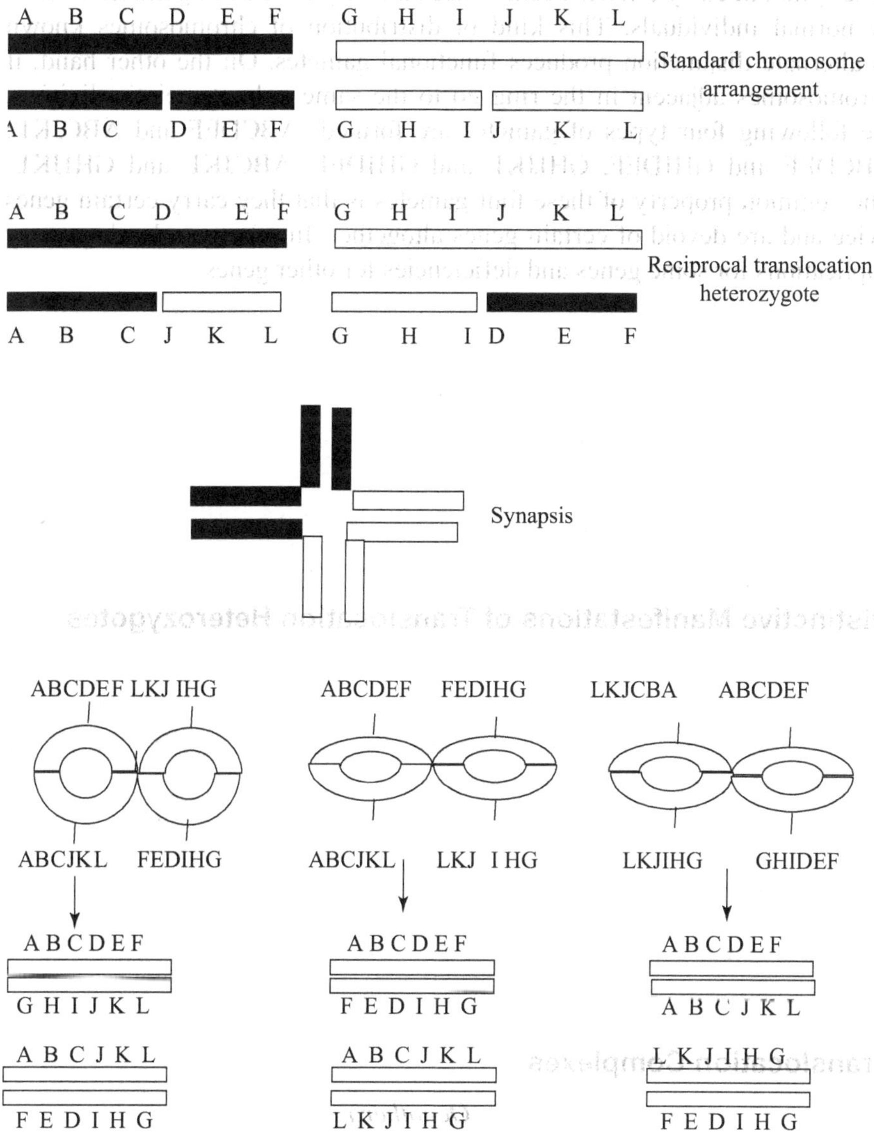

Fig. 36.1. Reduction division in translocation heterozygote.

There are several ways in which the chromosomes associated with a ring or chain may become distributed to the gametes formed as a result of meiosis. The two original chromosomes, ABCDEF and GHIJKL may go the same gamete and the translocation chromosomes, ABCJKL and GHIDEF may go to another gamete. It may be noted that in each of these gametes, every gene symbolized by a letter occurs once and only once as in gametes formed by normal individuals. This kind of distribution of chromosomes known as alternate disjunction produces functional gametes. On the other hand, if chromosomes adjacent in the ring go to the same pole at meiotic division, the following four types of gametes are formed: ABCDEF and ABCJKL; ABCDEF and GHIDEF; GHIJKL and GHIDEF; ABCJKL and GHIJKL. The common property of these four gametes is that they carry certain genes twice and are devoid of certain genes altogether. In other words, they carry duplications for some genes and deficiencies for other genes.

In most plants, pollen grains carrying deficiencies or duplications are as a general rule in-viable and are aborted. Such pollen grains appear empty and shriveled and can be clearly distinguished from pollen grains having normal chromosomal complements. Embryo sacs with deficiencies and duplications may also abort. Consequently, plants heterozygous for translocations may produce fewer seeds than normal plants and such plants appear semi-sterile. The semi-sterility of the translocation heterozygotes furnishes one of the simplest and most practical ways to detect translocations in maize.

Distinctive Manifestations of Translocation Heterozygotes

1. Semi-sterility is usually characteristic of translocation heterozygotes because of adjacent disjunctions.
2. Some genes which formerly assorted independently, will exhibit linkage relationships after translocation has occurred. A single reciprocal translocation will reduce the number of linkage groups by one.
3. The phenotypic expression of a gene may be modified when it is translocated to a new position in the genome. Position effects are particularly evident when genes in euchromatin (lightly staining areas usually containing genetic elements) are shifted near heterochromatin regions (darker staining areas presumably devoid of active genes).

Translocation Complexes

In the evening primrose of the genus *Oenothera* an unusual series of reciprocal translocations involving all 7 of its chromosome pairs has been reported. If we label each chromosome end with a different number, the normal set of 7 chromosomes would be 1-2, 3-4, 5-6, 7-8, 9-10, 11-12 and 13-14. A translocation set would be 2-3, 4-5, 6-7, 8-9, 10-11, 12-13 and 14-1. A multiple translocation heterozygote like this would form a ring of 14 chromosomes during meiosis. Different lethals in each of the two haploid sets 7 chromosomes enforce structural heterozygosity. Since only alternate segregation from the ring can form viable gametes, each group of 7 chromosomes behaves as though it were a single large linkage group with recombination confined to the pairing ends of each chromosome. Each set of 7 chromosomes which is inherited as a single unit is called a "Renner Complex".

Inversion

A rearrangement that reverses the order of a linear array of genes in a chromosome is known as inversion. An inversion that is entirely within one arm of a chromosome and does not include centromere is called as paracentric inversion. An inversion that includes centromere and hence involved both arms of a chromosome is termed as pericentric inversion. Inversions occur when a chromosome segment is detached, flipped around 180^0, and is reattached to the rest of the chromosome. As a result, the order of the segment's gene is reversed (Figs. 36.2 and 36.3).

Historically, soon after the discovery of linkage and recombination in *D. melanogaster*, it was noticed that some strains of this fly contain what were described as "C factors" which reduce or suppress the recombination of genes in certain chromosomes or a part of a chromosome in female heterozygote for such C factors. Sturtevant (1926) found that one of the C factors, which acted as suppressor of recombination in the right limb of the third chromosome, i. e from the centromere to the right end, specifically, from the loci of the genes *Dfd* and *p* to *ca* and *Mg* was an inversion of a section of this chromosome.

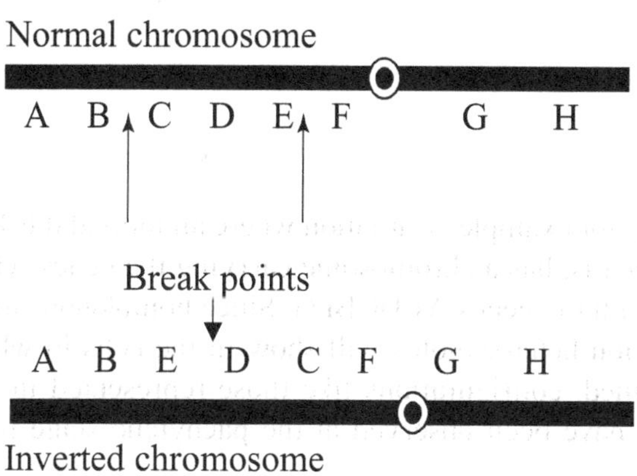

Fig. 36.2. Structure of an inversion. The chromosome has been broken at two points and the segment between them (containing regions C, D and E) has been inverted.

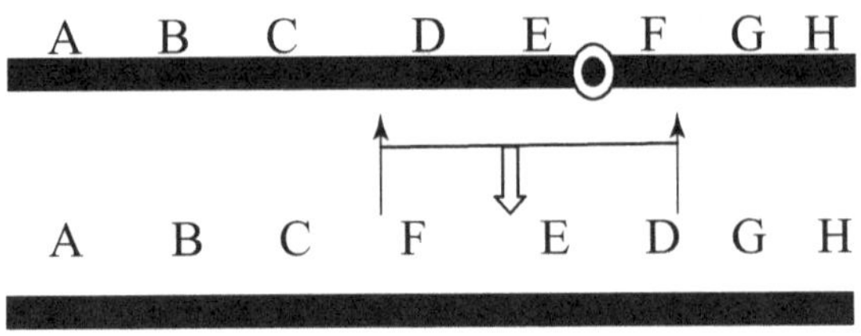

(a) Pericentric inversion-includes centromere

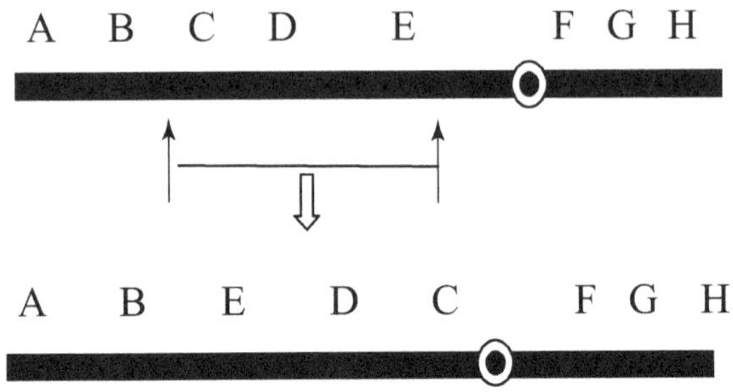

(b) Paracentric inversion-excludes centromere

Fig. 36.3. Pericentric and paracentric inversions. A pericentric inversion changes the size of the chromosome arms because centroemre is included within the inversion.

The reason why recombination is suppressed in inversion heterozygote will become clear when the cytological effects of inversions are considered. It may be noted that not only inversions but also heterozygous translocations and large duplications and deficiencies reduce the frequency of crossing over in the chromosomes involved. This is apparently caused by the difficulties the chromosomes encounter in establishing the meiotic pairing between the leptotene and the pachytene stages.

Figure 36.4. shows crossing over in heterozygote for a paracentric inversion (not including the centromere).

Let us consider, for example, a situation where an individal is heteozygous for an inversion, that is, has a chromosome carrying the genes ABCDEFG and a chromosome with the genes AEDCBFG. Since homologous loci are paired at meiosis, inversion heterozygotes will show in the cells in which a complete pairing is attained, configurations like those represented in Fig. 36.4. Such configurations have been observed at the pachytene stage in chromosomes

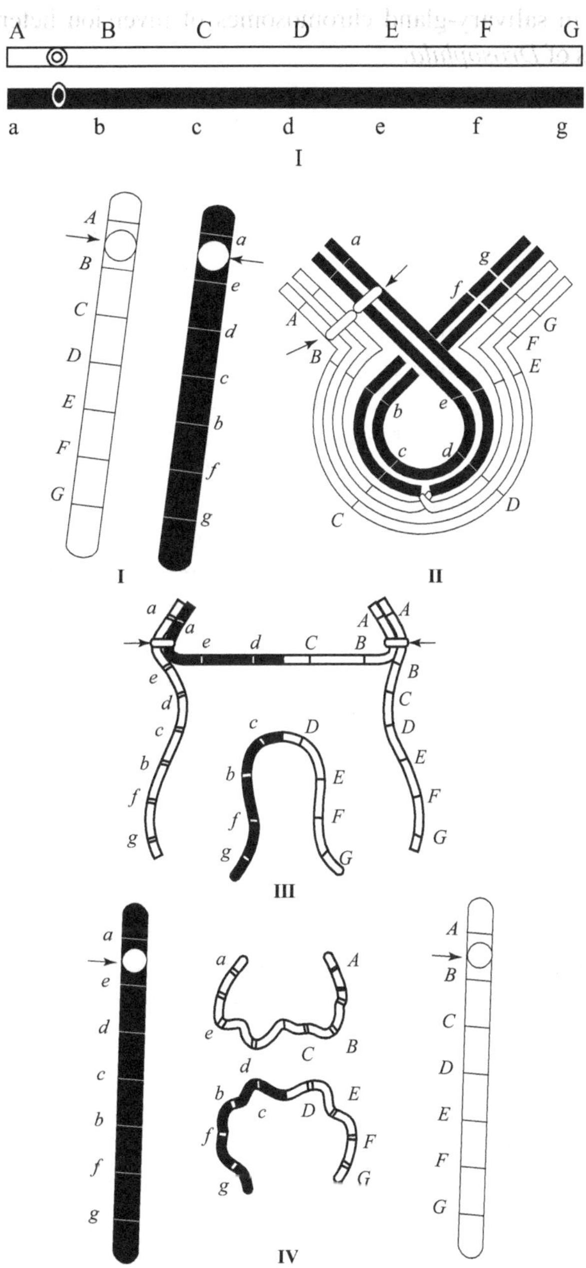

Fig.36.4. Crossing over in a heterozygote for a paracentric inversion (not including centromere). I, two chromosomes differing in a paracentric inverion, II, diplotene stage of meiosis showing a chiasma (X) inside the inverted section, III, anaphase of the first meiotic division showing a chromatid bridge and an accentric fragment, IV, the outcome of meiosis, two noncrossover chromosomes with normal gene complement and (in the middle) chromatids with two centromeres and with no centromere.

of maize and in salivary-gland chromosomes of inversion heterozygotes in several species of *Drosophila.*

Chiasata may become established in the paired inverted segments as shown in Fig. 36.4. The subsequent fate of crossover chromatids will be different depending upon whether the centromere lies within or outside the inverted segment. If the inversion does not include centromere (paracentric inversion), the meiotic anaphase will contain a chromatid connecting the two centromeres, called a chromatid bridge, and a chromatid lacking a cenromere altogether, called an acentric fragment. The formation of chromatid bridges and acentric fragments at the meiotic divisions has been observed by McClintock in maize and by Darlington, Stebbins and others in a series of other plants. The presence of such bridges and fragments furnishes a cytological method of detection of inversions which is used in organisms in which pachytene chromosomes or salivary-gland chromosomes are either absent or unfavourable for investigtion. Now neither the chromatid bridges nor the acentric fragments behave normally in the cell divisions and both are eventually lost.The only viable products of meiosis in paracentric inversion heterozygotes are the chromatids that underwent no crossing over within the inverted section. Sturtevant and Beadle (1936) have assumed that in oogenesis, one of these noncrossover chromatids becomes included in the egg nucleus whereas other chromosomes are eliminated in the polar bodies. The suppression of recombintion observed in inversion heterozygotes is explained by noninclsion of dicentric and acentric chromosomes in the egg nucleus.

Some inversions include the locus of the centromere and are called pericentric inversions (Fig. 36.5). Pericentric inversions sometimes lead to changes in the appearance of a chromosome at mitosis in somatic cells. For example, a chromosome with a median centromere usually appears as a more or less equal armed V-shaped body . If the two breaks giving rise to a pericentric inversion occur at different distances from the centromere in two arms, the resulting chromosome will have a more or less subterminal centromere and will appear hook-shaped or rod-shaped. Conversely, a pericentric inversion in a hook-shaped or rod-shaped chromosome may transform it into a V-shaped one with more or less equal arms.

Crossing over in a pericentric inversion does not result in production of chromatid bridges and acentric fragments at the first metaphase of meiosis. It does result, however, in formation of chromosomes having duplications for some genes and deficiencies for others (Fig. 36.5). Gametes that receive such chromosomes do not as a rule form viable zygotes. Recombination is therefore lowered in heterozygotes for pericentric inversions (Sinnott, Dunn and Dobzhansky, 1958).

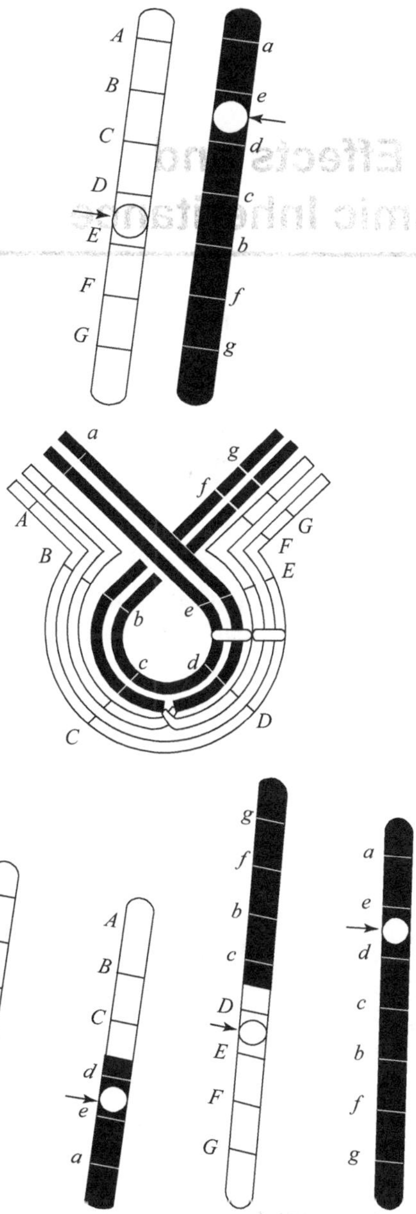

Fig. 36.5. Crossing over in a heterozygote for a pericentric inversion (including the centromere). Top: Two chromosomes differing in a pericentric inversion. Centre: Diplotene stage of meiosis showing a chisma (X) in the inverted section. Bottom: Four products of meiosis, two of which (the black and the white ones) are non-crossovers and carry normal complement of genes and the other two (partly black and partly white) are crossovers carrying some genes in excess and deficient for others. The centromeres are shown by open circles.

37

Maternal Effects and Cytoplasmic Inheritance

So far, we have seen the examples in which mode of inheritance of a particular character is tied to the behaviour of the nucleus or more specifically to the chromosomes. In other words, transmission of a character and its appearance in the individuals can be predicted from the knowledge of chromosome segregation and independent assortment. This analogy gets further strengthened on the ground that the DNA has been considered as the basic genetic material and practically all the cellular DNA is localized in the chromosomes. It has also been seen that DNA does not produce an organism alone but obviously depends upon an already existing medium in order to function. Thus, genotypic effects may considerably get modified by the environment. In the cell itself, one important source of environmental effect is the cytoplasm immediately surrounding the nucleus. The components of the cytoplasm may show wide variation between the individuals and therefore it cannot be ruled out that a genotype in one cytoplasm may function somewhat differently than the same genotype placed directly into another cytoplasm. It is well known fact that the cytoplasmic contribution of one of the parents in form of an egg is considerably larger than the cytoplasmic contribution of the other parent in form of sperm. For example, cross-fertilization between two strains capable of producing both sperm and eggs, leads to contribution of 'A' genotype and 'a' egg cytoplasm in one case and of 'B' genotype and 'b' egg cytoplasm in the other case. Thus, the cross-fertilized diploid zygotes have two possible genotypic-cytoplasmic combinations, namely, *AB-a*, and *AB-b* depending upon which strain provides the egg. If the egg is contributed by the A parent, then the genotype of the progeny shall be *AB-a*, and if the egg is contributed by the B parent, then the genotype of the progeny shall be *AB-b*. If each such cytoplasm produces a unique phenotype, then *AB-b* would have the same *b* phenotype as *BB-b* individuals. Such effects produced by the egg cytoplasm are termed as maternal effects.

Maternal Effect

A clear illustration of maternal effect has been reported by Caspari in the flour moth (*Ephestia kuhniella*) which has dark brown eyes and various pigmented larval parts which owe part of their colour to the presence of a pigment precursor, kynurenine. The appearance of kinurenine is controlled by a dominant gene *A*, which when homozygous (*aa*), causes the absence of kinurenine and consequent red eye colour and lack of larval pigmentation. Under this scenario, when a male heterozygous for *Aa* fertilizes an *aa* recessive female, then the offspring are expected to appear in a genetic ratio of 1 *Aa* : 1 aa and the larval phenotype will be in the ratio of 1 pigmented : 1 non-pigmented which will develop into dark-brown and red-eyed individuals, respectively, since *A*- causes dark brown eyes and *aa* causes red-eyes.

aa x Aa
Red-eyed female × Dark brown eyed male
↓
50% Aa (dark brown eye): 50% aa (red eye)

Aa X aa
Dark brown eyed female × Red-eyed male
↓
50% Aa: (all pigmented) 50% aa
↓ ↓
At maturity-dark brown red-eyed

The results were, however, quite different when the reciprocal crosses were made (Aa female x aa male). In this case, all early larvae appeared pigmented as if kynurenine, the precursor of the pigment were present in all the individuals although 50% of the individuals had the genotype *aa* and were without the kynurenine producing gene *A*. However, when the larvae matured, only half of them were dark brown-eyed and half were red-eyed. The explanation for these unexpected pigmented larvae is simple. The unexpected pigmented larvae were based on the egg cytoplasm contributed by the heterozygous mother (*Aa*). That is all eggs of the dark brown-eyed female contain kynurenine disregarding their own genotype and thus, they begin their development as pigmented larvae. However, half of the larvae with the genotype (*aa*) are unable to synthesize further kynurenine and ultimately these develop into red-eyed as the kynurenine initially present along with the egg is used up.

In some case, the maternal effect does not diminish during the development but lasts throughout the adult life. Such examples occur in case of body development related matters when the development is started in a particular direction depending upon the genotype of the mother and the same cannot be reversed. One striking example of this kind has been reported in a species of snail (*Limnaea peregra*). In this species, direction of coiling is determined by a single pair of genes. Snails can coil to the left (sinister) or right (dextral) and allele for dextral coiling (D) is dominant over the sinister coiling (d). The interesting part of this inheritance is that the direction of coiling of offspring is determined by the genotype of the maternal snail, even when this is completely different from the genotype of the offspring. Thus, a self-fertilized dextrally coiled heterozygote (Dd) always produces eggs that develop into dextrally coiled offspring even though some of them are homozygous recessive (dd) and should have been sinisterly coiled. These maternal effects last only one generation. This has been shown by Boycott, Diver and Garstang and as interpreted by Sturtevant (Fig. 37.1).

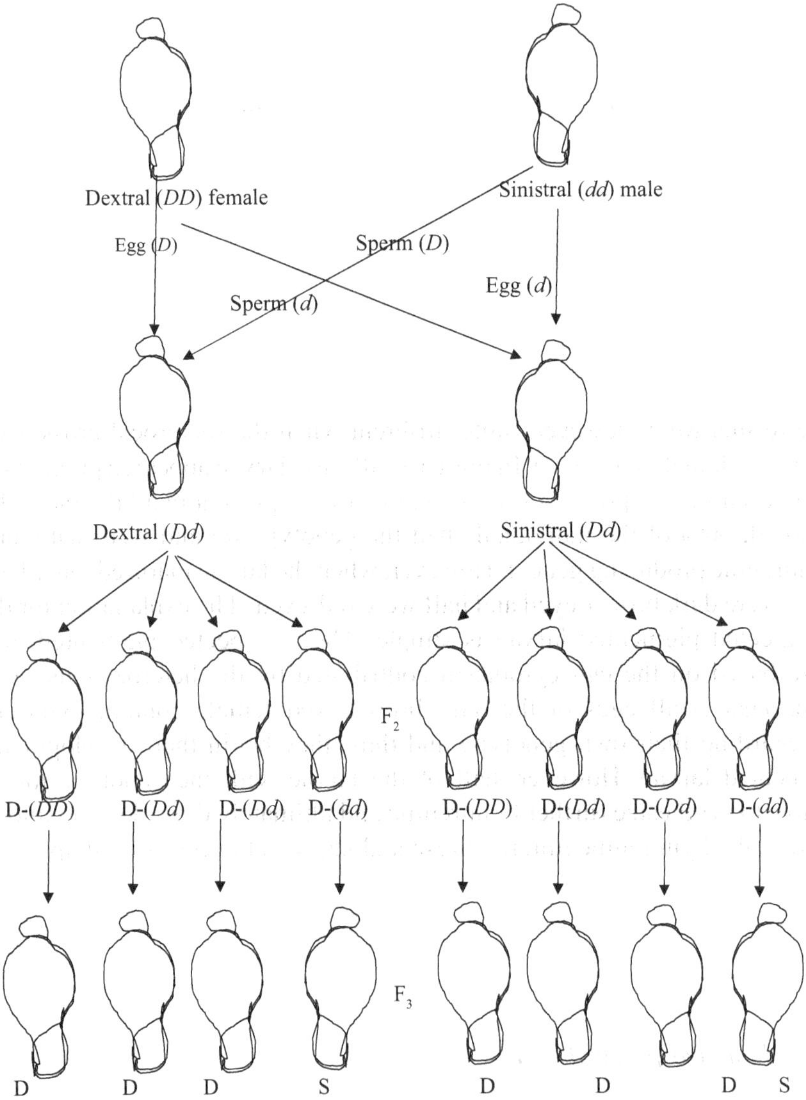

Fig. 37.1 The normal inheritance of dextral and sinistral coiling in Limnaea peregra. D, dextral phenotype; S, sinistral phenotype; D gene for dextrality; d gene for sinistrality (This species is hermaphrodite and may reproduce either by crossing or by self-fertilization.

It is obvious that the character of the coiling is, however, determined not by the individuals own genes but by those of its mother. Some of these snails that are themselves phenotypic dextral produce all sinistral offspring and such individuals by appropriate genetic tests may be shown to be homozygous for the recessive sinister gene. Their dextral character may have been determined by the presence of a dextral gene in their mothers. This shows that this is not the character of the mother but her genes that impress upon cytoplasm of all her eggs before maturation a certain type of pattern that finds expression during the early cleavage division of the egg. This phenomenon has been termed as "maternal determination".

Extra-Chromosomal Inheritance

The examples of maternal effect explained above indicate the presence of cytoplasmic factors that exist for a time independent of the nucleus and produce phenotypic effects of their own, such as kynurenine in *Ephistia.* It has also been seen that such a factor is not self-perpetuating and disappears unless replaced by the effect of an appropriate nuclear gene. Some cytoplasmic factors, however, show evidence for both self-perpetuation and independent transmission and hence these may be considered as genetic units fully equal to those in the chromosomes. Since they are located outside the chromosomes unlike the typical nuclear genes, they are called as extra-chromosomal and these factors have been termed as "plasmagenes" or "plasmons" or "plasmids" etc.

Generally, when crosses are made between the individuals carrying traits caused by extra-chromosomal factors, the inheritance pattern is not in accordance with the expected Mendelian ratios. The most common departure from chromosomal inheritance is indicated by the the differences in the reciprocal crosses. In contrast to maternal effect, however, the differences caused by extra-chromosomal factors do not usually disappear after one generation but may persist as long as the extra-chromosomal factor can perpetuate itself. Following are few examples of extra-chromosomal inheritance:

Inheritance of Chloroplast

The first example of inheritance of chloroplast was discovered by German Botanist Carl Correns in 1909 and Erwin Baur. Correns worked with variegating strains of the four o'clock (*Mirabilis jalapa*), a popular garden plant. He systematically made crosses between green and variegating strains. In these crosses, it was possible to use flowers from a white sector that had developed on variegating plant. Correns observed that the offspring of such crosses were always phenotypically identical to the tissue that produced the female gamete (Fig. 37.2).

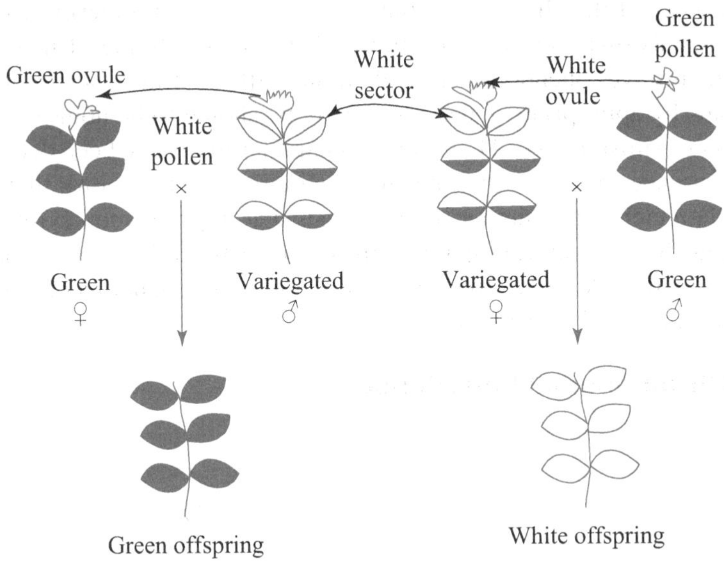

Fig. 37.2. Correns' experiment on the inheritance of leaf variegation in four o' clock plant.

Correns performed reciprocal crosses between green and variegated *Mirabilis* plants, using white sectors on the variegated plants as a source of gametes. From each cross, the offspring showed the same phenotypes as the female parent. Thus, in *Mirabilis* plant leaf colour exhibits strict maternal inheritance. In other words, a cross made by fertilizing ovules from green plants with pollen from a white sector on a variegated plant produced only green offspring. However, a cross made with ovules from a white sector on a variegated plant and pollen from a green plant produced only pure white offspring. This strict maternal inheritance could be explained if plant colour was controlled by factors that were transmitted through ovules but not through the pollen. Chloroplasts were the obvious candidates to have these factors. In *Mirabilis*, the chloroplasts are transmitted to the offspring through the female reproductive cells, but are largely, if not entirely, excluded from the male reproductive cells. When a plant inherits a mixture of pigmented and un-pigmented chloroplasts from the female parent, its tissues may variegate because the two types of chloroplasts sort themselves out during development.

Baur studied leaf variegation in another ornamental species, *Pelargonium zonale*. In this plant, crosses between green and variegating strains produced a mixture of offspring, some green, some variegating and some pure white. However, these different phenotypes were not recovered in Mendelian proportions. This non-Mendelian, bi-parental inheritance indicates that the colour of *Pelargonium* leaves is determined by a mixture of maternal and paternal factors located outside the nucleus, presumably in the chloroplasts. It therefore seems that in ***Pelargonium*** the chloroplasts are transmitted through the pollen as well as through the ovule.

The experiments of Correns and Baur have demonstrated that leaf variegation is inherited as a non-Mendelian trait, one that is almost certainly controlled by factors located in the chloroplasts. The discovery that chloroplasts contain DNA has made this hypothesis very attractive. Yet, in spite of the work of so many scientists, the molecular basis of leaf variegation is still unknown (Snustad *et al.*, 1997).

Cytoplasmic Male Sterility in Maize

In maize, male organs are in the tassel which develops only at the top and female reproductive organs are produced on the ears which develop at various positions along the stalk. In some strains of maize, the tassels fail to develop and these strains are called as male sterile. However, the male sterile plants can be easily crossed to other male-fertile strains. In the 1930s, M. Rhoades demonstrated that male-sterility in maize can be caused by a maternally inherited factor. Rhoades pollinated the ears of male-sterile strain with pollen from a male-fertile strain. All the progeny from this cross were male-sterile. Rhoades then backcrossed the male-sterile progeny to the male-fertile variety to see if male-sterility was caused by a dominant allele inherited from male-sterile strain. Because all the offspring were male-sterile, the dominant allele hypothesis was rejected. Rhoades continued to backcross the male-sterile offspring to male-fertile plants for several generations. At each step of backcrossing, he found that the male-sterile phenotype persisted. This showed that male-sterility was transmitted by a maternally inherited factor and evidently it must have been in the cytoplasm of the female reproductive cells. As the research moved forward, Rhoades discovered that dominant mutations in several nuclear genes could suppress the effect of this cytoplasmic factor. A plant that had the cytoplasmic factor plus one of these dominant mutations could be male-fertile. Such male-sterility restorer genes are called as restorer genes (*RfRf*). The male-sterile phenotype thus depended on an interaction between the cytoplasmic factor and products of several different nuclear genes. With one combination of nuclear genes, could cause male-sterility and with another combination, it could cause male-fertility signifying the fact that this male-sterility was cytoplasmic-genetic male-sterility.

Maize breeders have taken advantage of cytoplasmic male sterility to streamline the production of hybrid maize seed on commercial scale. However, this method have had its own share of bad luck in the past. In the early 1970s, overreliance on sterilizing cytoplasm of a particular strain of maize had a disastrous effect on the US corn crop. Plants that carried this cytoplasm proved to be especially susceptible to infection by a mutant race of southern corn blight (*Helminthosporium maydis*). In some parts of the United States, nearly 50% of the crop was destroyed. Fortunately, maize varieties that did not have the blight susceptible cytoplasm had been preserved and these varieties were pressed into breeding and US corn crop quickly got reestablished on firm footing.

What is responsible for the non-Mendelian inheritance of male-sterility in maize? Maize plants have both mitochondria and chloroplasts. Therefore, either organelle could be involved. A partial understanding of the basis of cytoplasmic male-sterility emerged in the 1970s and 1980s. During this time, it was possible to isolate and manipulate organelle DNA. Application of these techniques indicated that some maize male-sterile strains contained two small accessory DNA molecules, called as S1 and S2 in their mitochondria. These molecules appeared to be involved in the manifestation of male-sterility phenomenon in maize but the mechanism remained unknown at that time (Snustad *et al.*, 1997).

Antibiotic Resistance in *Chlamydomonas*

Photosynthetic green algae are an important part of biosphere. They are found in abundance in ocean and are the foundation of food chain. They are also found on the land. One of the terrestrial species, namely, *Chlamydomonas reinhardii* has been used extensively in genetical investigations. This unicellular, diploid organism has two mating types, designated as plus and minus. Cells of opposing mating types can fuse and form a diploid zygote which undergoes meiosis to produce four haploid offspring which divide mitotically to produce clones of vegetative cells. Genetic investigations have shown that the mating type is governed by a nuclear gene with two alleles, ***mt+*** (specifying plus mating types) and ***mt-*** (specifying minus mating type). Thus, out of four offspring produced from mt+ x mt- mating, two are mt+ type and two are mt- type.

There is a single large chloroplast and several mitochondria inside each cell of *Chlamydomonas*. Each of the four haploid offspring resulting from mt+ x mt- mating again has one chloroplast and several mitochondria. In 1954, Ruth Sager, an American geneticist discovered antibiotic resistance in *Chlamydomonas* follows a non-Mendelian inheritance. Sager discovered a mutant mt+ strain that was resistant to antibiotic streptomycin and the naturally occurring mt-strain was sensitive to this antibiotic. Sager mated a mutant+ (antibiotic resistant strain) with antibiotic sensitive strain. All the progeny from this cross were streptomycin resistant, although the mating type alleles segregated in a 1: 1 fashion. Thus, resistance to streptomycin in *Chlamydomonas* was found to be controlled by non-Mendelian factor. Sager then mated some of the resistant mt- cells to sensitive mt+ cells and this time she found that all the progeny were sensitive. Resistance or sensitivity to streptomycin was therefore always inherited through the cytoplasm of mt+ cell (Fig. 37.3).

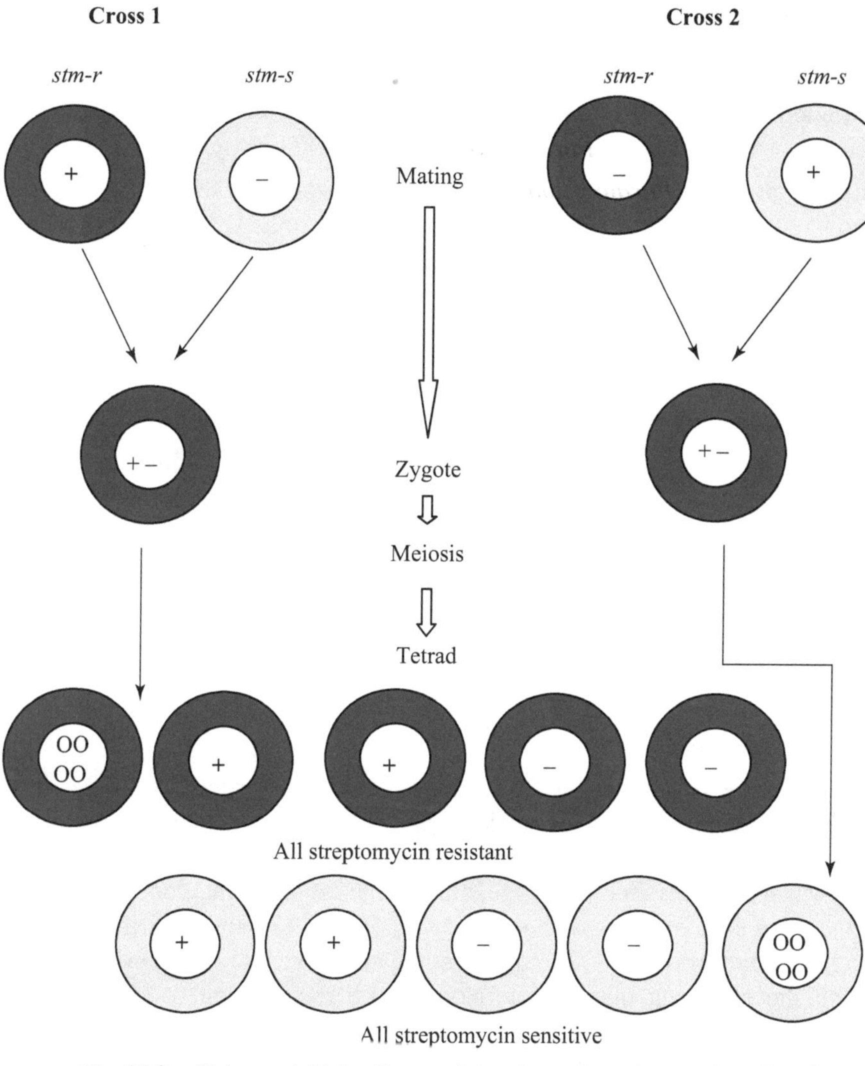

Fig. 37.3. Uniparental inheritance of streptomycin resistance (*stm-r*) and sensitivity (*stm-s*) in *Chlamydomonas*. All the offspring of cross have the same phenotype as the *mt*+ parent.

Infective Heredity in *Paramecia*

In 1943, Sonneborn showed the inheritance pattern of two alternative traits in strains of *Paramecium aurelia.* A strain No. 51 had the trait called 'killer' and had the property of destroying the other strains such as No. 47 which were called as 'sensitive'. The destruction of sensitive strains occurred through secretion of a toxic substance (paramecin) into the culture medium. In spite of this toxic relationship, conjugation could be accomplished between the killer and sensitive strains since *Paramecia* are resistant to killing during mating.

Conjugation in *Paramecium* proceeds through the establishment of a narrow cytoplasmic bridge between the two conjugants. During this process, there is mutual exchange of haploid nuclei across this bridge. The two resultant exconjugants are therefore diploid each bearing identical nuclear complement. However, during this process, very little cytoplasmic exchange normally occurs. Thus, both the exconjugants may be regarded as having maintained their original cytoplasm unchanged. When Sonneborn performed the killer-sensitive mating experiments, the exconjugants formed separate killer and sensitive clones depending upon the parent from whom they have been derived (Fig. 37.4A). Furthermore, when autogamy or self-fertilization was induced in these exconjugants, segregation for nuclear differences could not affect the appearance of killer trait. When conjugation was prolonged (Fig. 37.4B), massive transfer of cytoplasm took place between conjugants and under this case, killer trait could be transferred to No. 47 sensitive strain. It thus became evident that the killer trait was cytoplasmically inherited. This extra-chromosomal particle responsible for killer trait was given the name Kappa by Sonneborn.

Despite its cytoplasmic nature, the persistence of Kappa was soon found to be dependent upon at least one nuclear gene. This could be seen in the crosses between the killer strain No. 51 and a different sensitive strain No. 32. When there was no cytoplasmic exchange, only of the exconjugant clones possessed Kappa as in case of cross with sensitive strain No. 47. However, in the cross using sensitive strain No. 32, induced autogamy gave rise to a sensitive clone among the offspring of this heterozygous killer exconjugant (Fig. 37.5A). Similarly, when large amount cytoplasm was allowed to be exchanged, then both exconjugants showed the killer trait (Fig. 37.5B) and autogamy gave rise to sensitive clones. The killer trait or the Kappa particle was thereby shown to be maintaining itself in the presence of strain No. 51 or No. 47 nuclear gene called *K* but not in presence of a homozygous strain No. 32. Since both exconjugants from cross of strain No. 51 and 32 could maintain Kappa, the gene *K* was apparently dominant. Thus, the strain Nos. 51 and 47 could be designated as *KK* and the strain No. 32 as *kk*. Although present, however,

K could not by itself initiate the appearance of Kappa. That is the *KK* strain No. 47 remained sensitive until No. 51 cytoplasm was introduced. The presence of Kappa, in addition to the gene *K* was necessary for a strain to act as killer. When Kappa was removed from the killer strains by high temperature, x-rays and other techniques, these strains became sensitive. However, they could revert back to killer strains by ingesting chopped up killer animals. Thus, although relying upon nuclear genes for maintenance, Kappa seemed to be infective particle with a hereditary continuity of its own (Strickberger, 1968).

Normal conjugation (no cytoplasmic exchange)

Prolonged conjugation (cytoplasic exchange)

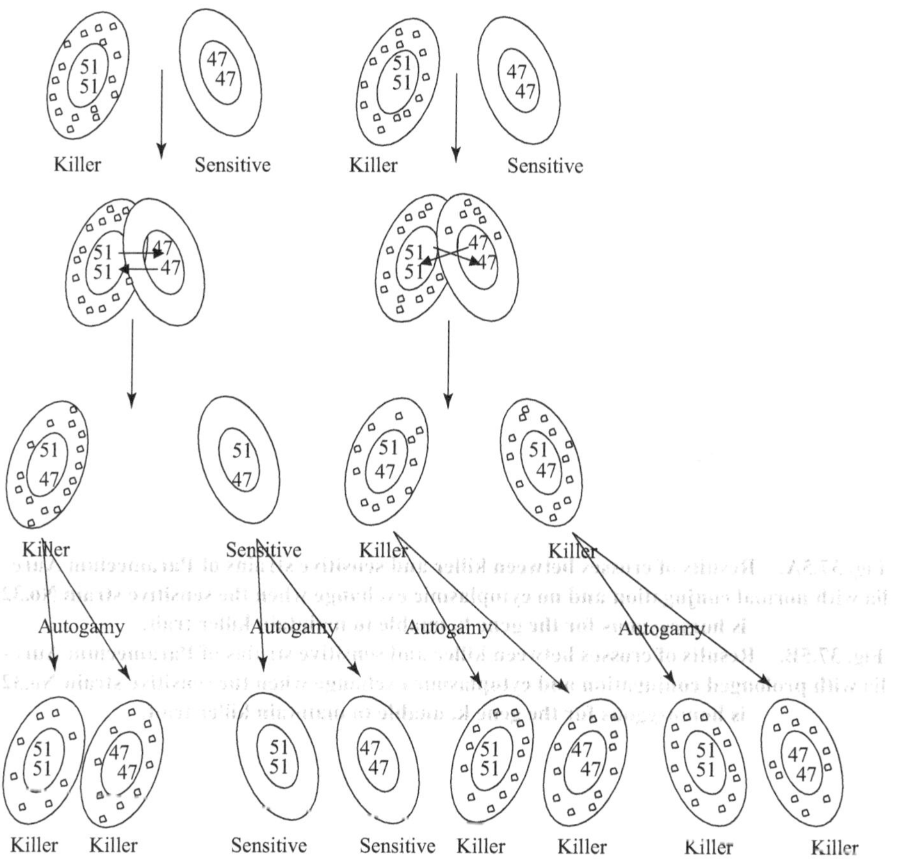

Fig. 37.4A. Results of crosses between killer strain No. 51 and sensitive strain No. 47 of Paramecium without cytoplasmic exchange.

Fig. 37.4B. Results of crosses between killer strain No. 51 and sensitive strain No. 47 of Paramecium with prolonged cytoplasmic exchange.

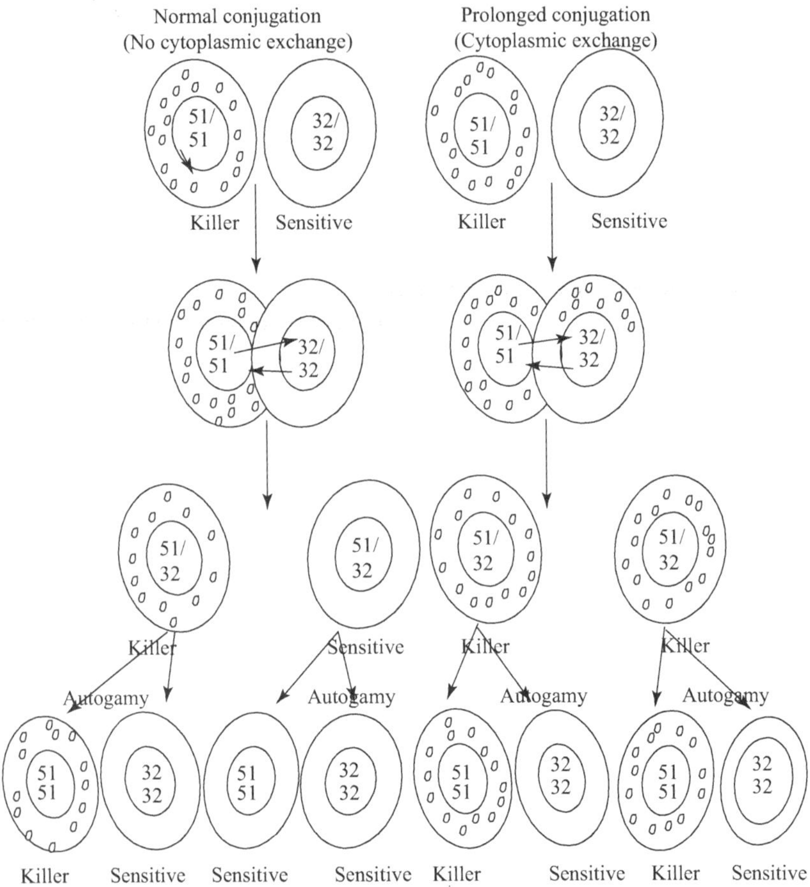

Fig. 37.5A. Results of crosses between killer and sensitive strains of Paramecium Aurelia with normal conjugation and no cytoplasmic exchange when the sensitive strain No.32 is homozygous for the gene k, unable to maintain killer trait.

Fig. 37.5B. Results of crosses between killer and sensitive strains of Paramecium Aurelia with prolonged conjugation and cytoplasmic exchange when the sensitive strain No.32 is homozygous for the gene k, unable to maintain killer trait.

38

DNA-The Chemical Basis of Heredity

In 1868, Johann Friedrich Miescher, a young Swish medical student got excited with an acidic substance that he isolated from pus cells obtained from bandages used to dress human wounds. To obtain large quantities of pus for his experiment, Miescher began purchasing used bandages from surgical clinics in Tuebingen, Germany. He separated pus cells from bandages and treated the cells with pepsin, a proteolytic enzyme that he isolated from stomach of pigs. After pepsin treatment, he recovered an acidic substance that was named as "nuclein" by him. This nuclein was unusual in the sense that it was rich in nitrogen and phosphorous, two elements known at that time to coexist only in certain types of fat. Miescher's paper on this investigation was published in 1871. At that time, importance of Miescher's nuclein was not fully appreciated. Even the existence of polynucleotide chains, the key component of the acidic material in Miescher's nuclein was not documented until 1940s. The role of nucleic acid in storing and transmitting genetic information remained un-established until 1944 and the double-helix structure of DNA was not discovered until 1953 by James Watson and Francis Crick. Even in 1953, many geneticists remained skeptical to accept the role of nucleic acid as the genetic information carrier rather than proteins as nucleic acids exhibited less structural variability than the proteins (Snustad *et al.*, 1997).

It is pertinent to recapitulate at this point of time that in 1865, Mendel has demonstrated that genes transmitted genetic information. Their patterns of transmission were studied extensively at that time but these classical genetic studies provided little insight into the molecular nature of genes. However, it was collectively realized that the genetic material disregarding its chemical nature must perform the following three essential functions in order to explain the on-going processes of inheritance and other related issues:

The genotypic function-replication: The genetic material must store genetic information and accurately transmit that information from parents to offspring, generation after generation.

The phenotypic function-gene expression: The genetic material must control the development of the phenotype of the organism. That is, the genetic material must dictate the growth and development of the organism from the single-celled zygote to the mature adult.

The evolutionary function-mutation: The genetic material must have capability to undergo change so that the organisms can adapt to the changing environment and evolution may proceed.

During the course of studies, it became clear that the chromosomes are composed of large organic molecules (macromolecules) called proteins and nucleic acids. The nucleic acids were found to be of two types, deoxyribonucleic acid (DNA) and ribonucleic acid (RNA). During 1940s and early 1950s, the results of some elegant experiments clearly proved that the genetic information is stored in the nucleic acids and not in the proteins. A few classical experiments to this effect are described in brief below:

Indirect Proofs That DNA Is the Chemical Basis of Heredity

Several indirect evidences suggest that the DNA is the chemical basis of heredity that it stores the genetic information and passes it on to the progeny generation after generation. A few such evidences are as follows:

- It is the DNA whose most portions are located in the chromosomes while RNA and proteins are also abundantly present in the cytoplasm.
- There is a precise correlation between the content of DNA per cell and number of sets chromosomes per cell. Most somatic diploid cells contain twice the amount of DNA as compared to that found in haploid germ cells (gametes).
- The molecular composition of DNA is the same with rare exceptions in all the different cells of an organism where the composition of both RNA and the proteins is highly variable from one cell type to another.
- DNA is more stable than the RNA and the proteins which are synthesized and degraded quite fast in the living organisms.

Although, these evidences are strong indications that the DNA is the real genetic material but they do not prove the direct role of DNA as the genetic material. The fact that the DNA is the genetic material has been proved through several experiments as listed below:

Direct Proof That DNA is The Genetic Material

Discovery of Transformation in Bacteria

Bacterial transformation, a type of recombination in bacteria was discovered by Frederick Griffith in 1928. Although this experiment on *Diplococcus pneumoniae* (pneumococcus) provided no evidence that DNA was directly involved in the transformation, it certainly set the stage for other follow-up discoveries demonstrating the direct role of DNA as the genetic material. In *Diplococcus pneumoniae*, thin capsule enclosing the cell wall is made up of substances called polysaccharides which are of specific types like Type II capsules and Type III capsules characterized by formation of specific antibodies. These capsules types are distinct properties of specific bacterial strains. Further, capsules are subject to variability in respect to their presence or absence. Some *Diplococcus pneumoniae* strains are without capsules and such non-capsulated bacteria have a rough appearance (R) when grown on culture plates in contrast to the encapsulated bacteria which give smooth appearance (S) on culture plates and are virulent. R and S colonies transmit their respective characters to future generations. All of these bacterial strains are sensitive to heat and are heat-killed.

In one of exceptional set of experiments, Griffith in 1928 demonstrated that heat-killed bacteria of one type could have hereditary influence on the bacteria of another type. In one experiment, he injected heat-killed bacteria of Type III-S into mice containing Type II-R and obtained virulent live cultures which were of Type III-S. Since injection of heat-killed Type III-S bacteria by itself does not result in any live bacterial culture, a change or transformation of Type II-R into Type III-S must have occurred through transfer of some active substance (Fig. 38.1).

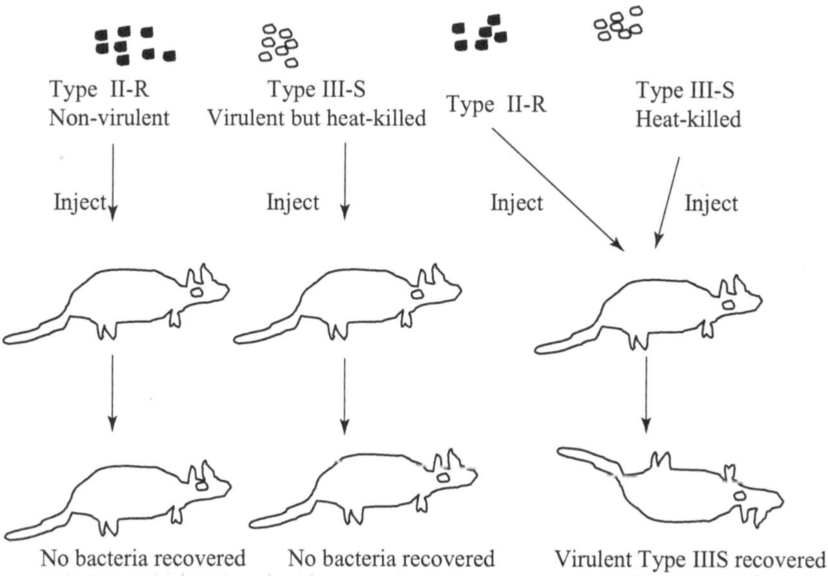

Fig. 38.1 Transformation experiment of Griffith in mice using *Diplococcus pneumoniae*.

Search for "transforming principle", the specific agent responsible for transformation began after the experiment by Griffith. In 1944, Avery,

MacLeod and McCarty showed that the "transforming principle" was entirely of DNA. Their experiment consisted of extraction of DNA from heat-killed cells of Type III-S and mixing this DNA directly with *in vitro* cultures of Type II-R. A serum was added whose antibodies react with the R cells and cause them to precipitate to the bottom. When transformation occurred, Type III-S cells, not being precipitated, grew diffusely throughout the medium (Fig. 38.2, Strickberger, 1968). Since 1944, several transforming experiments have been done with a variety of bacteria and in all the cases the "transforming principle" was identified to be DNA. Furthermore, like DNA the biological activity of the "transforming principle" was destroyed by DNA-specific enzyme, deoxyribonuclease but not by RNA-specific enzyme, ribonuclease or protein-specific enzyme trypsin or chemotrypsin. Labeling of transforming DNA with radio-active phosphorous results in incorporation of this radio-activity into recipient bacteria in the direct proportion to the number of the transformants achieved. The pure DNA nature of the "transforming principle" in these experiments is the strongest evidence that the DNA is the genetic material or the chemical basis of heredity.

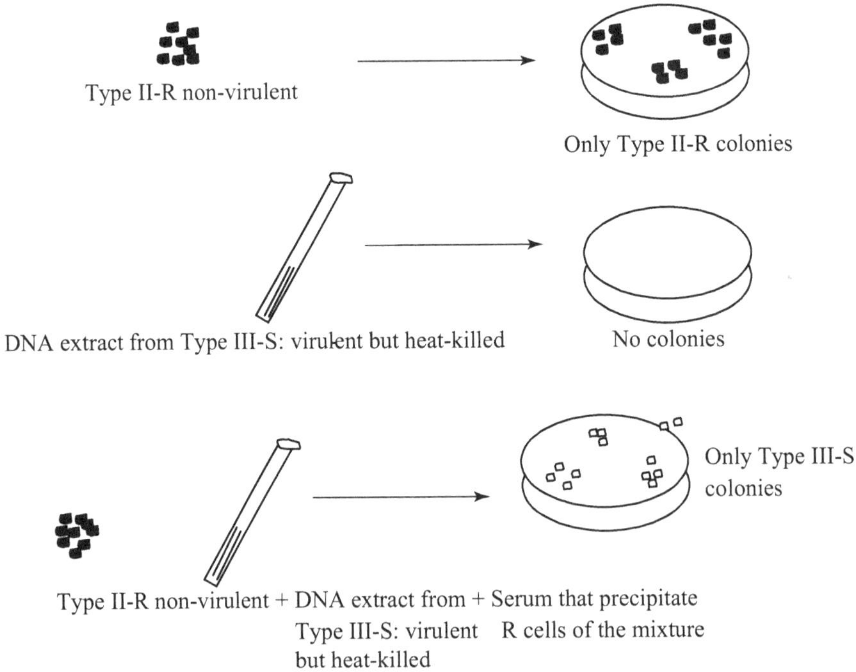

Fig. 38.2. Transformation experiment of Avery, MacLeod and MaCarty-1944.

Hershey and Chase Experiment-1952 on Viral DNA as The Genetic Material

Hershey and Chase (1952) showed that the production of new viral particles could happen if the protein envelopes of the virus remained outside the infected bacterial cell and this happened only when the viral DNA entered the bacterial cell. This experiment mainly consisted of two main steps (Fig. 38.3, Strickberger, 1968). First Hershey and Chase demonstrated that if bacteriophage cultures were labeled with radio-active phosphorous (^{32}P) or sulfur (^{35}S) and then ruptured in the presence of DNAse, the DNA contained all the ^{32}P and the protein ghosts had all the ^{35}S. Then they permitted phages labeled either with ^{32}P or ^{35}S to infect bacteria and shaved off the phage heads from the bacterial cell wall by the action of a Waring Blender. As only ^{32}P labeling could be recovered in the newly released phages after lysis of the bacteria, these experiments indicated that DNA was the genetic material transmitted between parents and offspring.

RNA As Genetic Material In Tobacco Mosaic Virus

Tobacco mosaic virus (TMV) is one of the several viruses having RNA instead of DNA as its sole nucleic acid. The RNA is surrounded by a protein coat which aids in infection of plant cells. The protein is specific for a particular strain and therefore there are different TMV strains each with a different protein envelope. Frankel-Conrat and others showed that the TMV protein could be separated from the RNA but the protein was not infective by itself. Only when the protein was recombined or reconstituted with the isolated RNA, only then the viral infectivity returned and new viral particles could be produced. The next question was whether chemically isolated TMV-RNA could be an infective agent on its own and thereby produce new viral particles. Experiments conducted by Gierer and Schramm in 1956 showed that this was the case as when they rubbed pure TMV-RNA into tobacco leaves, there was synthesis of new TMV and such synthesis could be prevented by ribonuclease enzyme (enzyme causing destruction of RNA). Further, experiments by Frankel-Conrat and Singer (1957) showed that the reconstituted TMV containing RNA from one strain and protein from another strain, always led to synthesis of viral particles whose protein was determined by the parental RNA rather than the parental protein (Fig. 38.4). Thus, in DNA containing organisms, DNA was the chemical basis of heredity but in the organism devoid of DNA but having RNA, the RNA functioned as the hereditary material.

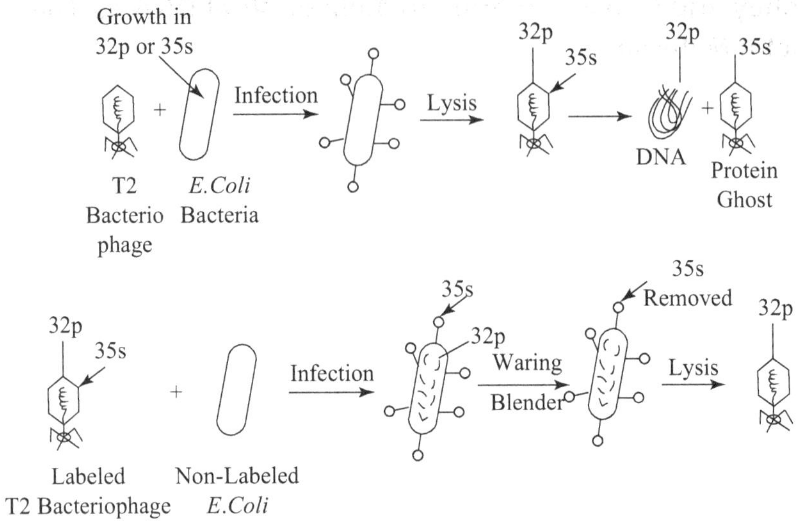

Fig 38.3 Hershey and Chase-1952 experiment.

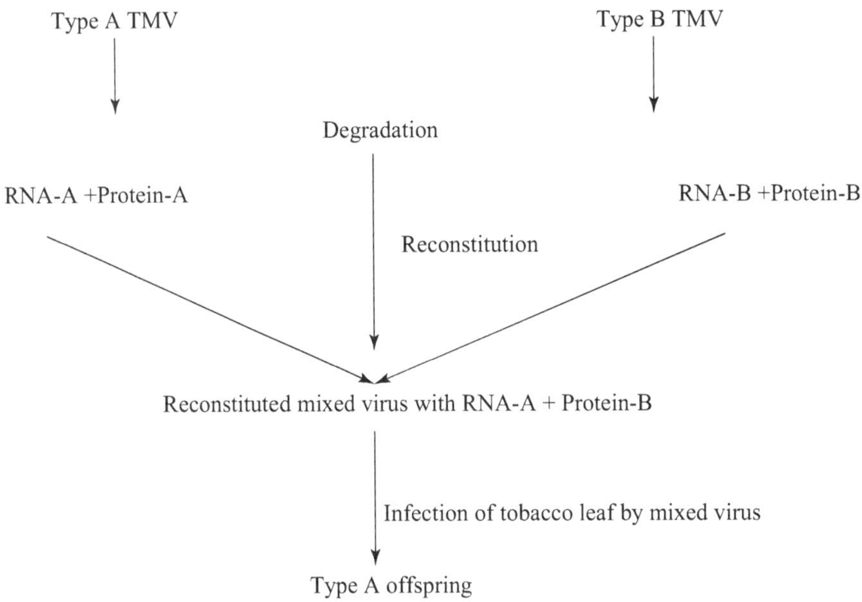

Fig. 38.4. Reconstitution of a mixed tobacco mosaic virus and demonstration that the type of offspring produced is determined by the nucleic acid (RNA in this case) and not by the protein.

39

Nucleic Acid

The search for the chemical basis of genetic material began about one and half century ago. Soon after discovery of Mendelian principles of segregation and independent assortment in 1865, Friedrich Miescher (1844-1895) published a method for separating cell nuclei from the cytoplasm and from these cell nuclei, he extracted an acid material termed as "nuclein" which had unusually large amount of phosphorous. Miescher concluded that this substance was unique and was just not comparable with any other group known at that point of time. Later on nuclein, now named as "nucleic acid" was found to be associated with various proteins and the combined entities were termed as "nucleoproteins".

Nucleotides

Nucleic acid was further shown to be sub-divisible into smaller units called "nucleotides". Each nucleotide in turn consists of a sugar, a phosphate group and a nitrogenous base. The sugar contains five carbons and is named as ribose. The other counterpart five carbon sugar lacking one oxygen atom is called as deoxyribose (Fig. 39.1).

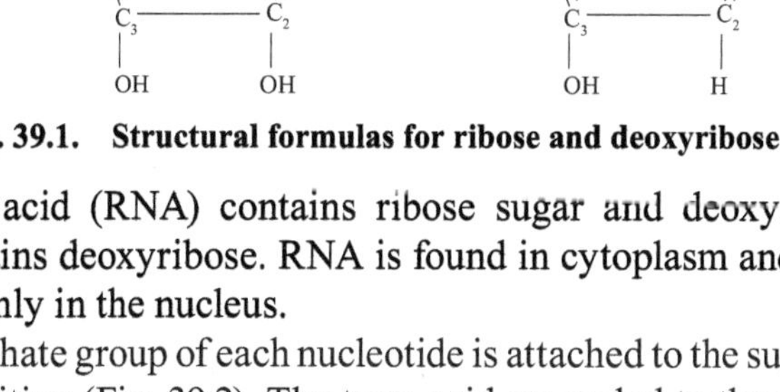

Fig. 39.1. Structural formulas for ribose and deoxyribose sugars.

Ribonucleic acid (RNA) contains ribose sugar and deoxyribonucleic acid (DNA) contains deoxyribose. RNA is found in cytoplasm and DNA with few exceptions only in the nucleus.

The phosphate group of each nucleotide is attached to the sugar at its number 5 carbon position (Fig. 39.2). The term acid appended to these chemicals was derived from the common representation of phosphate group as having a dissociable hydrogen ion (PO_4H^+).

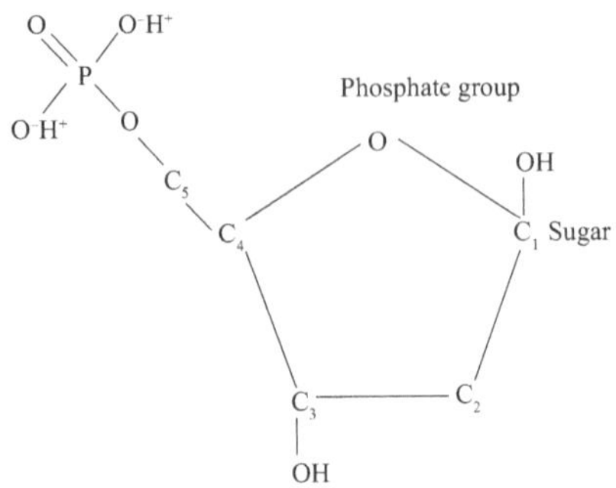

Fig. 39.2. Linkage between a phosphate and deoxyribonucleic acid (individual H atoms omitted).

Besides, 5 carbon sugar and phosphate groups, which are constant components of all nucleotides in a nucleic acid, a more variable nitrogen containing base is also present attached with the sugar molecule at 1 carbon position. The bases containing one carbon nitrogen ring are known as pyrimidines and the two ring bases are called as purines. Two main purines are adenine (A) and guanine (G). The pyrimidines include thymine (T), cytosine (C) and uracil (U). In addition to the differences in ring structures, the bases also show the presence of an amino group (NH_2) at number 6 position of one pyrimidine (cytosine) and one purine (adenine). A keto group (C=O) occurs at this same position in the other bases (thymine, uracil, guanine). The structural formulas of the common bases in the nucleic acid (DNA and RNA) are given in Fig. 39.3.

DNA contains adeine, guanine, thymine and cytosine. RNA contains adenine, guanine, cytosine and uracil. Depending upon the presence of a particular base, there are four different deoxyribonucleotides (deoxyribose sugar + phosphate + base). These are as follows:

- Deoxyadenilic acid or deoxyadenosine monophosphate-dAMP (deoxyribose + phosphate + adenine)
- Deoxythymidilic acid or deoxythymidine monophosphate-dTMP (deoxyribose + phosphate + thymine)
- Deoxyguanilic acid or deoxyguanosine monophosphate-dGMP ((deoxyribose + phosphate + guanine)
- Deoxycytidilic acid or deoxycytidine monophosphate-dCMP (deoxyribose + phosphate + cytosine)

RNA contains adenine, guanine, cytosine and uracil and there are four different ribonucleotides (ribose sugar + phosphate + base) as listed below:

(a) In RNA only (with rare exceptions);

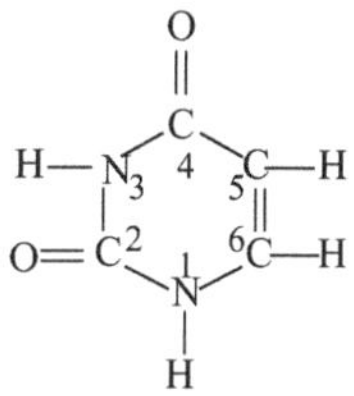

Uracil
(2,4-oxypyrimidine)

(b) In both RNA and DNA

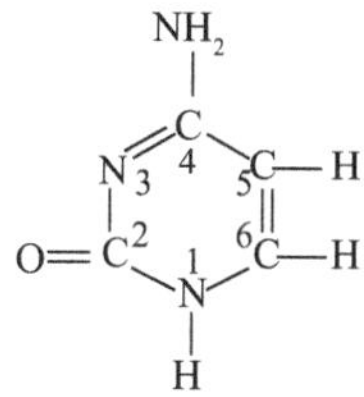

Cytosine
(4-amino-2-oxypyrimidine)

(c) In DNA only (with rare exceptions):

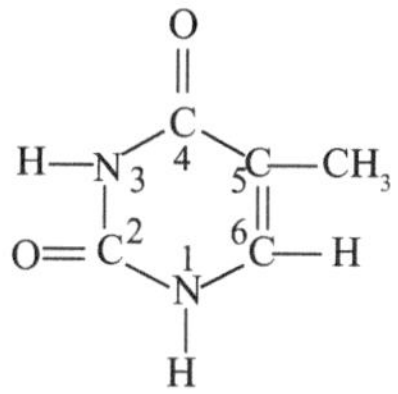

Thymine
(2, 4-oxy-5-methyl pyrimidine)

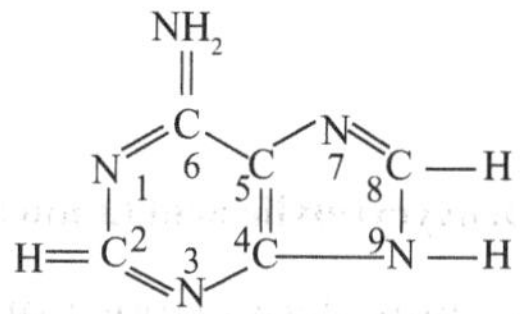

Adenine
(6-aminopurine)

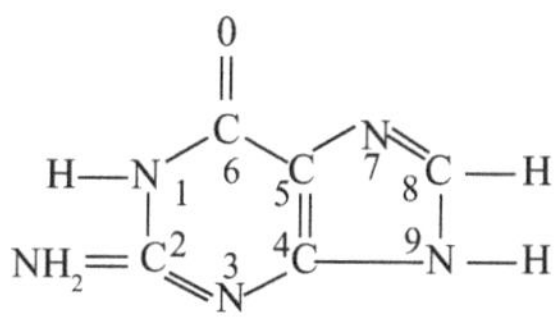

Guanine
(2-amino-6-oxypurine)

Fig. 39.3. Structures of common bases in nucleic acids.

- Riboadenilic acid or riboadenosine monophosphate-rAMP (ribose + phosphate + adenine)
- Ribouracilic acid or ribouracil monophosphate-rUMP (ribose + phosphate + uracil)
- Riboguanilic acid or riboguanosine monophosphate-rGMP ((ribose + phosphate + guanine)
- Ribocytidilic acid or ribocytidine monophosphate-rCMP (ribose + phosphate + cytosine)

Typical structure of a nucleotide, for example, deoxycytidilic acid is shown as follows (Fig. 39.4) where the base is attached at carbon 1 of the deoxyribose sugar and the phosphate group is attached at carbon 5 of the deoxyribose sugar.

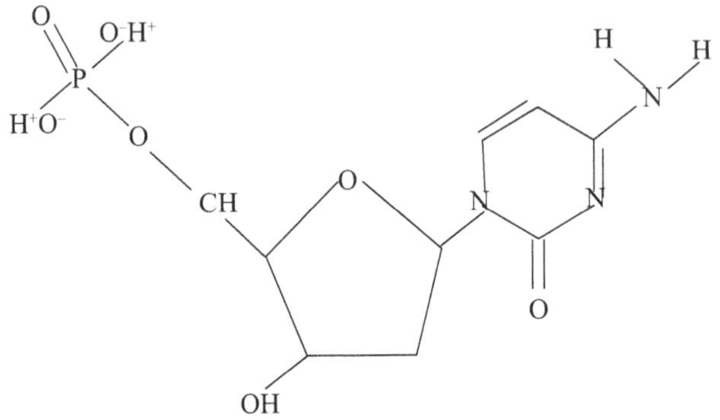

Fig. 39.4. Deoxycytidylic acid (a nucleotide in DNA).

Thus, it is clear that a base plus deoxyribose or ribose sugar plus phosphate group is called a nucleotide. The DNA molecule is thus a long polymer (i.e. a macromolecule composed of a number of identical subunits, monomers, covalently bonded) of thousands of nucleotide pairs, even in such simple organisms as virus.

In poly-nucleotides, such as DNA and RNA, these subunits are joined together in long chain. RNA usually exists as a single-stranded polymer that is composed of a long sequence of nucleotides. DNA is usually a double-stranded molecule. In the long chain of nucleotides, the phosphate group (PO_4) links adjacent sugars through a 3'-5' phosphodiester linkage, in one chain the linkages are polarized 3'-5', in the other chain, they are in reverse order (Fig. 39.5).

The adenine, guamnine, thymine and cytosine are classified into purines and pyrimidines. Purines only pair with pyrimidines and vice-versa thus producing a symmetrical double helix. Furthermore, the base pairs are linked in such a manner that the number of bonds between them is always maximized. For this reason, adenine (A) pairs only with thymine (T) and guanine (G) pairs only with cytosine (C).

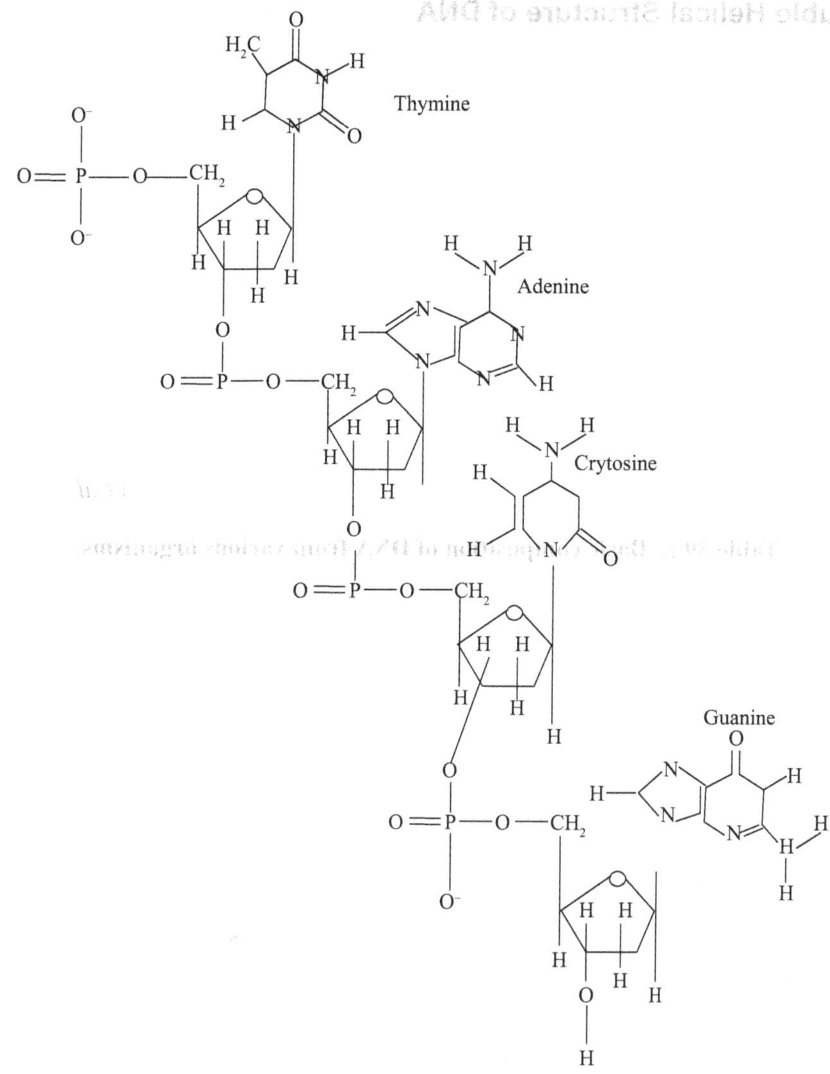

Fig. 39.5. Formation of poly-nucleotide chain by joining nucleotides with phosphodiester linkages. This poly-nucleotide has a 5' to 3' chemical polarity (top to bottom). Each phosphodiester linkage joins the 5' carbon of the 2'-deoxyribose in one nucleotide to the 3' carbon of 2'-deoxyribose in the adjacent nucleotide resulting in the chemical polarity of the poly-nucleotide chain.

The simplified version of structure of nucleic acid could be shown as follows:

```
      A    T      G        G         C          C          T         A       T
      |    |      |        |         |          |          |         |       |
------P------S-------P---------S----------P-------------S------------P-----------S---------P-------
```

P = Phosphate group, S = Sugar, A, T, G, C = Bases

Double Helical Structure of DNA

The correct structure of DNA was deduced in 1953 by James Watson and Francis Crick. Their double-helix model of DNA structure was based on two major evidences as follows:

When Erwin Chargaff and colleagues analyzed the composition of DNA sourced from many different organisms they could find that the concentration of thymine was always equal to the concentration of adenine and the concentration of cytosine was always equal to the concentration of guanine. These results indicated that the thymine and adenine as well as cytosine and guanine were present in DNA in some fixed interrelationship. Their data also showed that the total concentration of pyrimidines (thymine + cytosine) was always equal to the total concentration of purines (adenine + guanine). In contrast, the ratio of (thymine + adenine) and (cytosine + guanine) showed wide variation in DNAs of different species (Table 39.1, Snustad *et al.*, 1997).

Table 39.1. Basic composition of DNA from various organisms.

Species	Adenine (%)	Guanine (%)	Cytosine (%)	Thymine (%)	A+G/ T+C	A+ T/ G+C
Viruses						
Bacteriophage lamda	26.0	23.8	24.3	25.8	0.99	1.08
Bacteriophage T2	32.6	18.1	16.6	32.6	1.03	1.88
Herpes simplex	13.8	37.7	35.6	12.8	1.06	0.36
Vaccinia	31.5	18.0	19.0	31.5	0.98	1.70
Bacteria						
Escherichia coli	26.0	24.9	25.2	23.9	1.04	1.00
Micrococcus lysodeikticus	14.4	37.3	34.6	13.7	1.07	0.39
Ramibacterium ramosum	35.1	14.9	15.2	34.8	1.00	2.32
Fungi						
Neurospora crassa	23.0	27.1	26.6	23.3	1.00	1.04
Aspergillus niger	25.0	25.1	25.0	24.9	1.00	1.00
Saccharomyces cerevisiae	31.7	18.3	17.4	32.6	1.00	1.80
Higher eukaryotes						
Corn	25.6	24.5	24.6	25.3	1.00	1.04
Tobacco	29.3	23.5	16.5	30.7	1.12	1.50

Species	Adenine (%)	Guanine (%)	Cytosine (%)	Thymine (%)	A+G/ T+C	A+ T/ G+C
Drosophila	30.7	19.6	20.2	29.4	1.01	1.51
Human	30.2	19.9	19.6	30.3	1.01	1.53

When X-rays are focused through crystals of purified molecules, the rays are deflected by the atoms of the molecules in specific patterns, called diffraction patterns which provide information about the organization of the components of the molecules. These X-ray diffraction patterns can be recorded on X-ray-sensitive film. Watson and Crick used X-ray crystallographic data on DNA structure made available by Maurice Wilkins, Rosalind Franklin and co-workers. These data showed that DNA was a highly organized, two-stranded structure with repeating sub-structures spaced every 0.34 nanometer (1 nm $=10^{-9}$ meter) along the axis of the molecule.

On the basis of Chargaff's chemical data, Wilkins' and Franklin's X-ray diffraction data and inferences from model building, Watson and Crick proposed DNA exists as a right-handed double helix in which two poly-nucleotide chains are coiled about one another in a spiral. Each poly-nucleotide chain consists of a sequence of nucleotides linked together by phosphodiester bonds, joining adjacent deoxyribose moieties. The two poly-nucleotide strands are held together in their helical configuration by hydrogen bonding between bases in opposing strands. The resulting base pairs are stacked between two chains perpendicular to the axis of the molecule like the steps of a spiral staircase. The base pairing is specific. Adenine always paired with thymine and guanine always paired with cytosine. Specificity of base-pairing results from hydrogen bonding capacities of the bases in their normal configuration. In their normal structural configurations, adenine and thymine form two hydrogen bonds and guanine and cytosine form three hydrogen bonds. Hydrogen bonding is not possible between cytosine and adenine or thymine and guanine when they exist in their common structural states. Once the sequence of bases in one strand of DNA is known, the sequence of bases in the other strand is also known because of specific base pairing i.e. adenine (A) pairs only with thymine (T) and guanine (G) only with cytosine (C). The two strands of a DNA double helix are thus said to be complementary to each other. This property, the complementarity of the two strands of the double helix makes DNA uniquely suited to store and transmit genetic information from generation to generation.

The base pairs in DNA are stacked about 0.34 nm apart with 10 base pairs per turn (360 degree) of the double helix. The sugar-phosphate backbones of the two complementary strands are anti-parallel i.e. they have opposite chemical polarity. The phosphodiester bonds in one strand go from 3' carbon of one nucleotide to a 5' carbon of the nearby nucleotide, whereas those in the complementary strand go from a 5' carbon to a 3' carbon. This opposite polarity of the complementary strands of a DNA double helix plays an important role in DNA replication, transcription and recombination. Diagramme of the double-helical structure of DNA is given in Fig. 39.6. James Watson, Francis Crick and Maurice Wilkins were awarded jointly the Nobel Prize Physiology/Medicine in 1962 for their discovery concerning the molecular structure of nucleic acids and its significance for information transfer in the living material.

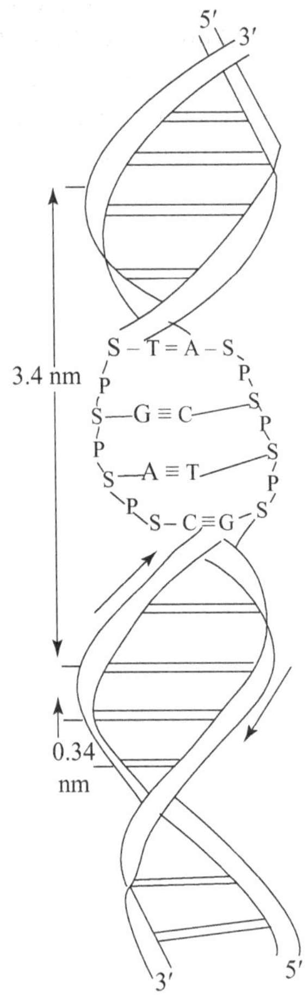

Fig. 39.6. Diagramme of Watson-Crick model of double-helix structure of DNA. S = Sugar, P = Phosphate, A = Adenine, T = Thymine, G = Guanine, C = Cytosine. There are two hydrogen bonds between A and T and three hydrogen bonds between g and C.

The Watson-Crick double helix structure depicted in Fig. 39.6 is called as B-DNA. B-DNA is the conformation that DNA takes under physiological conditions in aqueous solutions containing low concentrations of salts. The vast majority, if not all, of the DNA molecules present in the aqueous protoplasms of living cells exist in the B conformation. However, DNA is not static and exhibits considerable amount of conformational flexibility and the structures of DNA molecules change in response to changing environments. The exact conformation of a given DNA molecule or segment of DNA molecule will depend on the nature of the molecules with which it interacts. In fact, intracellular B-DNA appears to have an average of 10.4 nucleotide pairs per turn rather than 10 as shown in Fig. 39.6. In high concentrations of salts or in a partially dehydrated state, DNA exists as A-DNA which is right-handed helix like B-DNA but with 11 nucleotide pairs per turn. A-DNA is a shorter, thicker double helix with a diameter of 0.23 nm. DNA molecules almost certainly never exist as A-DNA *in vivo*. However, A-DNA conformation is is of importance because DNA-RNA heteroduplexes or RNA-RNA duplexes exist in a very similar structure *in vivo*. Additional right-handed helical forms of DNA designated as C, D, and E have been shown to exist under specific environmental conditions (dehydration, etc.). Certain DNA sequences have been shown to exist in a left-handed, double helical form called Z-DNA (Z for the zigzagged path of the sugar-phosphate backbones of the structure). Z-DNA was discovered by X-ray diffraction analysis of crystals formed by DNA oligomers containing alternating G: C and C: G base pairs. Z-DNA occurs in double helices that are G: C-rich and contain alternating purine and pyrimidine residues. In addition to its unique left-handed helical structure, Z-DNA differs from B-DNA in having 12 base pairs per turn, a diameter of 0.18 nm and a single deep grove.

40

DNA Replication

Watson and Crick based on their double helical DNA model, offered to explain DNA replication by a method where each single strand of DNA is template or mold for its complement and a new helix has one old strand and one newly synthesized strand. This type of DNA replication is called as semi-conservative as in this method one old strand is retained and one new complementary strand is synthesized. In contrast to semi-conservative mode of DNA replication, two more modes of DNA replication were proposed. One is conservative type where two new strands are synthesized in form of a double helix while the old double helix remains unchanged. The third type of DNA replication proposed has been dispersive mode of DNA replication which amounts to breaking down of double helical strands into small pieces along the length, their replication and random assemblage with newly synthesized pieces to form a patchwork single string of dispersed old and new pieces. These three modes of replication of DNA molecule are shown in Fig. 40.1.

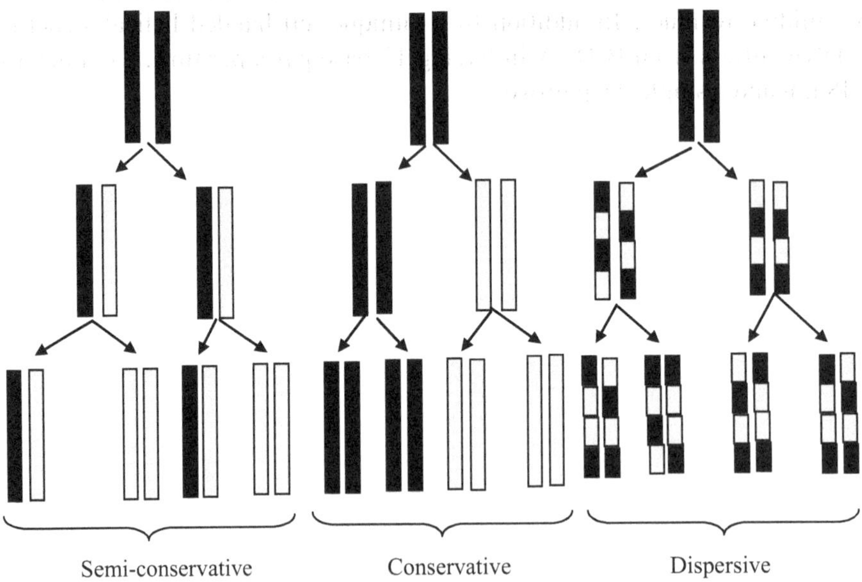

Fig. 40.1. Three modes of replication of DNA.

Semi-Conservative Mode of DNA Replication

This is the accepted mode of DNA replication based on experimental evidences. When Watson and Crick deduced the double-helix structure of DNA with its complementary base pairing, they could recognize that complementary base pairing specificity could provide a simple basis for the mechanism of DNA replication. In this proposition, the two complementary strands of double-helix get separated by breaking of hydrogen bonds between the base pairs and each parental strand directs the synthesis of a new complementary strand because of specific base pairing requirement. Each parental strand functions as a template for the synthesis of a new complementary strand where adenine in the parental strand serves as template via its hydrogen bonding potential for the incorporation of thymine in the nascent complementary strand. This mode of DNA replication is called as semi-conservative mode of DNA replication because each of the complementary strands of the parental double helix is conserved in the newly formed double-helix i.e. original double-helix is half conserved (Fig. 40.1). The hydrogen bonds linking base pairs are relatively weak bonds. At the time of replication, the two strands of the DNA separate along this line of weakness in zipper like fashion. Each single strand then becomes a template for the formation of another complementary strand by attracting free nucleotides to their complementary sites and linking the adjacent nucleotides under the direction of the enzyme DNA-polymerase.

Experimental Evidence for Semi-Conservative Mode of DNA Replication

The first critical test of Watson and Crick's proposal that DNA replicates by semi-conservative mechanism was performed by Matthew Meselson and Franklin Stahl in 1958. Their results showed that the chromosome (now known to contain a single double-helix of DNA) of the common colon bacillus *E. coli*, a prokaryote, replicates through semi-conservative mechanism of DNA replication. Meselson and Stahl grew *E. coli* cells for many generations in a medium in which heavy isotope of nitrogen, ^{15}N had been substituted for the normal, light isotope, ^{14}N. The purine and pyrimidine bases in DNA contain nitrogen and thus DNA of the *E. coli* cells grown on medium containing ^{15}N will have greater density (weight per unit volume) in comparison to the DNA of the cells grown on normal medium containing ^{14}N. The experimental results of Meselson and Stahl are depicted in Fig. 40.2. These results (heavy band in generation 0, hybrid DNA band in generation 1, half hybrid and half light band in generation 2 and one fourth hybrid and three fourth light bands in generation 3) show that these results are possible if the *E. coli.* chromosome replicates semi-conservatively.

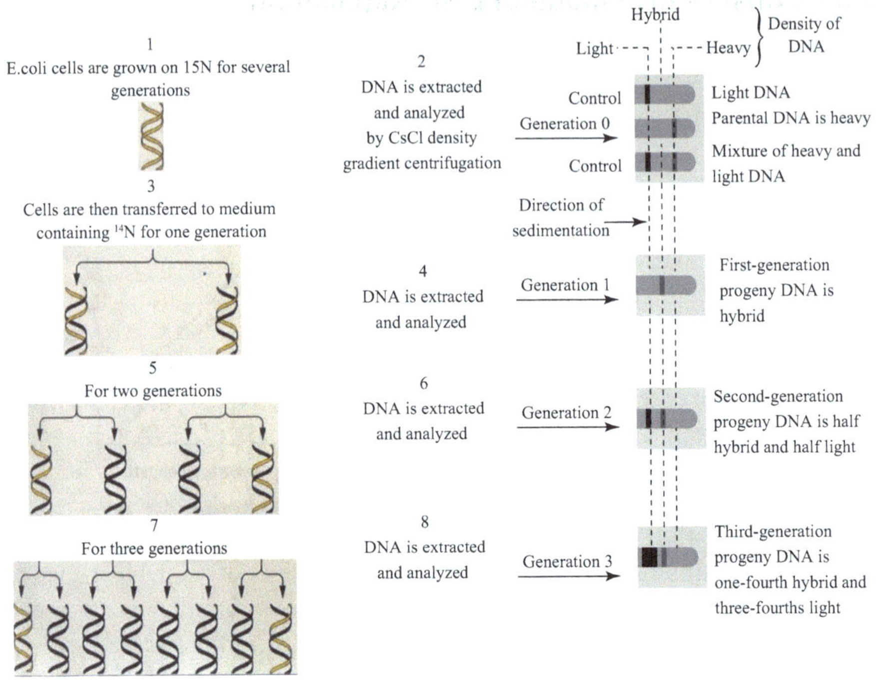

Fig. 40.2. Meselson and Stahl's experiment demonstrating that DNA replication is semi-conservation mechanism in *E. coli.*

DNA Synthesis

The *in vitro* synthesis of DNA was first accomplished by Arthur Kornberg and his co-workers in 1957. Kornberg received Nobel prize for this work in 1959. Kornberg and co-workers in 1957 isolated a polymerase enzyme from *Escherichia coli* bacteria that could be used for *in vitro* synthesis of DNA. Initially this enzyme was called as DNA polymerase or "Kornberg's enzyme" but now it is known as DNA polymerase I. The reaction mixture contained the polymerase enzyme; the four different kinds of 5'-triphosphates of each of the four deoxyribnucleotides-deoxyadenosine triphosphate (dATP), deoxythymidine triphosphate (dTTP), deoxyguanosine triphosphate (dGTP) and deoxycytidine triphosphate (dCTP), magnesium ions ($MgCl_2$), and the presence of some previously formed DNA called as primer DNA. Once all these necessary substances were present, DNA synthesis took place in amounts 20 times or more than the amount of initial DNA. This has been shown in Fig. 40.3 (Strickberger, 1968).

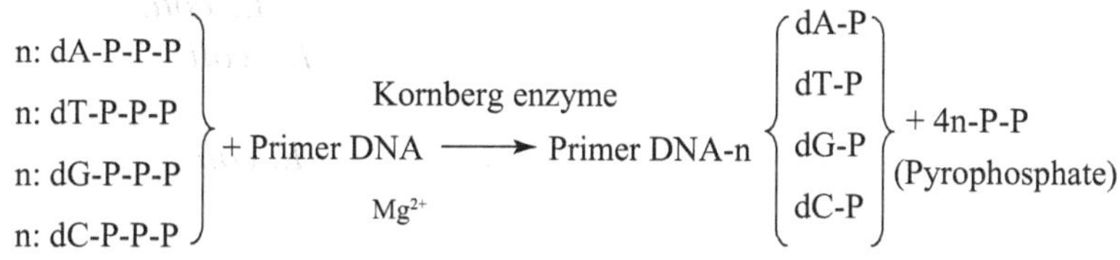

Fig. 40.3. Reaction system for the synthesis of DNA *in vitro* using the four different deoxribonucleotide triphosphate and primer DNA.

The pre-existing DNA must provide two essential components, one serving as a primer function and the other a template function. All DNA polymerases require a primer strand with a free 3'-hydroxyl. The primer strand is covalently extended by the addition of nucleotides (such as dTMP, derived from the incoming precursor dTTP). In addition, DNA polymerases require a template strand which determines the base sequence of the strand being synthesized. The new strand will be complementary to the template strand.

The Primer DNA

The primer DNA provides a terminus with a free 3-OH to which nucleotides are added during the synthesis of DNA. DNA polymerase I cannot initiate the synthesis of DNA chains *de novo.* It must get a free 3'-hydroxyl on a pre-existing DNA chain. DNA polymerase I catalyzes the formation of a phosphodiester bridge between the 3'-OH at the end of the primer DNA chain and 5'-phosphate of the incoming deoxyribonucleotide.

The Template DNA

Template DNA provides the nucleotide sequence that specifies the complementary sequence of the nascent DNA chain. DNA polymerase I needs a DNA template whose base dictates a DNA template by its base-pairing potential, the synthesis of a complementary base sequence in the strand being synthesized.

The reaction catalyzed by DNA polymerase I is a nucleophilic attack by the 3'-OH at the terminus of the primer strand on the nucleotide or interior phosphodiester atom of the nucleoside triphosphate precursor with the elimination of pyrophosphate. This reaction mechanism explains the absolute requirement of DNA polymerase I for a free 3'-OH group on the primer DNA strand that is being covalently extended and dictates that the direction of synthesis is always 5' 3' (Fig. 40.4, Snustad *et al.*, 1997).

The DNA polymerase I encoded by a gene *polA* is a single polypeptide with molecular weight of 109,000. Subsequently, it has been found that

DNA polymerase I is not the true "DNA replicase" in *E. coli.* It does not catalyze the semi-conservative replication of the *E. coli* chromosome. That particular function is performed by another enzyme. However, DNA polymerase I does perform important functions in the *E. coli* cells including a key role in the replication of the bacterial chromosome and a central role in repairing damaged DNA. DNA polymerase I has two other activities, both exonuclease activities. A nuclease is an enzyme that degrades nucleic acids. An exonuclease degrades nucleic acids starting at one or both ends, whereas an endonuclease cleaves nucleic acids at internal sites. DNA polymerase I has both 5'-3'exonuclease activity, which cuts back DNA strands starting at 5' termini and 3'-5' exonuclease activity which cleaves off mononucleotides from 3' termini of DNA strand. The 5'-3'exonuclease activity of DNA polymerase I normally excises small oligomers containing up to 10 nucleotides. Thus, DNA polymerase I has three enzymatic activities which are (i) a 5'-3'polymerase activity, (ii) a 5'-3'exonuclease activity, and (iii) a 3'-5'exonuclease activity.

The Complex Replication Apparatus

The results of DNA replication using autoradiography and electron microscopy indicated that the mechanism of DNA replication is much more complex than the researchers originally thought and proposed. The two daughter strands being synthesized at each replicating fork are extended in the same overall direction at the macromolecular level. Because of opposite polarity of the complementary strands, the synthesis of one of the two new strands occurs at 5'end of one strand (3'-5') and at the 3' end of the other strand (5'-3'). But it is also known that polymerases have an absolute requirement for a free 3' hydroxyl and they carry out only 5'-3' synthesis. Apparently, for many years, searches continued for polymerases that could catalyze 3'-5' synthesis of DNA. However, this type of polymerase could not be found. All experimental evidences showed that all DNA synthesis occurred in the 5'-3' direction. In background of these complex issues, DNA replication is considered much more complicated than scientists thought when the semi-conservative mode of DNA replication was proposed by Watson and Crick in 1953.

It should be absolutely clear by now that due to opposite polarity of both the strands of the DNA, one strand is extended in an overall 5'--3' direction and the other strand in 3'--5' direction but DNA polymerase can only carry forward synthesis in the 5'-- 3' direction. The resolution of this paradox could be realized with the demonstration that the synthesis of one strand of DNA is continuous whereas the synthesis of other strand is discontinuous. At the molecular level, the synthesis of the complementary strand of DNA occurs in opposite physical directions but both new strands are extended in the same 5'-- 3' chemical direction (Fig. 40.5).

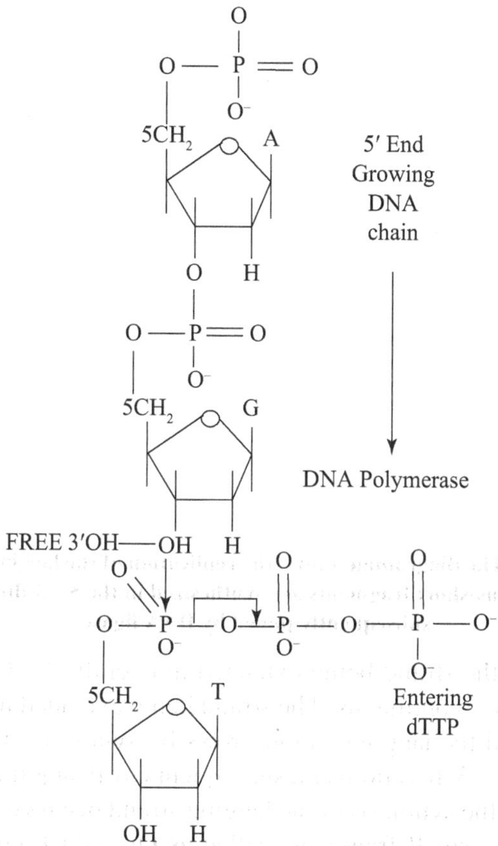

Fig. 40.4. Mechansim of action of DNA polymerase I. The diagramme shows the DNA polymerase –catalyzed addition of deoxythymidine monophosphate from the precursor deoxythymidine triphophate (dTPP) to the 3'end of the chain with release of pyrophosphate (P2O7).

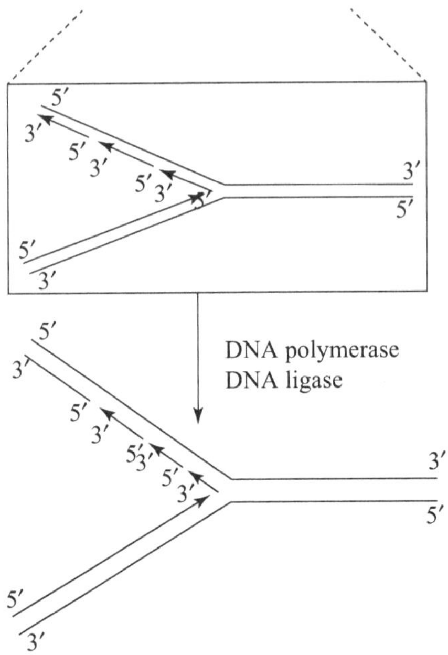

Fig.40.5. The diagramme shows the replication of the lagging strand is discontinuous-short fragments are synthesized in the 5'--3'direction and subsequently joined by DNA ligase.

The synthesis of the strand being extended in overall 5'—3' direction called the leading strand is continuous. The strand being extended in the overall 3'—5' direction called the lagging strand grows by synthesis of short fragments (synthesized in 5'—3'direction) and subsequent covalent joining of these short fragments. Thus, the synthesis of the lagging strand occurs by a discontinuous mechanism. These small fragments called as Okazaki fragments are joined together by DNA ligase.

Semi-conservative method of DNA replication requires that the two strands of a parental DNA molecule be separated during the synthesis of new complementary strands. Since a DNA double helix is a plectonemic coil with two strands that cannot be separated without untwisting them turn by turn. Naturally DNA replication requires an unwinding mechanism for the coiled strands. Given that each gyre or turn is about 10 nucleotide base pairs long, a DNA molecule must be rotated 360° once for each 10 replicated base pairs. This unwinding process is catalyzed by enzyme called DNA helicases. The major DNA helicase in *E. coli* is produced by the gene *dnaB* gene. DNA helicases unwind DNA molecules using energy derived from ATP. Once DNA strands are unwound by DNA helicase, they must remain in an extended single stranded form for replication. They are maintained in the extended single stranded state by a coating of single-stranded DNA-binding proteins (SSB proteins). Without the SSB protein coating, the complementary strands could renature or form intra-strand hairpin structures by hydrogen bonding between short segments of complementary or partially complementary nucleotide sequences. The required axes of rotation during the replication of circular DNA molecules are provided by the enzymes called DNA topoisomerases.

Most of the above experimental results on DNA replication are based on studies on *E. coli* and some of its viruses. However, enough information is available to conclude that most aspects of DNA replication are basically the same in both prokaryotes and eukaryotes including humans. RNA primers and Okazaki fragments are shorter in eukaryotes than in prokaryotes. However, leading and lagging strands replicate by continuous and discontinuous procedures respectively in eukaryotes as in case of prokaryotes. But there are a few unique aspects of eukaryotic DNA replication. For example, DNA synthesis takes place within a small portion of the cell cycle in eukaryotes not continuously as happens in case of prokaryotes. The giant DNA molecules present in eukaryotic chromosomes would take much too long to replicate if each chromosome contained a single origin. Thus, eukaryotic chromosomes contain multiple origins of replication. Rather than using two catalytic complexes of one DNA polymerase to replicate the leading and lagging strands at each replication fork, eukaryotic organisms utilize two different polymerases.

41

Transcription-The Transfer of Genetic Information

According to the central dogma of molecular biology, genetic information flows (i) from DNA to DNA during its transmission from generation to generation and (ii) from DNA to protein during its phenotypic expression in an organism (Fig. 41.1).

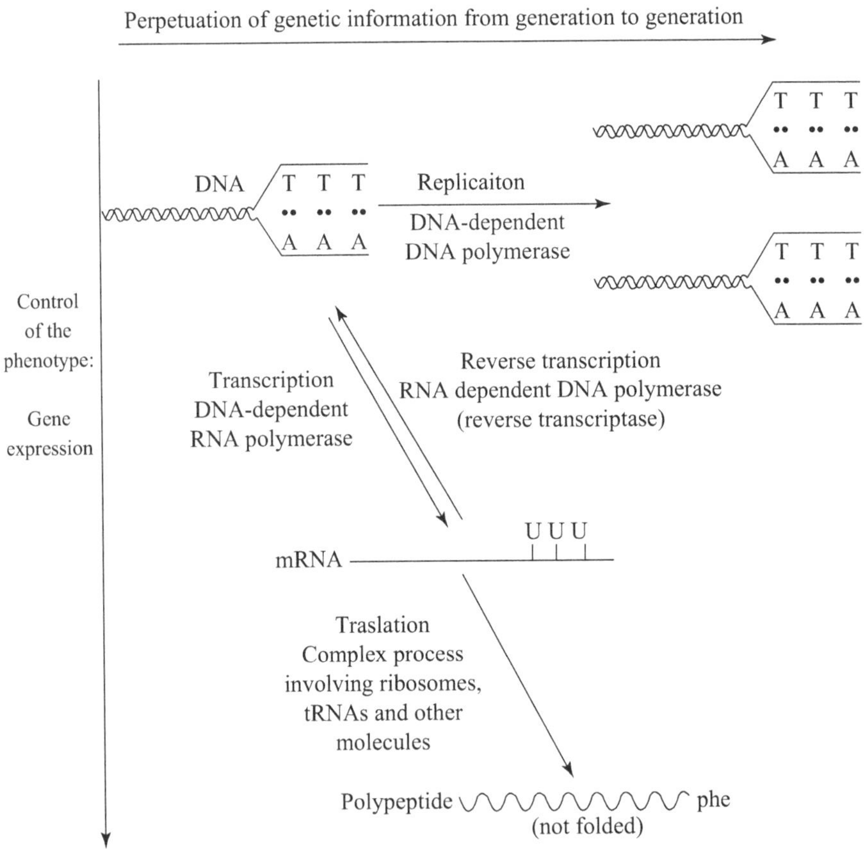

Fig. 41.1. The flow of genetic information according to the central dogma of molecular biology. Replication, transcription and translation occur in all organisms. Reverse transcription occurs only in cells infected with certain RNA viruses.

Transcription

Expression of genetic information involves two steps, namely, transcription and translation. In transcription, one strand of DNA of a gene is used as a template for synthesis of a complementary strand of RNA called as gene transcript. For example, DNA strand containing the nucleotide sequence as AAA is used as a template to produce the complementary sequence UUU in the RNA transcript. During translation, the sequence of nucleotides in the RNA transcript is converted into the sequence of amino acids in the polypeptide gene product. This conversion or translation is governed by genetic code, the specification of amino acids by nucleotide triplets known as codons in the gene transcript. For example, the UUU triplet in the RNA transcript specifies the amino acid phenylalanine in the polypeptide gene. Translation takes place on intricate macromolecular machines called ribosomes which are composed of three to five RNA molecules and 50-90 different proteins.

The RNA molecules that are translated on ribosomes are messenger RNAs (m-RNAs). In prokaryotes, the product of transcription, the primary transcript, usually is equivalent to the m-RNA molecule. This is shown in simple diagrammatic form in Fig. 41.2.

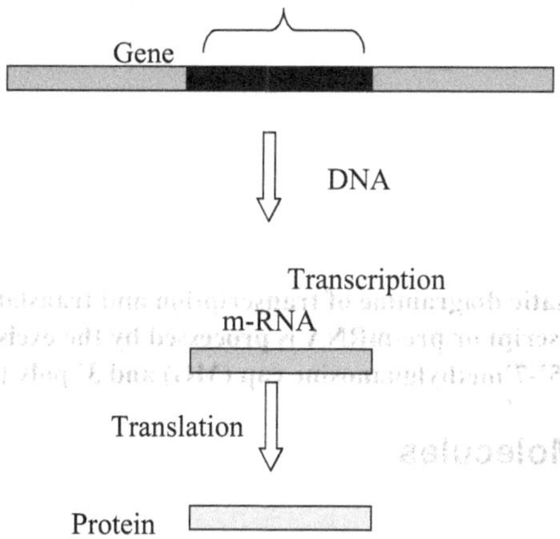

Fig. 41.2. Transcription and translation in prokaryotes.

In eukaryotes, primary transcript is often processed by the excision of specific sequences and modification of both termini before the primary transcript can be translated. Thus, in case of eukaryotes, the primary transcript of the gene is usually a precursor to the m-RNA and as such these precursor RNAs are called as pre-mRNAs. Most of the nuclear genes in higher organisms contain non-coding sequences which are called as introns that code the coding sequences or exons of the genes involved. The entire sequences of these split genes are transcribed into pre-mRNAs. The non-coding intron sequences are subsequently removed by splicing reactions which are carried out on macromolecular structures called spliceosomes (Fig. 41.3, Snustad *et al.*, 1997).

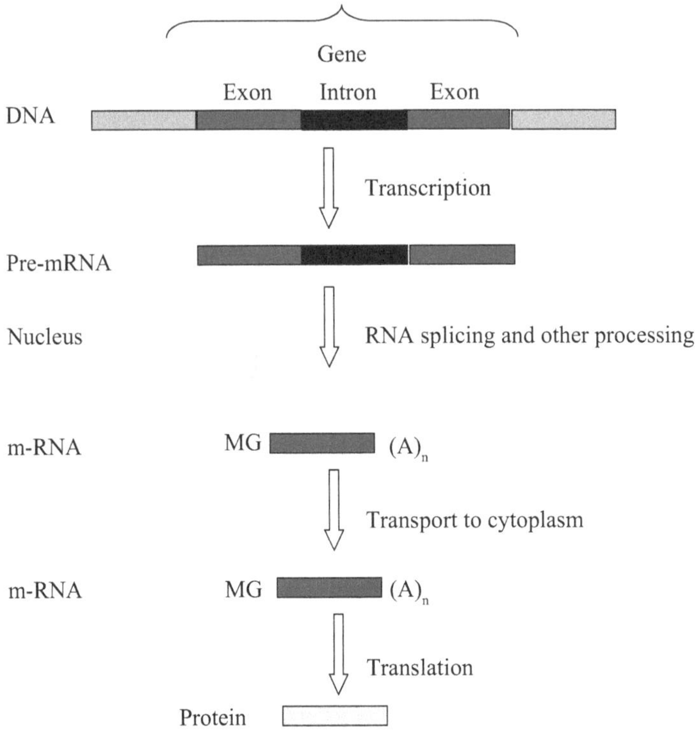

Fig. 41.3. Schematic diagramme of transcription and translation in eukaryotes. The primary transcript or pre-mRNA is processed by the excision of introns and addition of 5'-7'methylguanosine cap (MG) and 3' poly (A) tail (A)n.

Types of RNA Molecules

There are four different types of RNA molecules that play essential role in gene expression. The messenger RNA is the intermediary that carries genetic information from DNA to the ribosomes where the proteins are synthesized. The m-RNAs are needed to transfer genetic information from the nucleus to the cytoplasm. Transfer RNA (t-RNA) are small molecules of RNA that function as adaptors between amino acids and the triplet codons in the m-RNA during translation. There are from one to four t-RNAs for each of the 20 amino acids specified by genetic code. The t-RNA molecules are from 70 to 90 nucleotides long and fold into cloverleaf-shaped structures by intra-molecular base-pairing. In its folded form, each t-RNA has a specific amino acid attached at one end and a triplet nucleotide sequence called the anticodon at the other end. The recognition of nucleotide sequences in m-RNAs by t-RNAs is mediated by base-pairing between the anticodon in the t-RNA and the codon in the mRNA. Ribosomal RNAs (r-RNAs) are structural components of the ribosomes, the intricate machines that translate nucleotide sequences of m-RNA into amino acid sequences of polypeptides. The ribosomal RNAs interact with 50 different ribosomal proteins to produce the complex three-dimensional structures of the ribosomes. Prokaryotic ribosomes contain three r-RNAs, about 120, 1540 and 2900 nucleotides in length and about 50 proteins. Prokaryotic ribosomal RNAs are designated 5S, 16S and 23S RNAs based on their rate of sedimentation during sucrose gradient centrifugation. The four eukaryotic r-RNAs are designated as 5S, 5.8S, 18S and 28S and measure 120, 160, 1900 and 4700 nucleotides in length respectively. Smaller nuclear RNAs (sn-RNAs) are structural components of spliceosomes, the nuclear structures that excise introns from nuclear genes. There are five different smaller nuclear RNAs present in spliceosomes. They are small in size ranging from 100 to 215 nucleotides. They never leave the nucleus.

All four types of RNAs (m-RNA, t-RNA, r-RNA and sn-RNA) are produced by the process of transcription. Unlike m-RNAs, which specify polypeptides, the final products of t-RNA, r-RNA and sn-RNA genes are RNA molecules. Transfer RNA, ribosomal RNA and small nuclear RNA molecules are not translated.

In nutshell, "the central dogma of the molecular biology is that genetic information flows from DNA to DNA during chromosome replication and from DNA to protein during gene expression. The flow of information from DNA to protein occurs in two steps (i) transcription, DNA to RNA, and (ii) translation, RNA to protein. Transcription involves the synthesis of an RNA transcript complementary to one strand of DNA of a gene. Translation is the conversion of information stored in the sequence of nucleotides in the RNA transcript into the sequence of amino acids in the polypeptide gene product, according to the specification of genetic code" (Snustad *et al.*, 1997).

RNA Synthesis

RNA synthesis is similar to the synthesis of DNA except that (i) the precursors are ribonucleoside triphosphates instead of deoxyribonucleoside triphosphates, (ii) only one strand of DNA is transcribed for synthesis of a complementary RNA chain in any given region and (iii) RNA chains can be initiated *de novo* without any requirement for a pre-existing primer strand. The RNA molecule thus produced will be complementary to the DNA template strand and identical except that uridine residues replace thymidines, to the DNA nontemplate strand. If RNA molecule is an m-RNA, it will specify amino acids in the protein gene product. This specification is obtained by nucleotide triplet called as codons. Each codon specifies the incorporation of a single amino acid in a polypeptide chain. Therefore, m-RNA molecules are coding strands of RNA. They are also called as sense strands of RNA because their nucleotide sequences 'make sense' in the sense that they specify sequences of amino acids in the protein gene products. An m-RNA molecule that is complementary to an m-RNA is often called as antisense RNA. It is preferable to use template strand and non-template strand to refer to the transcribed and non-transcribed strands, respectively, of a gene. The synthesis of RNA chains occurs in 5'—3' direction as is the case with DNA with the addition of ribonucleotides to the 3'- hydroxyl group at the end of the chain. Simplified diagramme of RNA synthesis is given in Fig. 41.4.

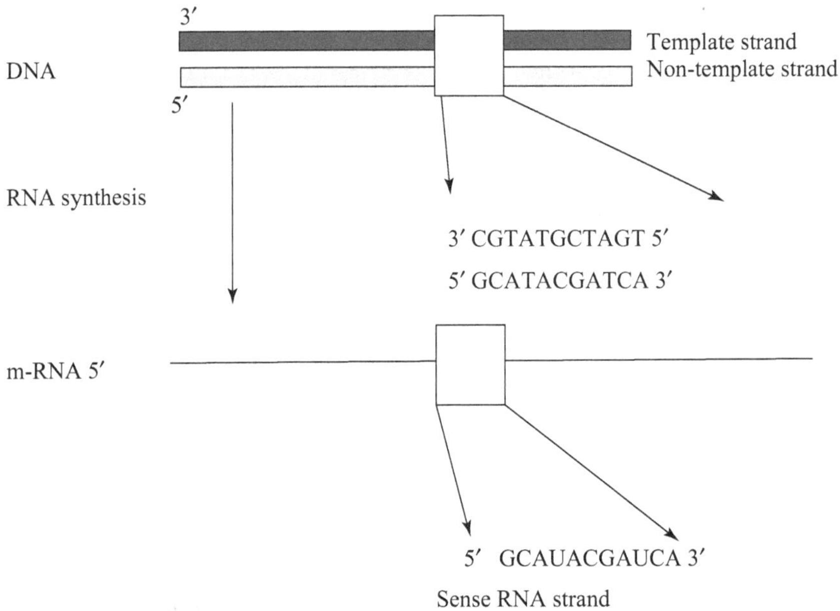

Fig. 41.4. RNA synthesis utilizing only one DNA strand of a gene as template.

The reaction involves a nucleophilic attack by the 3'-OH on the nucleotidyl (interior) phosphorous atom of the ribonucleoside triphosphate precursor with the elimination of pyrophosphate as happens in the DNA synthesis. This reaction is catalyzed by the enzyme DNA polymerase and is shown below:

$$n(RTP)\xrightarrow[\text{DNA polymerase}]{\text{DNA template}}(RMP)+n(pp)$$

where *n* is the number of moles of ribonucleotide triphosphate, (RTP), RMP is ribonucleoside monophosphate and pp is pyrophosphate.

RNA polymerases initiate transcription at specific nucleotide sequences called as promoters which are different in prokaryotes and eukaryotes. A single RNA polymerase carries out all transcription reactions in most prokaryotes in comparison to three different RNA polymerases present in eukaryotes. In this case, each polymerase is responsible for synthesis of a distinct class of RNAs. In principle, eukaryotic genes are present in the nucleus whereas polypeptides are synthesized in the cytoplasm. Messenger RNA molecules function as intermediaries that carry genetic information from DNA to the ribosomes, where proteins are synthesized. RNA synthesis catalyzed by RNA polymerases is similar to DNA synthesis in many respects. However, RNA synthesis occurs within a localized region of strand separation and only one strand of DNA functions as a template for RNA synthesis.

RNA Polymerases

The RNA polymerases are complex, multimeric proteins. In *E. coli*, RNA polymerase has a molecular weight of about 480,000 and consists of five polypeptides. Two of these five polypeptides are identical. Thus, as a net result, the enzyme contains four distinct polypeptides. The complete RNA polymerase molecule also known as holoenzyme has the composition as $\alpha_2 \beta \beta'\sigma$. The α subunits are involved in the assembly of the tetrameric core ($\alpha_2 \beta \beta'$) of the enzyme RNA polymerase. The β subunit contains the ribonucleoside triphosphate binding site and β' subunit harbours the DNA template binding region. The sigma (σ) factor is involved only in the initiation of transcription.

Initiation, Elongation and Termination of RNA Chains

Initiation of RNA chains involves three steps, namely, (i) binding of RNA polymerase holoenzyme to a promoter region in DNA, (ii) the localized unwinding of the two strands of DNA by RNA polymerase, providing a template strand free to base-pair with incoming ribonucleotides and (iii) the formation of phosphodiester bonds between the first few ribonucleotides in the nascent RNA chain. Elongation of RNA chain is catalyzed by the RNA polymerase core enzyme after the release of the σ subunit. The covalent extension of RNA chains takes place within the transcription bubble, a locally unwound segment of DNA. The RNA polymerase molecule contains both DNA unwinding and DNA rewinding activities. RNA polymerase continuously unwinds the DNA double helix ahead of the polymerization site and rewinds the complementary DNA strand behind the polymerization site as it moves along the double helix. The termination of RNA chain occurs when RNA polymerase encounters a termination signal. When this occurs, the transcription complex dissociates releasing the nascent DNA molecule.

There are two types of transcription terminators in *E. coli.* One type results in termination only in the presence of a protein called rho and such termination sequences are called as rho-dependent terminators. The other type results in the termination of transcription without the involvement of rho and such sequences are called rho-independent terminators.

Rho-independent terminators contain a G: C-rich region followed by six or more A: T base pairs, with A,s present in the template strand. The nucleotide sequence of G: C-rich region is such that regions of the single-stranded RNA can base-pair and form hairpin-like structures. The RNA hairpin structures form immediately after the synthesis of the participating regions of the RNA chain and retard the movement of RNA polymerase molecules along the DNA creating pauses in chain extension. Rho-dependent termination sequences are 50 to 90 base-pairs long, rich in C residues and largely devoid of G,S.

Transcription, Translation and m-RNA Degradation in Prokaryotes

In prokaryotes, the translation and degradation of an m-RNA molecule often begin before its synthesis/transcription is complete. Since m-RNA molecules are synthesized, translated and degraded in the 5' to 3' direction, all three processes occur simultaneously on the same RNA molecule. In prokaryotes, the polypeptide synthesizing apparatus is not separated by a nuclear wall or envelope from the site of m-RNA synthesis. All these activities occur in the same cell which has no clear-cut demarcation for the nucleus and the cytoplasm. Therefore, once the 5' end of an m-RNA has been synthesized, it can immediately be used as a template for polypeptide/protein synthesis. As a matter of fact, transcription and translation are tightly coupled in prokaryotes. Oscar Miller, Barbara Hamkalo and co-workers demonstrated this coupling between the two events in bacteria as it happened through electron microscopy techniques.

Transcription and RNA Processing in Eukaryotes

In general, overall process of RNA synthesis is similar in prokaryotes and eukaryotes but the process is more complex in eukaryotes where RNA is synthesized in nucleus and then transported to cytoplasm for occurrence of translation on the ribosomes. Prokaryotic m-RNAs often contain the coding regions of two or more genes and such m-RNAs are called multigenic. In contrast, the m-RNAs of eukaryotes contain the coding region of a single gene and are called as monogenic. However, there are evidences of m-RNA being multigenic in some eukaryotes like the small worm. Thus, eukaryotic m-RRNs could be monogenic or multigenic. Three different RNA polymerases are found in the eukaryotes and each enzyme catalyzes the transcription of a specific class of genes. All three eukaryotic RNA polymerases, designated as RNA polymerase I, II and III are more complex with 10 or more subunits than the *E. coli* polymerase. The key features of the three eukaryotic RNA polymerases are given in Table 41.1.

Table 41.1. Key features of the three RNA polymerases in eukaryotes.

Enzyme	Location	Products	Sensitivity to α-Amanitin
RNA polymerase I	Nucleus	Ribosomal RNAs, excluding 5S-r-RNA	No sensitivity
RNA polymerase II	Nucleus	Nuclear pre-m-RNAs	Complete sensitivity
RNA polymerase III	Nucleus	t-RNAs, 5S-r-RNA, and other smaller nuclear RNAs	Intermediate sensitivity

Additionally, in eukaryotes, majority of primary transcripts of genes undergoes three major modifications prior to their transport to the cytoplasm for translation i.e. formation of the polypeptide. The three modifications are as follows:

- 7-methyl guanosine caps are added to the 5'end of the primary transcript.
- Poly(A) tails are added to the 3'ends of the transcripts which are generated by cleavage rather than termination of chain extension.
- When present, noncoding intron sequences are spliced out of transcripts.

The 5'cap on most eukaryotic m-RNAs is a 7-methyl guanosine residue joined to the initial nucleoside of the transcript by a 5'-5'phosphate linkage. The 3'poly(A) tail is a polyadenosine tract having 20 to 200 nucleotides. In eukaryotes, the population of primary transcripts inside the nucleus is called heterogeneous nuclear RNA (hn-RNA) because of large variation in the sizes of the RNAs present. Major portions of these hn-RNAs are noncoding intron sequences and these are excised from the primary transcripts and degraded in the nucleus itself before moving to the cytoplasm. Also in eukaryotes, RNA transcripts are coated with RNA binding proteins during or immediately after their synthesis. These protein coats protect gene transcripts from degradation by ribonucleases that degrade RNA molecules during processing and transport to the cytoplasm. The average half-life of a gene transcript in eukaryotes is about five hrs as compared to about five minutes in *E. coli*.

Unlike prokaryotic organisms, eukaryotic RNA polymerases cannot initiate transcription by themselves. All three eukaryotic RNA polymerases require the assistance of protein transcription factors to start the synthesis of an RNA chain. These transcription factors bind to the promoter region in DNA and form an appropriate initiation complex before RNA polymerase binds and initiates transcription. Different promoters and transcription factors are used by the RNA polymerases I, II, and III. In all cases, the initiation of transcription involves the formation of a locally unwound segment of DNA providing a DNA strand that is free to function as a template for synthesis of a complementary strand of RNA. The formation of locally unwound segment of DNA required to initiate transcription involves the interaction of several transcription factors with specific *cis*-acting sequences in the promoter for the transcription unit. The promoters recognized by RNA polymerase II consist of short conserved-elements or modules located upstream from the transcription. The conserved element closest to the transcription start site is called the TATA box. It has consensus sequence TATAAAA (reading 5' to 3' on the non-template strand) and is centered at about position -30. The TATA box plays an important role in positioning the transcription start point. The second conserved element is called the CAAT box. It usually occurs near position -80 and has the consensus sequence GGCCAATCT. The initiation of transcription by RNA polymerase II needs the assistance of several basal transcription factors which must interact with promoters in the correct sequence to initiate transcription effectively. Each basal transcription factor is denoted as TFIIX (Transcription Factor for Polymerase II where X is a letter identifying the individual factor). Once eukaryotic RNA polymerases have been released from their initiation complexes, they catalyze RNA chain elongation by the same mechanism as the RNA polymerases of prokaryotes.

In brief, three different RNA polymerases are present in eukaryotes and each polymerase transcribes a distinct set of genes. Eukaryotic gene transcripts usually undergo three major modifications: (i) the addition of 7-methyl guanosine caps to 5' termini, (ii) the addition of poly(A) tails to 3' ends, and (iii) the excision of noncoding intron sequences. The information content of some eukaryotic transcripts is altered by RNA editing which changes the nucleotide sequences of transcripts prior to their translation.

42

Translation

In transcription, it has become clear as to how there is the transfer of genetic information stored in the sequence of nucleotides-pairs in DNA to the sequence of nucleotides in m-RNA molecules which carry that particular information from the nucleus to the sites of protein synthesis (ribosomes) in the cytoplasm. Further, there is the requirement of the process by which genetic information stored in the sequences of nucleotides in m-RNAs is used to specify the sequences of amino acids in the polypeptide gene products. This process known as translation takes place in the cytoplasm on complex workbenches called as ribosomes. With exception of water, proteins are by far the most prevalent component of living organisms in terms of total mass.

Proteins (= enzymes = genes) are composed of polypeptides and every polypeptide is encoded by a gene. Each polypeptide has a long sequence of amino acids linked together by covalent bonds. There are 20 different amino acids (glycine, alanine, valine, leucine, isoleucine, proline, phenylalanine, methionine, tryptophan, cysteine, serine, threonine, tyrosine, asparagine, glutamine, aspartic acid, glutamic acid, lysine, arginine, and histidine) present in most proteins. All amino acids except proline contain a free amino group (NH_2) and a free carboxyl group (COOH) as illustrated below:

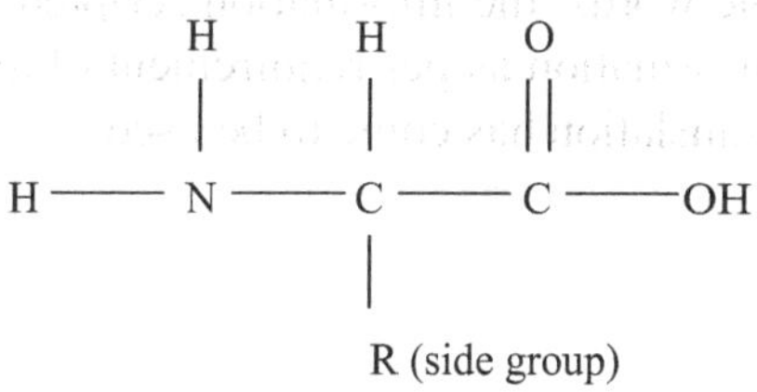

The amino acids differ from each other by the side groups represented by R. A peptide is a compound of two or more amino acids. Polypeptides length ranges from 51 amino acids in insulin to over 1000 amino acids in the silk protein fibroin. Amino acids in polypeptides are joined together by peptide bonds. Each peptide bond is formed by a reaction between the amino group of one amino acid and the carboxyl group of a second amino acid with the elimination of water molecule.

It should be obvious that most genes exert their effects on the phenotype of an organism through proteins which are larger macromolecules composed of polypeptides. Each polypeptide is a chain-like polymer assembled from different amino acids. The amino acid sequence of each polypeptide is specified by the nucleotide sequence of a gene. The vast functional diversity of proteins results from their complex three dimensional structures.

Translation-Protein Synthesis

The process by which genetic information stored in the sequence of nucleotides in an m-RNA is translated according to the specifications of the genetic code into sequence of amino acids in the polypeptide gene product is complex requiring the functions of a large number of macromolecules which are listed below:

- Over 50 polypeptides and from 3 to 5 RNA molecules present in each ribosome.
- At least 20 amino acids-activating enzymes.
- From 40 to 60 different t-RNA (transfer-RNA) molecules.
- Numerous soluble proteins involved in polypeptide chain initiation, elongation and termination.

The first step in gene expression, transcription involves the transfer of information stored in genes to messenger RNA (m-RNA) intermediaries which carry that information to the sites of polypeptide synthesis in the cytoplasm. Translation amounts to transfer of information in m-RNA molecules stored in form of nucleotide sequences into the sequences of amino acids in polypeptide gene product. In simple words, the information scripted in one language has been translated into information as per requirement of another language and that is how the word translation has come to be used.

Translation takes place on the ribosomes which are complex macromolecular structures located in the cytoplasm. The process of translation requires three types of RNAs, namely, m-RNA, t-RNA and r-RNA all of which are transcribed from DNA template i.e. chromosomal genes. The amino acids are attached to the correct t-RNA molecules by a set of activating enzymes called as aminoacyl-t-RNA synthetases. The nucleotide sequence of an m-RNA molecule is translated into appropriate amino acid sequences according to the dictionary of the genetic code. Each amino acid is specified by one or more codons which have three nucleotides, each. Of the 64 possible nucleotide triplets, 61 triplets specify amino acids and three specify polypeptide chain termination. The complementary nucleotide triplets of the t-RNA called as anticodons base-pair with the codons in m-RNA. Clearly codons are in the m-RNA and the anticodons are in the t-RNA (transfer-RNA). Ribosomes are considered to be something like work-benches complete with machines and tools required to make a polypeptide. The ribosomes (sites for protein synthesis) are non-specific which means that they can synthesize any polypeptide of any amino acid sequence encoded by a particular m-RNA molecule. Each m-RNA molecule is simultaneously translated by several ribosomes resulting in the formation of poly-ribosomes.

Ribosomes and Transfer RNA

Various experiments using autoradiography have shown that proteins are synthesized on the ribosomes which are distributed throughout the cells in prokaryotes and are located in the cytoplasm frequently on the extensive intracellular membrane network of the endoplasmic reticulum in eukaryotes. Composition-wise, ribosomes are half protein and half RNA. The ribosomes are composed of two subunits, one large and one small, which dissociate when translation of an m-RNA molecule is over. These subunits re-associate during the initiation of translation. Ribosome sizes are expressed in terms of their rate of sedimentation during sedimentation, in Svedberg units. *E. coli* ribosome has a molecular weight of 2.5×10^6 which in size term is 70S and dimension of about 20 nm x 25 nm. The ribosomes of eukaryotes are larger with about 80S. In *E. coli*, the small (30S) ribosomal subunit has a 16S (molecular weight about 6×10^5) RNA molecule plus 21 different polypeptides. The large subunit (50S) contains two RNA molecules (5S and 23S) plus 31 polypeptides. In mammalian ribosomes, the small subunit contains an 18S RNA molecule plus 33 polypeptides. The large subunit contains three RNA molecules of size of 5S, 5.8S and 28S plus 49 polypeptides.

The ribosomal RNA molecules like m-RNA molecules are transcribed from a DNA template. In eukaryotes, ribosomal RNA synthesis occurs in the nucleolus and is catalyzed by RNA polymerase I. The ribosomal RNA genes are present in tandomly duplicated arrays separated by intergenic spacer regions. The transcription of r-RNA genes produces precursors that are much larger than the RNA molecules found in ribosomes. The RNA precursors undergo post-transcriptional processing to generate mature r-RNA molecules. The synthesis and processing of (a) the 30S r-RNA precursor in *E. coli* and (b) that of the 45S r-RNA precursor in mammals is depicted in Fig. 42.1 (Snustad *et al.*, 1997). In addition to the post-transcriptional cleavages of r-RNA precursors, several nucleotides in r-RNAs are methylated after transcription which protects r-RNA molecules from degradation by ribonucleases. There are multiple copies of genes for r-RNA in genomes of all organisms.

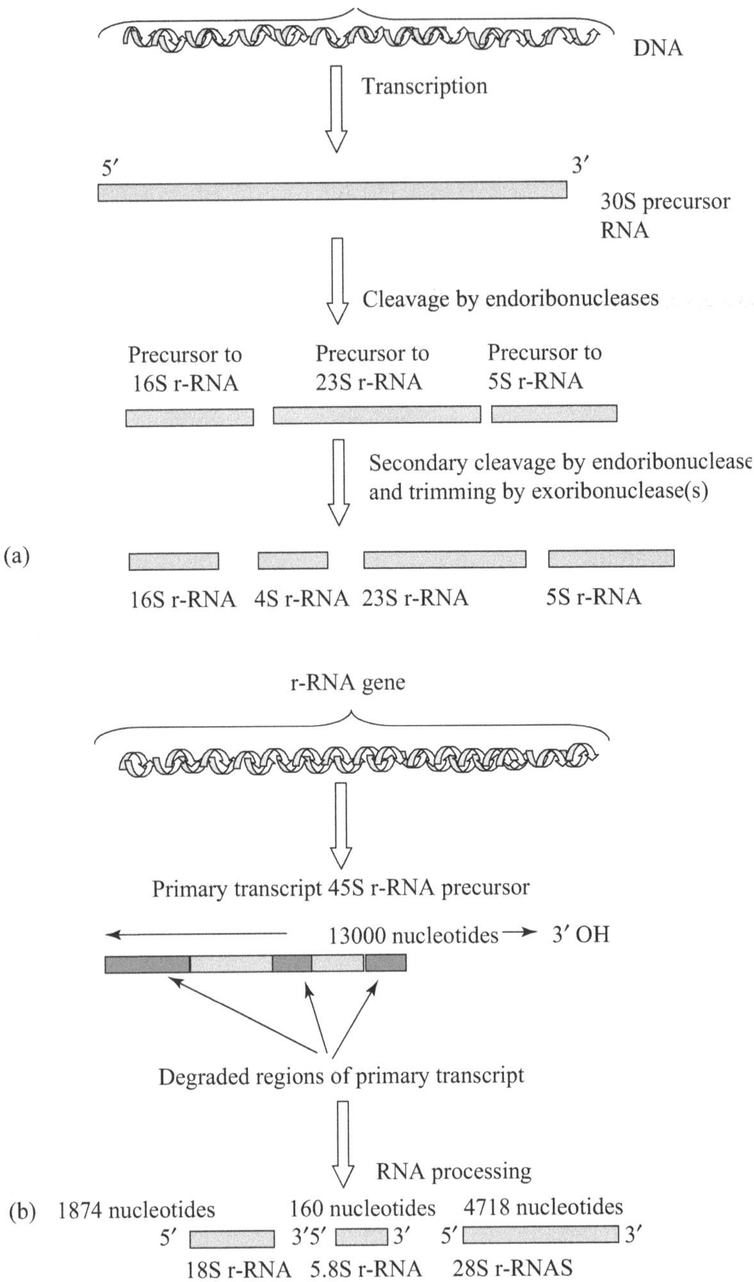

Fig.42.1. Synthesis and processing of (a) the 30S r-RNA precursor in *E. coli* and (b) the 45S r-RNA precursor in mammalians.

The amino acids are attached to the t-RNAs by high energy bonds between the carboxyl groups of the amino acids and the 3'-hydroxyl termini of the RNAs. The t-RNAs are activated or charged with amino acids in a two step process where both reactions are catalyzed by the same enzyme, aminoacyl-tRNA synthetase for each of the 20 amino acids. The first reaction involves the activation of amino acid using energy from ATP (adenosine triphosphate) and the second step involves reaction with appropriate t-RNA. Both the steps are as follows:

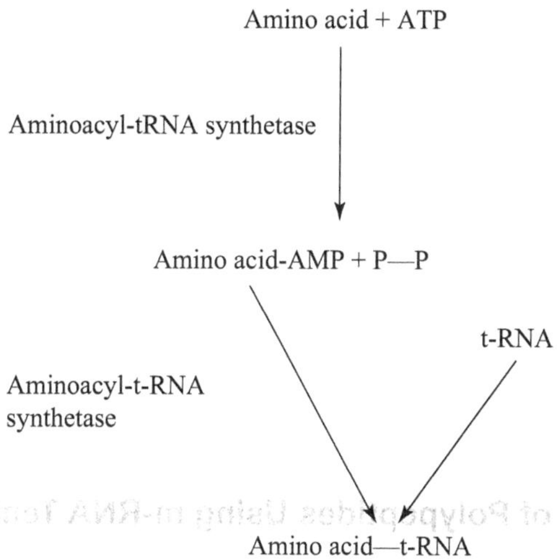

Amino acid—t-RNAs function as the substrate for polypeptide synthesis on the work-bench, ribosomes where each activated t-RNA recognizes the correct m-RNA codon based on base-pairing principle (A pairs with U in place of T and G pairs with C). The t-RNAs are transcribed from nuclear genes. The t-RNAs are transcribed initially in the form of larger precursor molecules that undergo post-transcriptional processing involving cleavage, trimming, methylation and so on. The mature t-RNA molecules are equipped with several nucleosides that are not present in the primary t-RNA gene transcript. These unusual nucleosides such as inosine, pseudouridine, dihydrouridine, 1-mehylguanosine, etc are produced by post-transcriptional enzyme catalyzed modifications of the four nucleosides incorporated into RNA during transcription.

t-RNAs are usually small having 70-80 nucleotides and therefore have been amenable to structural analysis. Robert W. Holley and colleagues published the complete nucleotide sequence and clover leaf structure of alanine t-RNA of yeast in 1965. Holley shared the 1968 Nobel Prize in Physiology and Medicine for this pioneering work. Since then many t-RNAs have been sequenced and the yeast alanine t-RNA gene has been synthesized *in vitro* using mononucleotides by Indian American Har Gobind Khorana and co-workers. The anticodon of each t-RNA occurs within a loop (non-hydrogen-bonded region) near the middle of the molecule. The nucleotide sequence and cloverleaf model of the alanine t-RNA of *S. cerevisiae* is shown in Fig. 42.2. The t-RNA molecules contain great deal of specificity in terms of correct anticodon sequence so as to respond to the right codons, recognition by the correct aminoacyl t-RNA synthetases and so that they are activated with correct amino acids and binding to the appropriate sites on the ribosomes to carry out their adaptor function. Three t-RNA binding sites are present on each ribosome. The *A* or aminoacyl site binds the incoming aminoacyl t-RNA, the t-RNA carrying the next amino acid to be added to the growing polypeptide chain. The *P* or peptidyl site binds the t-RNA to which the growing polypeptide is attached. The *E* or exit site binds the departing uncharged t-RNA. The specificity for aminoacyl t-RNA binding in these sites on the ribosomes is provided by the m-RNA codons that make up part of the binding sites. As the ribosome moves along an m-RNA, the specificity for amnoacyl t-RNA binding in the *A*, *P* and *E* sites changes as different m-RNA codons move into register in the binding sites. The ribosomal binding sites by themselves (minus m-RNA) are thus capable of binding any aminoacyl t-RNA.

The Synthesis of Polypeptides Using m-RNA Templates (Translation)

The m-RNA molecules provide the specifications for amino acid sequences of the polypeptide gene products. The ribosomes provide many of the macromolecular components needed for the process of translation to happen. The t-RNAs provide the adaptor molecules needed to incorporate amino acids into polypeptides in response to codons in the m-RNAs. The translation of the sequence of nucleotides in an m-RNA molecule into the sequence of amino acids in the resultant polypeptide product includes the following three steps, (i) polypeptide chain initiation, (ii) polypeptide chain elongation and (iii) polypeptide chain termination.

The initiation of translation all events that precede the formation of a peptide bond between the first two amino acids of the new polypeptide chain. In *E. coli*, the chain initiation process involves the 30S subunit of the ribosome, a special initiator t-RNA, an m-RNA molecule, three soluble protein factors, IF-1, IF-2 and IF-3 and one molecule of GTP. The translation occurs on 70S ribosomes. However, these ribosomes dissociate into their 30S and 50S subunits after completion of synthesis of each of the polypeptide chains. In the first stage of initiation of translation, a free 30S subunit interacts with an m-RNA molecule and the initiation factors. The 50S subunit joins the complex to form 70S ribosome in the final step of the initiation process. The synthesis of polypeptides is initiated by a special t-RNA designated as t-RNA$_f^{Met}$ meaning thereby that all polypeptides begin with methionine during their synthesis. The amino terminal methionine is subsequently cleaved from many polypeptides. A distinct methionine t-RNA (t-RNA^{Met}) responds to the internal methionine codons. However, both methionine t-RNAs have the same anticodon and both respond to the same codon (AUG) for methionine.

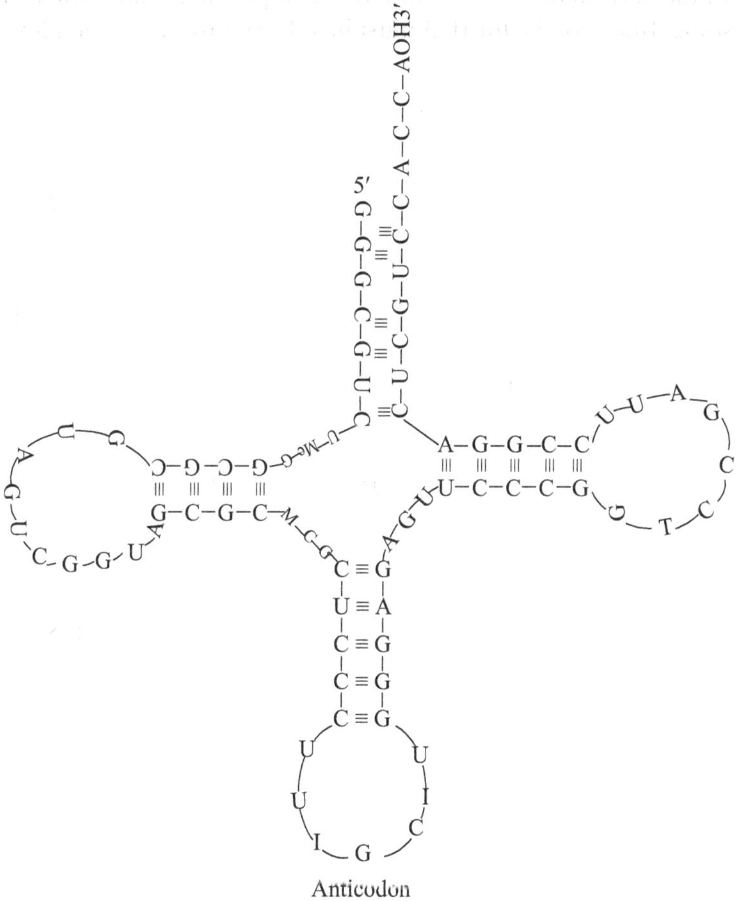

Fig. 42.2. Nucleotide sequence and cloverleaf configuration of the alanine t-RNA of S. cerevisiae.

The polypeptide chain initiation begins with the formation of two complexes, namely, (i) one contains initiation factor IF-2 and methionyl t-RNA$_f^{Met}$ and (ii) the other complex contains an m-RNA molecule, a 30S ribosomal subunit and initiation factor IF-3 which plays a crucial role to begin the process of initiation. Further prokaryotic m-RNAs contain a conserved polypurine tract, consensus AGGAGG, located about seven nucleotides upstream from the AUG, initiation codon. This conserved hexamer, named as Shine-Dalgarno sequence after scientists who discovered it, is complementary to a sequence near the 5' terminus of the 16S ribosomal RNA. IF-2/methionyl t-RNA_f^{Met} complex and m-RNA/30S subunit/IF-3 complex then combine with each other and with initiation factor IF-1 and one molecule of GTP to form the complete 30S initiation complex. The final step in the initiation of translation is the addition of the 50S subunit to 30S initiation complex to produce the complete 70S ribosome. Initiation factor IF-3 must be released from the complex.

The initiation of translation is more complex in eukaryotes involving several soluble initiation factors. However, overall process is similar except that (i) the amino group of methionine on the initiator t-RNA is not formylated as in prokaryotes and (ii) the initiation complex forms at 5' terminus of the m-RNA, not at the Shine-Dalgarno/AUG translation start site as in *E. coli.*

The process of peptide chain elongation is almost same in both prokaryotes and eukaryotes. The addition of each amino acid to the growing polypeptide involves three steps, namely, (i) binding of an aminoacyl t-RNA to the *A* site of the ribosome, (ii) transfer of growing polypeptide chain from the t-RNA in the *P* site to the t-RNA in the *A* site by formation of a new peptide bond and (iii) translocation of the ribosome along the m-RNA to position the next codon in the *A* site. During step 3, the nascent polypeptide –t-RNA and the uncharged t-RNA are translocated from the *A* and *P* sites to the *P* and *E* sites respectively. These steps are repeated in a cyclic manner throughout the elongation process of the polypeptide chain.

Polypeptide chain elongation undergoes termination when any of three chain-termination codons (UAA, UAG and UGA) enters the *A* site on the ribosome. These three 'stop codons' are recognized by soluble proteins called release factors (RFs).

The entire complex process can be captured in brief as follows: Genetic information carried in the sequence of nucleotides in an m-RNA molecule is translated into a sequence of amino acids in polypeptide gene product by intricate macromolecular machines called ribosomes. The translation process is complex, requiring the participation of many different RNAs and protein molecules. Transfer RNA molecules serve as adaptors, mediating the interaction between the amino acids and codons in messenger RNA. The process of translation involves the initiation, elongation and termination of polypeptide chains and is governed by the specifications of the genetic code.

43

Genetic Code

Biochemical reactions are mediated by enzymes and all enzymes are proteins. Proteins are polymers of subunits (monomers) called amino acids which are often termed as 'residues'. Each amino acid has an amino group (NH_2) at one end and a carboxyl group (COOH) at the opposite end. 20 different kinds of amino acids occur naturally in proteins. Each enzyme consists of a certain number of amino acids in a precisely ordered sequence. The blueprint for making proteins is coded in the nucleotide sequence of DNA. The number of nucleotides which code for an amino acid is termed a codon. There are 20 common amino acids but only four different nucleotides. Obviously a singlet code (one nucleotide coding for one amino acid) could code only for four amino acids. A coding sequence of two nucleotides for one amino acid or a doublet code would produce $4^2 = 16$ possible coding combinations or codons and this arrangement is not sufficient to take care of 20 amino acids. A codon size of three nucleotides for one amino acid or triplet code seems more likely as it produces $4^3 = 64$ possible codons as illustrated in Table 43.1 (Strickberger, 1968).

Table 43.1. Possible coding combinations using different-sized DNA nucleotide sequences (on the level of m-RNA translation into protein, the m-RNA codon sequences will be complementary to those shown here, for example ATC in DNA) = UAG in m-RNA).

Singlet code	Doublet code				Triplet code			
A	AA	AG	AC	AT	AAA	AAG	AAC	AAT
G	GA	GG	GC	GT	AGA	AGG	AGC	AGT
C	CA	CG	CC	CT	ACA	ACG	ACC	ACT
T	TA	TG	TC	TT	ATA	ATG	ATC	ATT
					GAA	GAG	GAC	GAT
					GGA	GGG	GGC	GGT
					GCA	GCG	GCC	GCT
					GTA	GTG	GTC	GTT
					CAA	CAG	CAC	CAT
					CGA	CGG	CGC	CGT
					CCA	CCG	CCC	CCT
					CTA	CTG	CTC	CTT
					TAA	TAG	TAC	TAT
					TGA	TGG	TGC	TGT
					TCA	TCG	TCC	TCT
					TTA	TTG	TTC	TTT

With this arrangement of stipulating minimum of three nucleotides coding for one amino acid, 64 codes are available for 20 amino acids indicating 44 codons possibly to be superfluous. This excess of seemingly unnecessary codons would even be larger if coding ratios were to be 4: 1 ($4^4 = 256$ codons) or 5: 1 ($4^5 = 1024$ codons). To account for excess of codons beyond 20, one may have to suppose that more than one codon can code for a particular amino acid. For example, if each kind of amino acid were to be coded by three different possible codons, 60 possible codons will be accounted for by 20 amino acids. A code in which there is more than one codon for the same amino acid, that particular code is called as degenerate. It is also possible that some or all of the codons in excess of 20 do not code for any amino acid and such codons are termed as nonsense codons. However, all these problems are now fully solved and properties of genetic code are fully understood.

Disregarding the codon size at the moment, the next problem to be encountered will be to decide whether the code could be overlapping or non-overlapping where each of the situations will have different consequences as shown in Fig. 43.1.

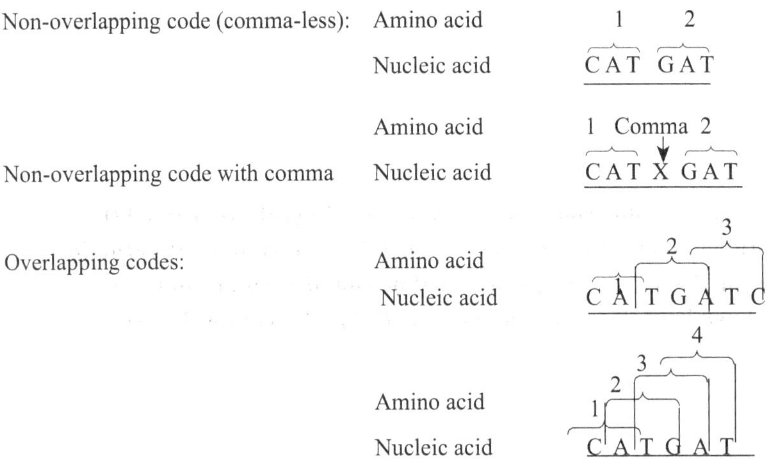

Fig. 43.1. Overlapping and non-overlapping triplet codons and their consequences.

Based on experimental evidences and the resultant amino acids, it has been concluded that triplet codons are comma-less and non-overlapping.

Properties of Genetic Code

Main features of genetic code were worked out in the 1960s. Cracking the genetic code has been one of the most exciting eras in the history of science with new information appearing almost daily. By the mid-1960s, the genetic code was largely solved. The most important properties of genetic code are as follows (Snustad *et al.*, 1997):

- **The genetic code is composed of nucleotide triplets:** Three nucleotides in m-RNA specify one amino acid in the polypeptide product. Thus, each codon contains three nucleotides.
- **The genetic code is non-overlapping:** Each nucleotide in m-RNA belongs to just one codon except in rare cases where genes overlap.
- **The genetic code is comma-free:** There are no commas or other forms of punctuation within the coding regions of m-RNA molecules. During translation, codons are read consecutively.
- **The genetic code is degenerate:** All but two of the amino acids are specified by more than one codon.
- **The genetic code is ordered:** Multiple codons for a given amino acid and codons for amino acids with similar chemical properties are closely related, usually differing by a single nucleotide.
- **The genetic code contains start and stop codons:** Specific codons are used to initiate and to terminate peptide chains.
- **The genetic code is nearly universal:** With minor exceptions, the codons have the same meaning in all living organisms from viruses to humans.

Some Common Terms Used in Describing the Genetic Code

Several common terms are used while dealing with genetic code. Definitions of some common terms describing genetic codes are given in Table 43.2 (Strickberger, 1968).

Table 43.2. Definitions of some common terms in describing genetic codes.

Term	Meaning
Code letter	Nucleotides, e g. AUGC (in m-RNA) and ATGC (in DNA)
Code Word or Codon	Sequence of nucleotides specifying an amino acid, for example, UUU = phenylalanine
Anticodon	Sequence of nucleotides on t-RNA that complements the codon, for example, AAA = anticodon for phenylalanine
Genetic Code or Coding Dictionary	A table of all code words or codons that specify amino acids
Word Size	The number of letters in a code word, for example, three letters in a triplet code (this is the same as coding ratio in a non-overlapping code)
Non-overlapping Code	When only as many amino acids are coded as there are code words in end-to-end sequence, for example, (triplet code) UUUCCC = phenylalanine (UUU) + proline (CCC)
Overlapping Code	When more amino acids are coded for than there are code words in end-to-end sequence, for example, UUUCCC = phenylalanine (UUU) + serine (UCC) + proline (CCC)

Term	Meaning
Degenerate Code	When there is more than one codon for a particular amino acid, for example, UUU, UUC = phenylalanine
Ambiguous Code	When one codon can code for more than one amino acid, for example, GGA = glycine, glutamic acid
Comma-less Code	When there are no intermediary nucleotides (spacers)between words, for example, UUUCCC = two amino acids in triplet non-overlapping code
Reading Frame	The particular nucleotide sequence that starts at a specific point and is then partitioned into codons. The reading frame may be "shifted" by removing or adding nucleotides, thereby causing a new sequence of codons to be "read"
Sense Word	A codon that specifies an amino acid normally present at that position in a protein
Missense Mutation	A change in nucleotide sequence either by deletion, insertion, or substitution resulting in the appearance of a codon that produces a different amino acid in a particular protein, for example, UUU (phenylalanine) UGU (cysteine)
Nonsense Mutation	A codon that does not produce an amino acid, for example, UAG
Universality of Code	Utilization of same genetic code in all organisms, for example, UUU = phenylalanine in bacteria, mouse, man and tobacco

Genetic Codes

By combining the results of tri-nucleotide binding assays and in vitro translation experiments performed with synthetic m-RNAs, Nirenberg, Severo Ochoa, Khorana and others were able to decipher the meaning of all 64 triplet codons. These codons assignments are now firmly established, supported by definitive data from both *in vitro* and *in vivo* studies. The genetic codes for each of amino acids are presented in Table 43.3 (Snustad *et al*., 1997).

Experimental Evidences for Coding Dictionary

In 1961, Nirenberg and Matthei reported the first experiment showing that a particular RNA sequence produced a particular amino acid sequence. They utilized a cell-free system extracted from *E. coli* bacteria containing ribosomes, radioactively labeled amino acids, transfer RNA and necessary enzymes and ATP as source of energy for protein synthesis. In the absence of m-RNA, there was no synthesis of observable protein indicating the critical role of m-RNA in protein synthesis. They then added RNA from tobacco mosaic virus (TMV) and showed an increase in amino acid incorporation into proteins. In other words, m-RNA was necessary for protein synthesis and an *E. coli* system could use TMV-RNA to synthesize TMV-like proteins. However, their most remarkable and unique discovery occurred when they

Table 43.3. Genetic code.

First (5') letter	Second letter: U		C		A		G		Third (3') Letter
U	UUU	Phe	UCU	Ser	UAU	Try	UGU	Cys	U
	UUC		UCC		UAC		UGC		C
	UUA	Leu	UCA		UAA	Ochre (terminator)	UGA	Opal	A
	UUG		UCG		UAG	Amber (terminator)	UGG	Trp	G
C	CUU	Leu	CCU	Pro	CAU	His	CGU	Arg	U
	CUC		CCC		CAC		CGC		C
	CUA		CCA		CAA	Gln	CGA		A
	CUG		CCG		CAG		CGG		G
A	AUU	Ilu	ACU	Thr	AAU	Asn	AGU	Ser	U
	AUC		ACC		AAC		AGC		C
	AUA		ACA		AAA	Lys	AGA	Arg	A
	AUG	Met (initiator)	ACG		AAG		AGG		G
G	GUU	Val	GCU	Ala	GAU	Asp	GGU	Gly	U
	GUC		GCC		GAC		GGC		C
	GUA		GCA		GAA	Glu	GGA		A
	GUG		GCG		GAG		GGG		G

NB: Each triplet nucleotide sequence or codon refers to the nucleotide sequence in m-RNA (not DNA) that specifies the incorporation of the indicated amino acid or polypeptide chain termination. The table is based on the data of Nirenberg and co-workers and Khorana and co-workers.

added the synthetic polynucleotide, poly-U to this system and recovered a polypeptide whose amino acids were all phenylalanine in spite of the presence of other amino acids in the original mixture.

These results obtained by Nirenberg and Matthei clearly demonstrated that phenylalanine or more precisely the anticodon of a phenylalanine carrying t-RNA was coded by a sequence of uracil bases (UUU codon of m-RNA = AAA anticodon of t-RNAPhe). It became immediately evident that other synthetic RNA messengers (m-RNAs) introduced into the system could be used to identify the code for a variety of amino acids. Thus poly-A was found to be the code for lysine and poly-C for proline. Poly-G was also tried but this could not attach to ribosomes, since it remained in solution as multi-stranded helixes caused by hydrogen bonding between its guanine bases. Aside from specifying the coding sequences of these three amino acids, these results also indicated that the code was comma-less, without special intervening nucleotides between codons. The argument in this favour is as follows: If the code has specific commas, the incorporation of phenylalanine into poly-U means that the commas are all uracil. However, the incorporation of lysine into poly-A and of proline into poly-C in turn means that the commas are made of adenine and of cytosine. This explanation that commas are made of sometimes of uracil, sometimes of adenine and sometimes of cytosine is highly inconsistent and hence rejected.

Within a short period of time after this discovery, synthetic ribonucleotides containing mixed bases were also tried and evaluated both by Nirenberg's group and by Ochoa's group. Poly-UC for example potentially contained only eight different codons.

Poly-UC = UUU, UUC, UCU, UCC, CUU, CUC, CCU, CCC (resulting from all possible combinations of U and C in triplet configuration).

Some codon assignments then could be made by varying the relative amounts of the two bases and observing which amino acids were preferentially incorporated. In this particular experiment, poly-UC caused the polypeptide incorporation of leucine and serine in addition to the expected incorporation of phenylalanine (UUU) and proline (CCC). Modifying the proportion of bases in poly-UC so that it contained more uracil than cytosine caused more serine to be incorporated into the polypeptide indicating that serine may be caused by codons having more of U than C and accordingly serine may either be due to UUU, CUU or UCU. Further determinations with other nucleotide polymers enabled other amino acids to be characterized by their general code, for example, poly-UG caused the polypeptide incorporation of cysteine, glycine, tryptophan, and valine. Likewise, poly-UA led to incorporation of asparagine, isoleucine, leucine, lysine and tyrosine. As in serine, however, the precise nucleotide sequence within the codon of most amino acids could not be determined through these experiments.

The real breakthrough in assigning specific codons to the specific amino acids as given in Table 43.3 came through the experiments of Nirenberg and Leder, and Khorana and his co-workers in 1965 by using ribonucleotides whose base sequences were known. In Nirenberg-Leder technique, a "species of m-RNA molecules containing three nucleotides in known order is placed into a cell-free system containing ribosomes mixed with different activated t-RNA molecules bearing amino acids. In this system, the m-RNA is too short to enable protein synthesis but nevertheless causes binding of amino acid charged t-RNA molecules to the ribosomes. The amino acid bound by a particular trinucleotide m-RNA sequence is detected by radioactive labeling of only one kind of amino acid in the mixture and then observing whether there is pronounced incorporation of this radioactivity into the ribosomal fraction that is specially isolated on a separate filter. UUU trinucleotdes, for example, will cause radioactivity of the ribosomal fraction only when phenylalanine is the radioactive amino acid in the mixture.

A major breakthrough in deciphering the genetic code occurred when Har Gobind Khorana (1968 Physiology/Medicine Nobel Prize recipient with Marshall Nirenberg and Robert Holley) and colleagues developed a procedure by which copolymers with known repeating di-, tri- and tetranucleotide sequences could be synthesized. Because the codon sequence in these repeating copolypers were fixed, their use as synthetic m-RNAs yielded results that were more easily interpreted than those obtained with random copolymers. For example, an m-RNA molecule with a repeating UG dinucleotide sequence will direct the synthesis of polypeptide containing cysteine and valine residues as shown below:

5'-UGU GUG UGU GUG UGU GUG-3'

H_2N—Cys—Val—Cys—Val—Cys—Val—COOH

Because the triplets UGU and GUG alternate in the poly(UG)n, these two codons must specify cysteine and valine. However, it is yet not clear as to which codon specifies which amino acid. In contrast, a repeating UUG trinucleotide polymer directed the synthesis of a mixture of poly-leucine, poly-cysteine and poly-valine.

Synthetic m-RNA

5'-UUG UUG UUG UUG UUG UUG UUG

Reading frame 1

Reading frame 2

Reading frame 3

Polynucleotide products

Polyleucine	Leu—Leu—Leu—Leu—Leu—Leu—Leu
Polycystein	Cys—Cys—Cys—Cys—Cys—Cys—Cys—Cys
Polyvaline	Val—Val—Val—Val—Val—Val—Val

Because the initiation of translation occurs at random in these *in vitro* systems, some ribosomes will translate these polymers as UUG, UUG, UUG…… and so on while others will translate them as UGU, UGU, UGU……and still others as GUU, GUU, GUU…….Thus these three codons must specify leucine, cysteine, and valine. Becaue the polymer (UG)n product was a cysteine-valine copolymer, one of the new codons in poly(UUG)n either UUG or GUU must be a leucine codon. By analyzing the amino acids incorporated in response to many different repeating di-, tri, and tetranucleotide copolymers, Har Gobind Khorana and co-workers were able to assign many of the codons to specific amino acids. For example, UGU = cysteine, UUG = leucine and GUU = valine.

This simple binding assay allowed Nirenberg, Leder and others to test the ability of each of the 64 possible trinucleotides to activate ribosomes so that specific aminoacyl t-RNA would be bound. Not all of the trinuleotides worked. In some case, the level of binding was very low, yielding ambiguous results. However, in most case, a given trinucleotide caused one specific aminoacyl t-RNA to be bound to the activated ribosomes. Based on the data of Nirenberg and co-workers and Khorana and co-workers, the information has been complied as given in Table 43.3 by various authors. These results have been supported by the results of tri-nucleotide binding experiments and have been subsequently verified by direct comparison of the nucleotide sequences of genes and the amino acid sequences of their polypeptide products.

Initiation and Termination Codons

The genetic code also provides for punctuation of the genetic information at the level of translation. In prokaryotes and eukaryotes, the codon AUG is used as chain initiation codon (Table 43.3) for polypeptide chain formation. In rare cases, GUG is used as an initiation codon. In both the cases, the initiation codon is recognized by an initiator t-RNA (t-RNA_f^{Met} in prokaryotes and t-RNA_i^{Met} in eukaryotes). Three codons, namely, UAA, UAG and UGA specify polypeptide chain termination. These codons are recognized by protein release factors rather than by t-RNAs. Prokaryotes contain two release factors, RF-1 (terminates polypeptide in response to codons UAA and UAG) and RF-2 (terminates polypeptide in resonse to codons UAA and UGA). In eukaryotes, there is a single release factor that recognizes all three termination codons.

Recognition of Codons by t-RNAs-The Wobble Hypothesis

The hydrogen bonding between the bases in the anti-codons of t-RNAs and the codons of m-RNA appears to follow strict base-pairing rules only for the first two bases of the codon. The base-pairing involving the third base of the codon seems to be less stringent allowing what Crick has called 'wobble' at this site. Table 43.4 shows the base-pairing predicted by Crick's wobble hypothesis.

Table 43.4. Base-pairing between the 5' base of the anticodon of t-RNAs and 3' base of codon of m-RNAs according to the wobble hypothesis.

Base in anticodon	Base in codon
G	U or C
C	G
A	U
U	A or G
I	A, U or C

The wobble hypothesis predicted the existence of at least two t-RNAs for each amino acid with codons that exhibit complete degeneracy and this has been proven to be true. The wobble hypothesis also predicted the occurrence of three t-RNAs for the six serine codons. Three serine t-RNAs have been characterized: (i) t-RNA^{Ser1} (anticodon ACG) binds to codon UCU and UCC, (ii) t-RNA^{ser2} (anticodon AGU) binds to codons UCA and UCG, (iii) t-RNA^{Ser3} (anticodon UCG) binds to codons AGU and AGC. These specificities have been verified by trinucleotide-stimulated binding of purified aminoacyl t-RNAs to ribosomes in vitro. Further, there are several tRNAs that contain the base inosine which is a ribonuclotide made from the purine hypoxanthine. According to Crick's wobble hypothesis, when inosine is present at the 5' end of anticodon (the wobble position), it could base-pair with adenine, uracil or cytosine in the codon.

In nutshell, the wobble hypothesis explains how a single t-RNA can respond to two or more degenerate codons. According to this hypothesis, the pairing between the third base of a codon and the first base of an anticodon is less stringent than normal permitting wobble at this site. Some suppressor mutations alter the anticodons of t-RNAs so that the mutant t-RNAs recognize chain termination codons and insert amino acids in response to their presence in m-RNA molecules.

Comparisons of the nucleotide sequences of genes with the amino acid sequences of their polypeptide products have verified the codon assignments deduced from *in vitro* studies.

Har Gobind Khorana –1968 Nobel Laureate with Marshall W. Nirenberg and Robert W. Holley in Physiology and Medicine for Breakthrough Research on Deciphering Genetic Code and Hugely Popular Among Students and Researchers on Molecular Biology and Biochemistry in India

Biography

9 January 1922-
9 November 2011

Har Gobind Khorana was an Indian-born-American biochemist. While on the faculty of the University of Wisconsin–Madison, he shared the 1968 Nobel Prize for Physiology or Medicine with Marshall W. Nirenberg and Robert W. Holley for research that showed the order of nucleotides in nucleic acids, which carry the genetic code of the cell and control the cell's synthesis of proteins. Khorana and Nirenberg were also awarded the Louisa Gross Horwitz Prize from Columbia University in the same year.

Born in British India, Khorana served on the faculties of three universities in North America. He became a naturalized citizen of the United States in 1966, and received the National Medal of Science in 1987. ❑

Khorana was born to Krishna Devi Khorana and Ganpat Rai Khorana, in Raipur, a village in Punjab, undivided India (now in present-day Pakistan) in a Hindu family. The date of his birth is mentioned as 9 January 1922. He was the youngest of five children. His father was a patwari, a village agricultural taxation clerk in the British Indian government. In his autobiography, Khorana wrote this summary: "Although poor, my father was dedicated to educating his children and we were practically the only literate family in the village inhabited by about 100 people. The first four years of his education were provided under a tree, a spot that was, in effect, the only school in the village.

He attended D.A.V. High School in Multan, in West Punjab. Later, he studied at the Punjab University in Lahore, with the assistance of scholarships, where he obtained a bachelor's degree in 1943 and a Master of Science degree in 1945.

Khorana lived in India until 1945, when he moved to England to study organic chemistry at the University of Liverpool on a Government of India Fellowship. He received his PhD in 1948 advised by Roger J. S. Beer. The following year, he pursued postdoctoral studies with Professor Vladimir Prelog at ETH Zurich in Switzerland. He worked for nearly a year on alkaloid chemistry.

During a brief period in 1949, he was unable to find a job in India quickly and returned to England on a fellowship to work with George Wallace Kenner and Alexander R. Todd on peptides and nucleotides. He stayed in Cambridge from 1950 until 1952.

He moved to Vancouver, British Columbia, with his family in 1952 after accepting a position with the British Columbia Research Council at University of British Columbia. Khorana was excited by the prospect of starting his own lab, a colleague later recalled. His mentor later said that the Council had few facilities at the time but gave the researcher "all the freedom in the world". His work in British Columbia was on "nucleic acids and synthesis of many important bio-molecules" according to the American Chemical Society.

In 1960 Khorana accepted a position as co-director of the Institute for enzyme research at the Institute for Enzyme Research at the University of Wisconsin at Madison. He became a professor of biochemistry in 1962 and was named Conrad A. Elvehjem Professor of Life Sciences at Wisconsin–Madison. While at Wisconsin, "he helped decipher the mechanisms by which RNA codes for the synthesis of proteins" and began to work on synthesizing functional genes according to the American Chemical Society. During his tenure at this University, he completed the work that led to sharing the 1968-Nobel Prize in Physiology and Medicine along with Nirenberg and Holley for deciphering genetic code. The Nobel web site states that it was "for their interpretation of the genetic code and its function in protein synthesis". Har Gobind Khorana's role is stated as follows: he "made important contributions to this field by building different RNA chains with the help of enzymes. Using these enzymes, he was able to produce proteins. The amino acid sequences of these proteins then solved the rest of the puzzle."

He became a US citizen in 1966. Beginning in 1970, Khorana was the Alfred P. Sloan Professor of Biology and Chemistry at the Massachusetts Institute of Technology and later, a member of the Board of Scientific Governors at The Scripps Research Institute. He retired from MIT in 2007. Har Gobind Khorana married Esther Elizabeth Sibler in 1952. They had met in Switzerland and had three children, Julia Elizabeth, Emily Anne, and Dave Roy.

Research

Ribonucleic acid (RNA) with two repeating units (UCUCUCU → UCU CUC UCU) produced two alternating amino acids. This, combined with the Nirenberg and Leder experiment, showed that UCU genetically codes for serine and CUC codes for leucine. RNAs with three repeating units (UACUACUA → UAC UAC UAC, or ACU ACU ACU, or CUA CUA CUA) produced three different strings of amino acids. RNAs with four repeating units including UAG, UAA, or UGA, produced only dipeptides and tripeptides thus revealing that UAG, UAA and UGA are stop codons.

Their Nobel lecture was delivered on 12 December 1968. Khorana was the first scientist to chemically synthesize oligonucleotides. This achievement, in the 1970s, was also the world's first synthetic gene; in later years, the process has become widespread. Subsequent scientists referred to his research while advancing genome editing with the CRISPR/Cas9 system.

He extended the above to long DNA polymers using non-aqueous chemistry and assembled these into the first synthetic gene, using polymerase and ligase enzymes that link pieces of DNA together, as well as methods that anticipated the invention of polymerase chain reaction (PCR). These custom-designed pieces of artificial genes are widely used in biology labs for sequencing, cloning and engineering new plants and animals, and are integral to the expanding use of DNA analysis to understand gene-based human disease as well as human evolution. Khorana's invention(s) have become automated and commercialized so that anyone now can order a synthetic oligonucleotide or a gene from any of a number of companies. One merely needs to send the genetic sequence to one of the companies to receive an oligonucleotide with the desired sequence.

After the middle of the 1970s, his lab studied the biochemistry of bacteriorhodopsin, a membrane protein that converts light energy into chemical energy by creating a proton gradient. Later, his lab went on to study the structurally related visual pigment known as rhodopsin.

A summary of his work was provided by a former colleague at the University of Wisconsin: "Khorana was an early practitioner, and perhaps a founding father, of the field of chemical biology. He brought the power of chemical synthesis to bear on deciphering the genetic code, relying on different combinations of trinucleotides."

Awards and Honors

1958: Merck Award from Chemical Institute of Canada.

1959: Fellow Chemical Institute of Canada.

1960: Gold Medal, Professional Institute of the Public Service of Canada.

1967: Dannie-Heinneman Preis, Goettingen Academy of Germany.

1968: Nobel Prize for Medicine/Physiology (shared with M. Nirenberg and R. W. Holley).

1968: ACS Award for Creative Work in Synthetic Organic Chemistry.

1968: Lasker Foundation Award for Basic Medical Research.

1968: Distinguished Service Award, Watmull Foundation, Honolulu, Hawaii.

1971: American Academy of Achievement Award, Philadelphia, PA.

1972: Padma Vibhushan President Award, India.

1974: Willard Gibbs Medal of the Chicago Section of the American Chemical Society.

1980: Gairdner Foundation Annual Award, Toronto, Canada.

1986: Ellis Island Medal of Honour.

1987: Paul Kayser International Award of Merit in Retina Research.

2000: MIT School of Science Distinguished Service Award.

Har Gobind Khorana died on 9 November 2011, in Concord, Massachusetts, at the age of 89. In his obituary, the Washington Post provided this summary of the man: "Dr. Khorana was known for a modest, ingratiating manner. He tended to shun publicity, making many of his most important scientific announcements at departmental seminars and in scientific publications".

44

Gene Regulation

Regulator Genes

Bacterial cells are capable of producing many kinds of enzymes in relatively large quantities but the kinds and amount of enzymes actually produced at any given time are usually those which are required at that particular time in response to the prevailing environment. Inducible enzymes are ordinarily absent or present in very small quantities until a specific compound appears. This compound is called as inducer. A typical example of this is production of β galactosidase (an enzyme that splits lactose into galactose and glucose, Fig. 44.1) once the lactose is introduced into the medium on which *E. coli* cells are grown. Here lactose acts as inducer for the enzyme β galactosidase.

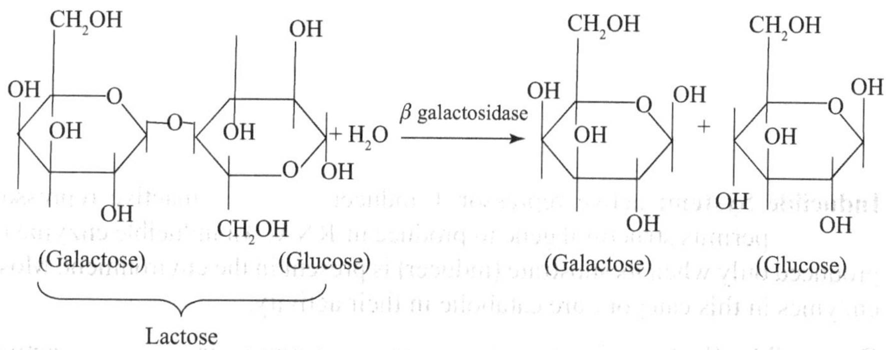

Fig. 44.1. Hydrolysis of sugar lactose into galactose and glucose under enzyme β galactosidase.

Along with the induction of β galactosidase, a second enzyme called permease is produced which permits entry of lactose into the cells and this lactose gets split up into the component sugars, galactose and glucose by the enzymc β galactosidase. A third enzyme, acetylase also appears upon induction of lactose. All three of these enzymes are affected by genes that map within the *lac* locus in the sequence as *z* (β galactosidase)-*y* (permease)-*ac* (acetylase). The induction and repression of gene expression can be accomplished by the same mechanism, with one relatively minor modification. This mechanism known as operon model was developed in 1961 by Francois Jacob and Jacques

Monod to explain the regulation of genes coding the enzymes required for the lactose utilization in *E. coli.* Jacob and Monod, 1965 Nobel Prize recipients, proposed that the transcription of one gene or a set of contiguous structural genes (genes specifying the amino acid sequences of enzymes or structural proteins) is regulated by two controlling elements. One of these elements, the regulator or repressor gene, encodes a protein called the repressor. Under the appropriate conditions, the repressor binds to the second element, the operator. The operator is always contiguous with structural genes whose expression is regulated by the operator.

When the repressor is bound to the operator, transcription of the structural genes cannot occur. Transcription is initiated at RNA polymerase-binding sites called promoters located just upstream (5') from the coding regions of the structural genes. Promoters were unknown at the time of Jacob and Monod's proposal, but they have since been shown to be essential components of operons. When repressor is bound to the operator, it sterically prevents RNA polymerse from transcribing the structural genes in the operon. Operator genes are usually located between the promoters and structural genes that they regulate. The complete contiguous unit including the structural gene or genes, the operator and the promoter is called operon. The diagrammatic representation of this system is shown in Fig. 44.2 (Strickberger, 1968). Both repressible systems and inducible systems control gene expression through repressors (negative control), the former producing an inactive repressor that is inactivated by a substance called co-repressor, the latter producing an active repressor that is inactivated by an inducer. This is shown as follows:

Inducible System: active repressor + inducer ———→ inactive repressor ———→ permits structural gene to produce m-RNA. An inducible enzyme is produced only when its substrate (inducer) is present in the environment. Most enzymes in this category are catabolic in their activity.

Repressible System: inactive repressor + co-repressor ———→ active repressor ———→ prevents structural gene from producing m-RNA. A repressible enzyme is normally present in the cell but ceases to be synthesized when high concentrations of its end products are present. The end product is called co-repressor. A regulator gene produces a substance called aporepressor which unites with co-repressor to form a functional repressor molecule. The repressor inhibits synthesis of m-RNA by all genes specifying enzymes in the synthetic pathway. Most repressible enzymes are found in anabolic pathways.

Genes may be broadly classified into two types, (i) genes for structure and (ii) genes for function. Structural genes encode the specifications for the synthesis of some types of chemical substances. The substances may be (i) an enzyme or (ii) non-enzymatic structural protein (e. g. actin or myosin filaments of muscle cells, hemoglobin red blood cells, etc.), (iii) a protein called repressor which prevents the activity of some other gene. Genes which produce repressor substances are called regulatory genes. Operator genes which are closely linked to one or more structural genes produce no detectable substance. An operator controls the synthetic activity of the immediately adjacent structural genes. An operator plus the structural genes under its control is called operon. Perhaps the operator locus initiates transcription of adjacent DNA cistrons into m-RNA molecules coding for one or more polypeptides. An active operator allows transcription to occur. If repressor substance is present, it may combine with the operator site, thereby preventing transcription of the structural genes of the operon into m-RNA.

Stansfield (1969) has presented this aspect in a simplified manner as follows: An example of the way an inductive system is presumed to function is provided by lactose system of bacteria *E. coli.* Catabolism of the sugar lactose requires the cooperation of two enzymes, beta galactoside permease which makes the cell permeable to the substrate (lactose) and beta galactosidase, an intracellular enzyme which hydrolyzes the disaccharide (lactose) into the simple sugars, namely, galactose and glucose.

The operon consists of operator (*o*), the gene for galactosidae (*z*) and the gene for permease (*y*). A regulator gene (*i*) is located near this operon. Some of the alleles which have been found at each of these loci and their functions are listed as follows:

Operator Alleles

o^+ = This operator "turns on" the structural genes in its own operon, namely, y^+ and z^+ alleles in the same segment of the DNA (cis-position) can produce the enzymes. This operator is sensitive to repressor substance i. e. it will "turn off" the synthetic activity of the structural genes in its operon if repressor is present.

o^c = A constitutive operator which is insensitive to repressor substance and permanently "turns on" the structural genes in its own operon.

o^o = This is a defective operator which permanently "turns off" the structural genes in its own operon.

Galactosidase Alleles

z^+ = Makes beta galactosidase if its operon is "turned on".

z^{-} = Makes a modified enzymatically inactive substance called C_z protein.

Permease Alleles

y^+ = Makes beta galactosidase permease if its operon is "turned on".

y^- = No detectable product is formed regardless of the state of the operator.

Regulator Alleles

i^+ = Makes a diffusible repressor substance which inhibits synthetic activity in any o^+ operon in the absence of lactose. In the presence of lactose, repressor substance is inactivated.

i^- = A defective regulator unable to produce active repressor substance.

i^s = Suppressor which is insensitive to the presence of lactose and inactivates any o^+ operon.

Briefly, in bacteria, genes with related functions frequently occur in coordinated regulated units called operons. Each operon contains a set of contiguous structural genes, a promoter (the binding site for RNA polymerase) and an operator (the binding site for a regulatory protein called a repressor). When repressor is bound to the operator, RNA polymerase cannot transcribe the structural genes in the operon. When the operator is free of repressor, RNA polymerse can transcribe the operon.

Gene Regulation in Eukaryotes

In eukaryotes, gene expression is spatially temporally regulated. Some genes such as those encoding the alpha and beta tubulins in *Arabidopsis* are expressed in a tissue-specific manner. Other genes such as those encoding the alpha and beta globins in vertebrates are expressed in a specific temporal pattern during development. Eukaryotic gene expression can be regulated at the transcriptional, processing or translational levels. Transcriptional regulation involves protein-DNA interaction, processing regulation involves the alternate splicing of primary gene transcripts, and translational regulation involves m-RNA stability. Eukaryotic gene expression can be induced by environmental factors such as heat and light and by signaling molecules such as hormones and growth factors. Hormone-induced gene expression is mediated by proteins that interact with the hormones. Some of these hormone receptors act directly as transcription factors by binding to DNA sequences in the vicinity of a gene. Others control transcription indirectly through a signal transduction pathway that targets transcription factors in a gene.

Cancer causing genes or oncogenes were initially discovered in the genomes of retroviruses. These genes encode a variety of different proteins including

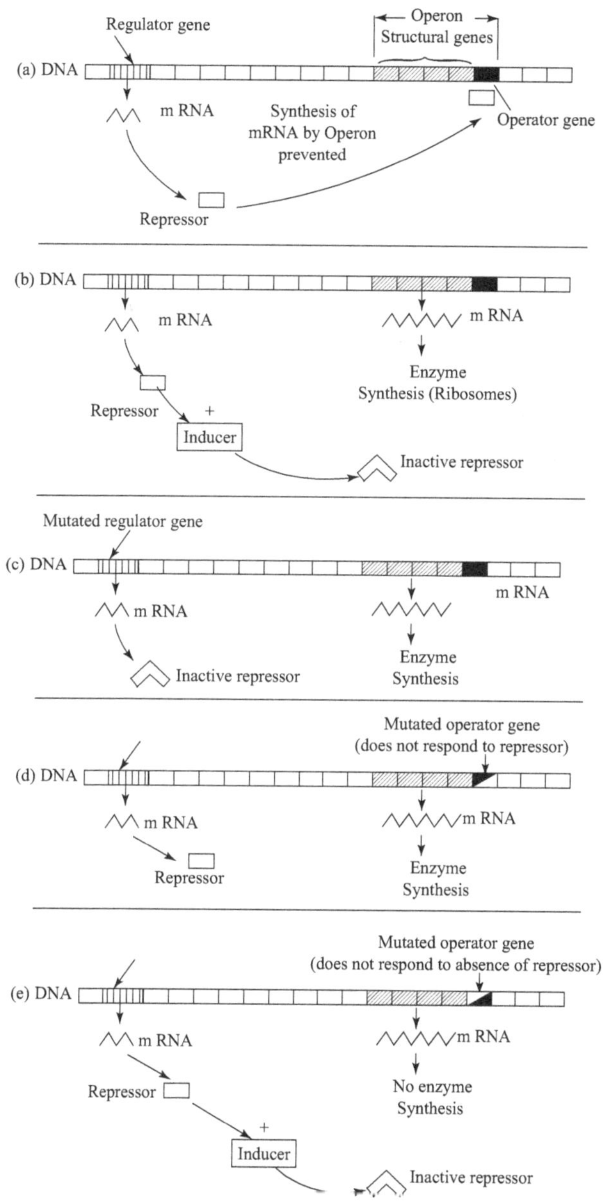

Fig. 44.2. Regulation of enzyme synthesis. (a) Regulation by activity of a repressor which prevents enzyme synthesis by structural genes of the operon, (b) Repressor activity removed by an inducer which permits enzyme synthesis to proceed, (c and d) Mutation of the regulator gene or the operator gene allowing enzyme synthesis to proceed even in the absence of an inducer, (e) Mutation of the operator gene preventing function of the structural genes even in presence of inducer.

growth factors, growth-factor receptors, tyrosine kinases and transcription factors. Later, these genes were found to have normal cellular homologs, but unlike their wild counterparts, these cellular oncogenes possess introns. The gene regulation processes have been extended to the cancers. Some cancers are caused by mutant proteins that function as dominant activators of cell division. Mutations that produce these proteins have been identified by transferring cultured cells with DNA isolated from tumor cells. Other cancers such as Burkitt's lymphoma and chronic myelogenous leukemia are associated with chromosome translocations that alter the expression of a particular cellular oncogene. Still other cancers such as retinoblastoma are associated with deletions of genes that suppress tumor formation. Such a deletion may predispose an individual to develop a cancer if a spontaneous somatic mutation eliminates the other copy of the gene (Snustad *et al*., 1997).

45

The Gene Concept

The concept of gene has undergone several refinements since its discovery in 1866 by Mendel. Most genes encode one polypeptide and can be operationally defined by the complementation test. The complementation or cis-trans test provides an operational definition of the gene. It is used to determine whether mutations are in the same gene or in different genes. Intragenic complementation may occur when a protein is a multimer containing at least two copies of one gene product. The concept of gene has evolved from a bead on a string, not divisible by recombination or mutation to a sequence of nucleotide pairs in DNA encoding one polypeptide chain. The unit of genetic material not divisible by recombination or mutation is the single nucleotide pair. Historically several experimental developments have occurred to facilitate understanding of the structure and function of gene and these are briefly described in a chronological order as follows to have better understanding of the gene concept as it evolved. However, before discussing evidences supporting various gene concepts, it is better to summarize important stages in the evolution of the gene concept.

Evolution of the Concept of the Gene

The gene theory of inheritance began with the publication of Mendel's classical paper in 1866. After seven years of experimental work in the monastery garden at Bruenn, Austria (today, Brno, a city in the South Moravian Region of the Czech Republic), he presented the results of his experiments, together with the generalizations we now know as "Mendel's laws" at two meetings of the Natural History Society of the then Bruenn in the spring of 1865. The results and the theory were published in the annual proceedings of the society, which appeared and were distributed to libraries in Europe and America in 1866. It is safe to say that no one who heard Mendel's paper and no one who read it in the nineteenth century appreciated its significance and it remained neglected until 1900 when the law of segregation was rediscovered almost simultaneously by three different investigators who had reported results like Mendel. These three, namely, De Vries in Holland, Correns in Germany, and Tschermak in Austria found Mendel's forgotten paper and proclaimed its importance. Mendel's gene (not so named then) was the "character" or "constant factor" that controlled one specific phenotypic trait, such as form of seed, colour of cotyledons, colour of seed coat and flower, form of pod, colour of pod, position of flowers, and length of stem. At the time of discovery of Mendel's work, several inherited diseases in humans were being investigated. English physician, Sir Archibald E. Garrod first recognized in 1902 that homozygosity for a recessive mutant allele was responsible for a defect in the normal process of metabolism. His concept of gene amounts to "one mutant gene-one metabolic path". This concept was refined as "one gene-one enzyme" concept by G. W. Beadle and E. L. Tatum in 1941. Since many enzymes contain two or more different polypeptides, each encoded by a separate gene, the "one gene-one enzyme" concept got modified to "one gene-one polypeptide"

Prior to 1940, genes were compared to beads on a string where recombination could occur between the genes but not within the gene. The gene was then considered both as a "functional unit" and a "structural unit". Gene as a functional unit meant that one gene controlled one phenotype and gene as a structural unit meant that the gene cannot be subdivided by recombination or mutation. Based on a few developments occurring subsequently, it was established that the "nucleotide pair" had replaced gene as the basic unit of structure, not sub-divisible by recombination or mutation.

Cis-trans/complementation test was developed by E. B. Lewis in early 1940s for test of functional allelism in *Drosophila*. This test was subsequently exploited by S. Benzer to define experimentally the gene in bacteriophage T4. Two similar or identical mutant phenotypes isolated from unrelated parental material may or may not be result of allelic modification at the same gene locus, as is generally the case. This may be verified by observing the phenotype of the F_1 hybrid between the two. The F_1 hybrid will have the same phenotype as one of the mutant parents when the expression is determined by the alleles of the same gene. In case the alleles are not identical, the F_1 hybrid will segregate the two parental forms in the F_2 generation yielding a mono-hybrid ratio. If, on the other hand, the F_1 hybrid between the two mutants does not have the mutant phenotype, but is normal wild type, the mutant forms, even though they are phenotypically identical, are considered to be determined by the alleles of two separate genes and a di-hybrid ratio is to be expected in the F_2 generation. This is clarified further as follows.

Let us consider the breeding behaviour of two dwarf mutant phenotypes. Two dwarf phenotypes which have arisen independently in un-related pedigrees may be the result of a recessive mutation to an allele conditioning dwarfism at a single gene locus, *d* or they may be the expression of alleles at two gene loci, *d* and *dw*. In the former case, the genotypes of the two dwarfs may be identical, both being d_1d_1. In case two different mutant alleles of *d* locus are involved, then the mutant genotypes could be d_1d_1 and d_2d_2 respectively. If two loci, *d* and *dw* are involved, the genotypes will be *dd* ± ± and ± ± *dwdw* where the symbol ± represents the wild type alleles at the *dw* and *d* loci, respectively. Given these genotypic constitutions, the breeding results expected from hybrids between the two dwarfs, according to whether one locus or two loci are involved, can be illustrated as follows (Williams, 1964):

Dwarfs due to identical recessive alleles:

Dwarf (A) d_1d_1 x Dwarf (B) d_1d_1 = F_1 hybrid *d*1*d*1 (non-segregating F_1 hybrid).

Dwarfs due to two different recessive alleles of one gene:

Dwarf (A) d_1d_1 x Dwarf (B) d_2d_2 = F_1 hybrid dwarf d_1d_2 (F_1 hybrid segregates dwarfs, d_1d_1, d_1d_2, d_2d_2).

Dwarf due to recessive alleles at two different loci:

Dwarf (A) *dd*++ x Dwarf (B) ++*dwdw* = F_1 hybrid normal *d* + *dw* + (F_1 hybrid segregates 9 normal : 7 dwarfs).

In the cross between recessive non-allelic genes, the phenotype of the hybrid (F_1) between two dwarfs is normal because the expression of the *d* and *dw* alleles at each of the two loci is masked by the presence of the dominant + allele. In other two combinations, the F_1 hybrid is dwarf. But where two different alleles of a single gene locus are involved, as in example 2, the hybrid may be one of three phenotypes depending on the dominance relationship of d_1 and d_2. If d_1 or d_2 are dominant, the phenotype of the hybrid will be either d_1d_1 or as d_2d_2. In case there is no dominance, then the hybrid will be intermediate between d_1d_1 and d_2d_2. It should be clear that the test for allelism among recessives determines difference between genes and alleles. Genes complement one another in the production of normal phenotype where as alleles fail to do this. It may be stated that whenever mutant phenotypes show a complementary effect in conditioning a normal F_1 phenotype, the controlling alleles belong to different genes. If they are non-complementary, they represent alleles of the same gene. However, it has been shown in several cases later on that the phenomenon of complementation has been described between large numbers of alleles of the single gene in certain organism. But this does not essentially alter the basic test of allelism given here. The complementation test allows a geneticist to determine whether two independent mutations occurred in the same gene or in two different genes. Moreover, the gene as defined by complementation test is a perfect fit to "one gene-one polypeptide" concept. Charles Yanofsky, Sydney Brenner and their collaborators in the 1960s demonstrated that the gene and its polypeptide products were collinear structures with a direct correlation between the sequence of nucleotide pairs in the gene and sequence of amino acids in the polypeptide. However, this simple concept of gene as a continuous sequence of nucleotide pairs specifying a collinear sequence of amino acids in the polypeptide gene product was short-lived as overlapping genes and genes within genes were discovered in the late 1960s. Coding sequences of eukaryotic genes were found to be interrupted by non-coding intron sequences in late 1970s. Further some genes, for example genes encoding immunoglobulins have been shown to be stored in germ-line chromosomes as "short gene segments" which are assembled into mature, functional genes during development (Snustad *et al.*,1997). Despite these few examples not fitting to normally accepted gene concept, it is safer to conclude that the concept of the gene has undergone many refinements since its discovery by Mendel in 1866. In general, most genes encode one polypeptide and can be operationally defined by the complementation test.

Following are the key historical developments that have led to the evolution of concept of gene function and gene structure:

- Evolution of the Concept of the Gene: Function:
- Mendel's Constant Factors Controlling Phenotypic Traits:

Mendel explained the transmission of traits from one generation to another and explained this based on the transmission of the "factors" while explaining his experimental results from crosses made in garden pea which were started in 1857. Those factors are now known as genes. These factors or the genes controlled specific phenotypic traits such as flower colour, seed colour, and seed shape. Those "factors" were the basic unit of function, the units of the genetic information that governed one specific aspect of the phenotype. This definition of gene as the "basic unit of function" is almost universally accepted by the present day geneticists/plant breeders. There has been no change in this concept of gene as the "basic unit of function" since its discovery by Mendel in 1866. However, the concept of molecular structure of this basic unit of function has undergone several changes and refinement since rediscovery of Mendel's work in 1900. The phenotype of an organism is the product of the action of all the genes superimposed with the effect of environment. The relationship between the gene and the phenotype is shown in Fig. 45.1 (Snustad *et al.*, 1997).

Garrod's Discovery of One Mutant Gene-One Metabolic Path

Sir Garrod was engaged in studying several congenital metabolic diseases in humans when Mendel's laws were discovered in 1900. One such metabolic inherited disorder was alkaptonuria. This disorder is easily detected as urine of the affected person becomes black upon exposure to air. The substance responsible for this disorder is alkapton or homogentisic acid which is an intermediate product in the degradation of the aromatic amino acids tyrosine and phenylalanine as shown in Fig. 45.2.

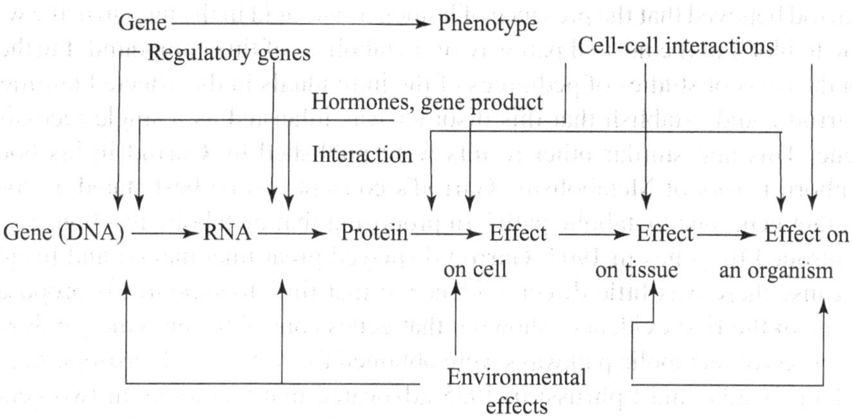

Fig. 45.1. Complex pathway between gene and phenotype.

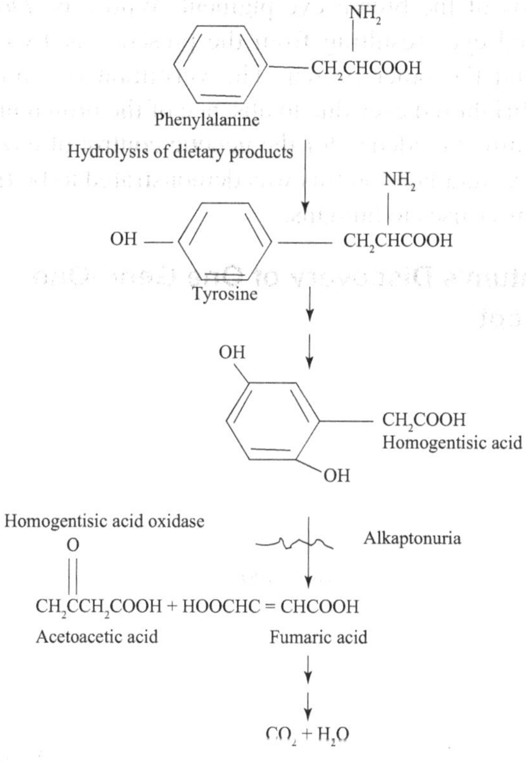

Fig. 45.2. Alkaptonuria in humans results from a block in phenylalanine catabolism caused by a mutation in the gene encoding the enzyme homogentisic acid oxidase. In absence of this enzyme, there is accumulation of the substrate homogentisic acid in tissues and the urine.

Garrod believed that the presence of homogentisic acid in the human urine was due to block in the normal pathway of metabolism of this compound. Further, on the basis of studies of pedigrees of the individuals in the affected families, Garrod could establish that this disorder was inherited as a single recessive gene. This and similar other results were published by Garrod in his book "Inborn Errors of Metabolism" Garrod's concept can be best stated as "one mutant gene-one metabolic path". In proposing that metabolic reactions were controlled by genes in 1902, Garrod displayed great imagination and insight because there was little direct evidence at that time to support his proposal. Some of the first evidences showing that genes control the enzyme-catalyzed reactions of metabolic pathways were obtained from studies of *Drosophila* eye colour. Beadle and Ephrussi in 1935 advocated that mutations in two genes controlling eye colour in *Drosophila* resulted in blocks at two different steps in the biosynthesis of the brown eye pigment. Wild type *Drosophila* (fruit flies) had dark red eyes resulting from the presence of two eye pigments, one bright red and the other brown. The vermilion (*v*) and cinnabar (*cn*) mutant flies had bright red eyes due to absence of the brown pigments. These studies provided direct evidence for the genetic control of enzyme catalyzed metabolic pathways and later on this was demonstrated to be true in all types of organisms from viruses to humans.

Beadle and Tatum's Discovery of One Gene-One Enzyme Concept

Beadle and Tatum in 1942 demonstrated genetic control of biochemical reactions in *Neurospora* that led to the "one gene-one enzyme" hypothesis. *Neurospora crassa* can grow on medium containing only (i) inorganic salts, (ii) a simple sugar and (iii) one vitamin, biotin. The growth medium containing only these components is called "minimal medium". Beadle and Tatum hypothesized that all other essential metabolites like purines, pyramidines, vitamins and amino acids must be synthesized *de novo*. Further, they reasoned out that the biosynthesis of these growth factors must be under the genetic control as there have been evidences to this in the earlier investigations by others. Then they thought that mutants could be produced which may not be able to grow on the minimal medium without the additional growth factor requirements.

Beadle and Tatum tested this prediction by producing *Neurospora crassa* mutants and screening them for new growth factor requirements. In order to select strains with mutation in only one gene, they studied only mutant strains that yielded a 1: 1 mutant to wild type progeny ratio when crossed with wild type. Beadle and Tatum identified mutant strains that grew in the presence of vitamins and they further investigated the ability of these vitamins-requiring strains to grow on media supplemented with each of the vitamins separately.

In this way, Beadle and Tatum demonstrated that each mutation resulted in a requirement for one growth factor. Further, they demonstrated in several cases that one mutation resulted in the loss of one enzyme activity. Thus "one gene-one enzyme" concept became a central tenet of molecular biology and Beadle and Tatum were awarded Nobel Prize for this brilliant work in 1958. Beadle acknowledged the original contribution of Garrod to arrive on this proposition which is clear from his Nobel Prize acceptance speech as quoted below (Snustad *et al.*, 1997):

"In this long, roundabout way, we had discovered what Garrod had seen so clearly so many years before. By now we knew of his work and were aware that we had added little if any thing new in principle. …….Thus, we were able to demonstrate that what Garrod had shown for a few genes and a few chemical reactions in man was true for many genes and many reactions in Neurospora".

One Gene-One Polypeptide

Subsequent to the work of Beadle and Tatum, it was shown that there were many enzymes and structural proteins that were heteromultimeric that is they contained two or more different polypeptide chains where each polypeptide was encoded by a separate gene. For example, in *E. coli*, it has been show that the enzyme tryptophan synthetase is composed of two α polypeptides encoded by the *trpA* gene and two *β* polypeptides encoded by *trpB* gene. In view of this example and several other similar cases, the "one gene-one enzyme" concept has been modified to "one gene-one polypeptide" concept.

Evolution of the Concept of the Gene: Structure

This section deals with the structure of the gene in contrast to the earlier concept where the focus has been on the function of the gene.

Pre-1940 Beads on String Concept

Prior to 1940, the genes were considered as beads on string to draw a parallel that the genes were located on the chromosomes in a manner something like similar to beads on string. Recombination was assumed to occur only between the genes and not within the genes as per this structural concept of the gene. According to this "beads-on-a string" concept, the gene was considered as the basic unit of genetic information defined by three criteria, namely, (i) function, (ii) recombination and (iii) mutation. More specifically, the gene was considered as:

Unit of Function: This implied that the gene controlled the inheritance of one "character" of one attribute of the phenotype.

Unit of Structure: This was operationally defined in two ways:

By Recombination: Considered as unit of genetic information not sub-divisible by recombination.

By Mutation: Considered as the smallest unit of genetic material capable of independent mutation.

Initially the geneticists thought that all these three criteria defined the same basic unit of inheritance that is the "gene". Now it is well understood that gene remains the unit of function or the unit of genetic information controlling the synthesis of one polypeptide chain or in some cases one RNA molecule but structurally, the gene is the sequence of nucleotide pairs in the DNA. Each nucleotide pair cannot be called a gene. Thus, the geneticists have retained the original definition of gene as the unit of function but have discarded beads-on-a-string concept and have accepted that structurally gene is sub-divisible by recombination and mutation.

Discovery of Recombination Within the Gene

C. P. Oliver in 1940 published the first evidence demonstrating that recombination could occur within a gene. Oliver studied two mutations in *Drosophila*, lz^s (spectacle eye) and lz^g (glassy eye) which were thought to be alleles, that is different forms of the same gene. Data prior to 1940 indicated that these two genes mapped on the same locus on X-chromosome. They had similar effect on phenotype of the eye and heterozygous females, lz^s/lz^g had lozenge instead of wild type eyes. However, when lz^s/lz^g females were crossed with either lz^s or lz^g males and large numbers of progeny were examined, wild type progeny occurred with a frequency of about 0.2 percent. This high frequency of recovery of wild types was explained as if recombination has occurred within the earlier known gene and this meant recombination between the alleles lz^s and lz^g which was confirmed using different sets of outside markers. These results strongly indicated that the lz^s and lz^g mutations were located at different and distinct sites within the lozenge locus and the *lz*+ chromosome was produced by crossing over between the two sites as shown in the Fig. 45.3 (Snustad *et al.*, 1997).

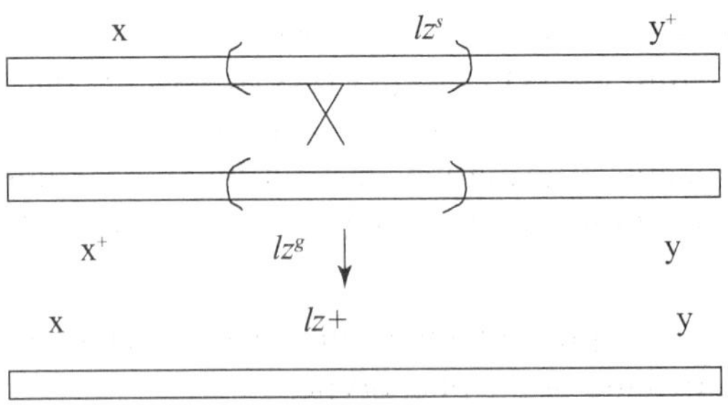

Fig. 45.3. Crossing over within lozenge locus.

The conclusive proof of the crossing over within the lozenge locus as shown in Fig. 45.3 required the recovery of *lz*s-*lz*g double mutant with the reciprocal combination of outside markers as shown below:

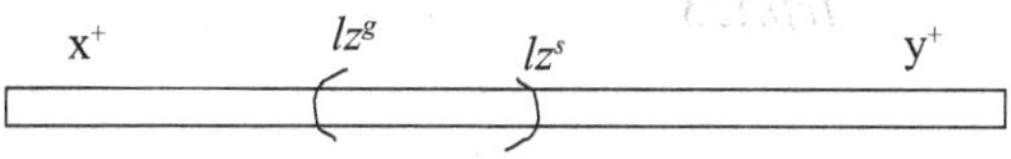

The identification of both products, the wild type and double mutant chromosomes produced by crossing over within the lozenge gene was first accomplished by M. M. Green who happened to be one of Oliver's students.

These results clearly demonstrated that the structurally gene was more complex than a bead on a string. It became obvious that the gene was divisible containing sites that were separable by crossing over.

Recombination Between Adjacent Nucleotide Pairs

The results obtained by Oliver, Lewis and Green in *Drosophila* indicated that mutable sites that are separable by recombination can exist within a single gene. S. Benzer extended this picture of gene by demonstrating the existence of 199 distinct sites of mutation that were separable by recombination within the *rIIA* gene of the bacteriophage T4. Benzer's picture of gene as a sequence of nucleotide pairs capable of mutating at many different sites was soon verified by the results of many investigators in several different organisms. Based on these, it was thought that the smallest unit of genetic material capable of mutation might be the single nucleotide pair that recombination might occur between adjacent nucleotide pairs, whether between or within the genes. Recombination between the adjacent nucleotide pairs of a gene was first demonstrated by Charles Yanofsky for the *trpA* gene encoding α polypeptide of tryptophan synthetase in *E. coli* in 1972. This enzyme is a tetramer containing two α polypeptides and two *β* polypeptides. This is involved in the final step in the biosynthesis of the amino acid tryptophan.

Yanofsky and co-workers isolated large number of tryptophan auxotrophs with mutations in the *trpA* gene. The wild type *trpA* gene encodes an α polypeptide that is 268 amino acids long. Yanofsky and colleagues determined the complete amino acid sequence of the wild type α polypeptide and amino acid substitutions that had occurred in several mutant forms of the tryptophan synthetase α polypeptide. These scientists mapped the mutations within the *trpA* gene using two and three-factor crosses and compared the map positions with the locations of the amino acid-substitutions in the mutant polypeptide. They correlated genetic and biochemical data for *trpA* gene and tryptophan synthetase α polypeptide and demonstrated that recombination can occur between mutations that alter the same amino acid. Mutations at *trpA23* and *trpA46* both are responsible for substitution of another amino acid for example arginine in case the of *trpA23* and glutamic acid in the case of *trpA46* for the glycine present at position 211 of the wild-type tryptophan synthetase α polypeptide. Thus glycine at position 211 in wild type is getting replaced by arginine in the mutant *trpA23* and by glutamic acid in the mutant *trpA46*. However, these two closely related mutations were found to be occurring at two different mutable sites meaning thereby that these two sites A23 and A46 are separable by recombination. With some more experiments and information, Yanofsky and colleagues were able to determine which of the glycine, arginine and glutamic acid codons were present in the *trpA* mRNA at the position encoding amino acid residue 211 of the α polypeptides present in *trp+*, *trpA23* and *trpA46* cells, respectively. The data presented by Yanofsky demonstrated that the mutational events that produced the *trpA23* and *trpA46* alleles were G: C to A: T transition at adjacent nucleotide pairs. The *trp+* cells produced by recombination between chromosomes carrying mutations *A*23 and *A*46 showed that recombination had occurred between adjacent nucleotide pairs in the *trpA* gene which has been shown in Fig. 45.4 (Snustad *et al.*, 1997).

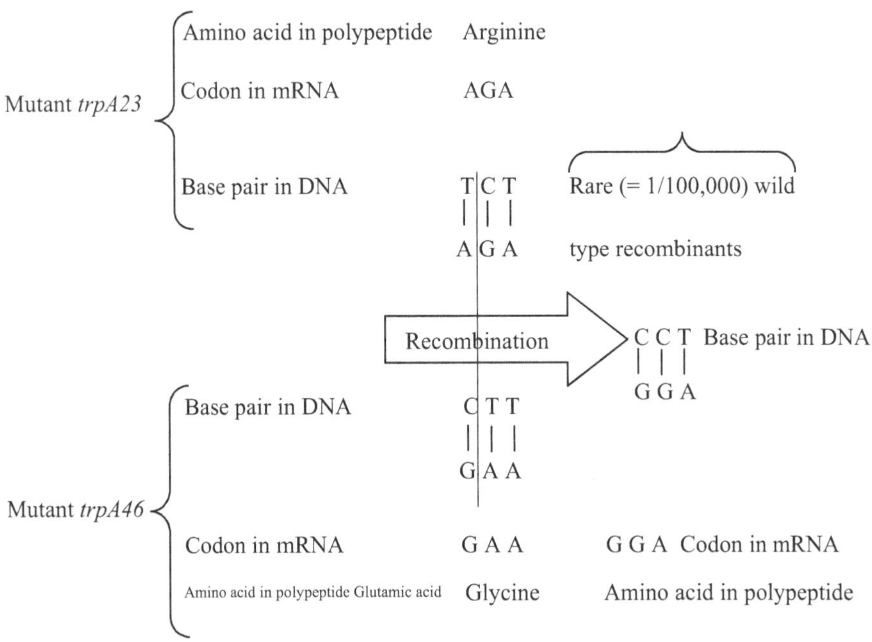

Fig. 45.4. Recombination between mutations at adjacent nucleotide pairs in the *trpA* gene of *E. coli.* Mutations *A*23 and *A46* both result in an amino acid substitution at position 211 of tryptophan synthetase α polypeptide. Wild type bacteria has a glycine residue at this position of the α polypeptide. *A*23 causes a glycine to arginine substitution and *A*46 causes a glycine to glutamic acid substitution.

Collinearity Between the Coding Sequence of a Gene and Its Polypeptide Product

It is now well established that the nucleotide pair sequences of the coding regions of the structural genes and the amino acid sequences of the corresponding polypeptides are co-llinear i.e. the first three base-pairs of the coding sequence of a gene determine the first amino acid of the polypeptide and likewise the next three base-pairs specify the second amino acid and so in a collinear fashion. It is also known that in the higher eukaryotes, the coding regions of the most of the genes are interrupted by the noncoding introns. However, this interruption by the introns in the genes does not invalidate the concept of collinearity between the base-pairs and the amino acids. This interruption by introns simply means that there is no direct correlation in physical distances between the positions of base-pair coding triplets in a gene and the position of amino acids in the resultant polypeptide specified by the particular gene.

The first strong evidence regarding collinearity between a gene and its polypeptide product was the outcome from studies of Charles Yanofsky and co-workers on the *E. coli* gene encoding the α subunit of the enzyme tryptophan synthetase which catalyzes the final step in the biosynthesis of amino acid tryptophan. This enzyme has two α polypeptides encoded by the *trpA* gene and two *β* polypeptides encoded by the ***trpB*** gene. Charles Yanofsky and co-workers demonstrated that there was a direct correlation between the map position of mutations in the *trpA* gene and the positions of the resultant amino acid substitutions in the tryptophan synthetase α polypeptide as shown in Fig. 45.5 (Snustad *et al*., 1997). Almost at the same time, S. Benzer and colleagues showed a similar collinearity between the positions of mutation in the gene of bacteriophage T4 that encodes the major structural protein of the phage head and the position of amino acids in the polypeptide affected by these mutations. Brenner and colleagues studied *amber* (UAG, chain termination codon) mutations and demonstrated a direct correlation between the length of the polypeptide fragment that has been produced and the position of the mutation within the gene. By now, collinearity has been established between many genes and their protein (polypeptide) products in organisms ranging from viruses to humans. Thus, collinearity between the coding regions of genes and amino acid sequences of their polypeptide products is a universal or nearly universal characteristics of "gene-protein" relationships.

To conclude, the concept of the gene has evolved from a bead on a string, not divisible by recombination or mutation, to a sequence of nucleotide pairs in DNA encoding one polypeptide chain. The unit of genetic material not divisible by recombination or mutation is the single nucleotide pair.

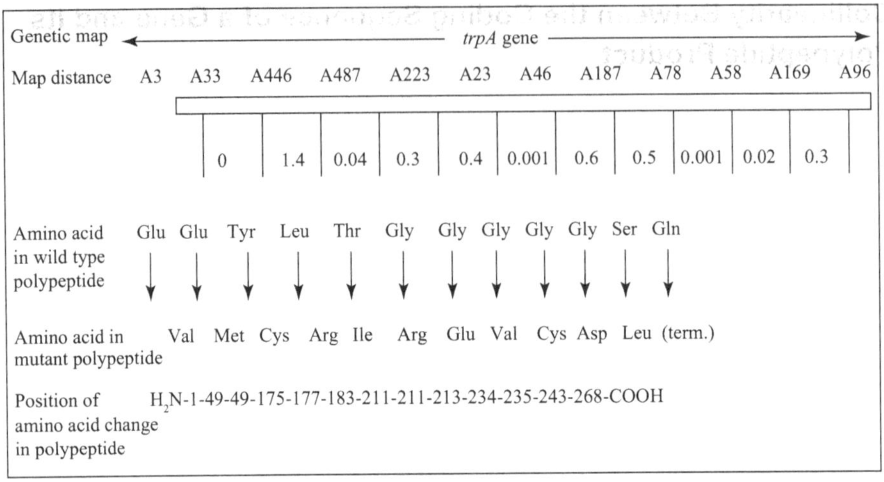

Fig. 45.5. Collinearity between *trpA* gene and its polypeptide product, the α polypeptide of tryptophan synthetase in *E. coli.*

A Genetic Definition of Gene Based on the Complementation Test

Let us recall that the elementary concept of gene meant that the genes were discrete elements analogous to beads on a string. The chemical structure of gene has been fully unraveled and it is clear that a gene is a sequence of nucleotides. Initially a gene was assumed to be a genetic unit by three criteria, namely, (i) the unit of physiological function leading to phenotypic expression, (ii) the unit of structure which cannot be sub-divided by crossing over and (iii) the unit which can be changed by mutation. Now it is well known that these three criteria generally do not define the same amount of genetic material. Each of these criteria may indicate a different genetic element.

A gene can be defined based on its function. The entire structure of an enzyme may be specified by a segment of DNA. Well known concept that one gene produces an enzyme, may hold true for some, but not for all enzymes. As explained earlier, the enzyme tryptophan synthetase has been found to consist of two structurally different protein chains (α and β) where each chain is produced by an adjacent segment of DNA. Thus "one gene-one enzyme" concept has now been refined to "one gene-one polypeptide". The portion of DNA specifying a single polypeptide chain is termed a "cistron" and is synonymous with the unit of gene function.

There are many positions or sites within a cistron where mutations can occur. It has been demonstrated in case of hemoglobin that a change in a single nucleotide pair can result in a mutant phenotype. A "muton" is the smallest unit of genetic material that when changed (mutated) produces a phenotypic effect. A "muton" may be delimited to a single nucleotide. Different forms of a mutationally defined gene are called homoalleles.

A muton within a cistron may render its product defective. A fully active wild type cistron may be produced by recombination between two defective cistrons containing mutations at two different sites. The smallest unit of recombination (the recon) involves two adjacent nucleotides. Recombinationally separable forms of a gene within a cistron are referred to as heteroalleles.

Two mutants can be assigned to the same or different cistrons on the basis of "complementation of cis-trans" test. Two mutants in the same cistron will not complement each other to produce wild type or normal phenotype when in trans position. A normal phenotype indicating complementation can be produced by two mutants if they belong to different cistrons regardless of their linkage relationship i. e. whether in cis or trans position. This is illustrated in Fig. 45.6 (Stansfield, 1969).

The unit which comes closest to the concept of the classical Mendelian gene is the cistron. The cistron is a relatively large segment of genetic material when compared with the muton or recon. A generalized definition of gene based on cistron, recon and muton is "The gene of function is that sequence of nucleotides (providing numerous sites for intragenic mutation and/or recombination) which specifies the amino acid sequence of a specific polypeptide chain". Seymour Benzer introduced the term cistron to refer to the unit of function operationally defined by the cis-trans test. However, today, most geneticists consider the terms gene and cistron to be synonymous. However, gene is more commonly used rather than cistron.

The gene is operationally defined as the unit of function by the complementation or trans-test, which is used to determine whether mutations are in the same gene or different genes as explained in Fig. 45.6. The complementation test is one of the three basic tools of genetics. The other two genetic tools are recombination and mutation. The information provided by complementation test is totally distinct from that obtained from recombination analysis. The results of complementation test indicate whether mutations are allelic, whereas results of recombination analysis demonstrate whether mutations are linked and if so, provides estimates of map distance on a chromosome.

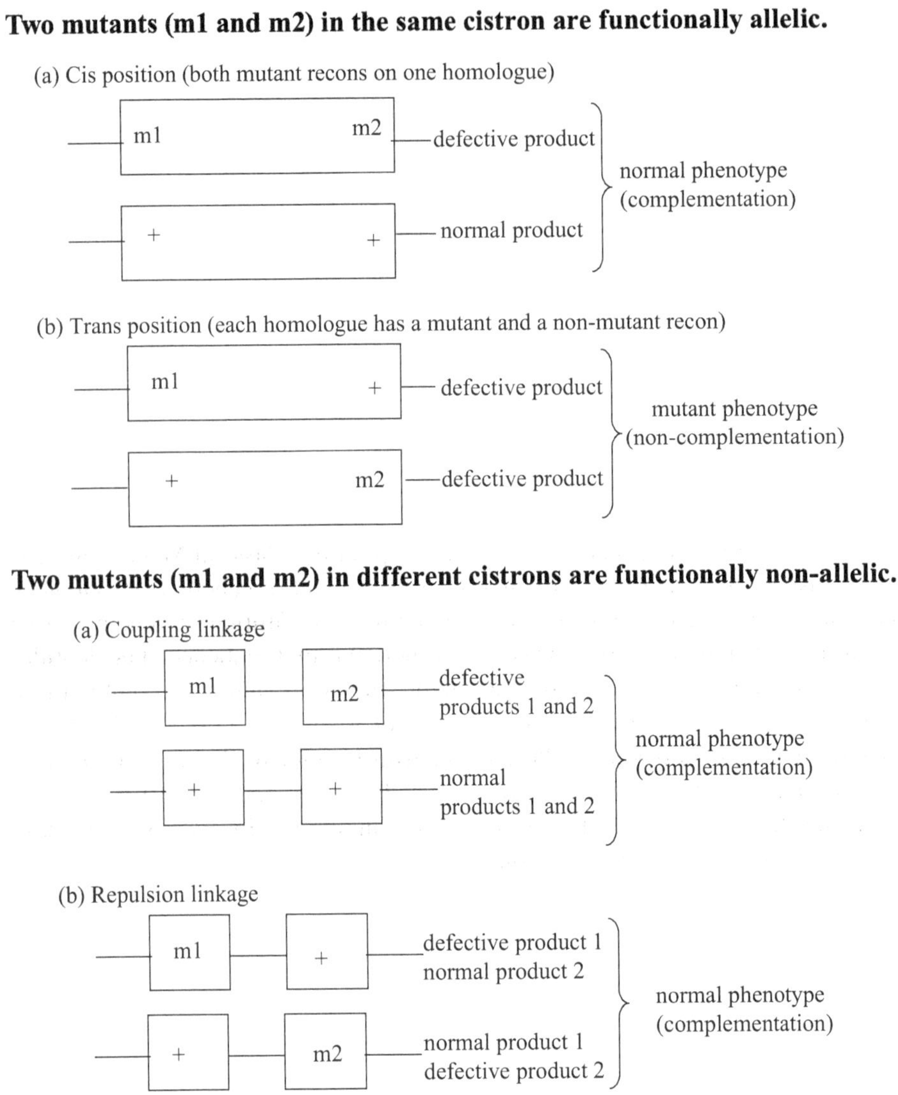

Fig. 45.6. Cis-trans or complementation test.

References

I-PLANT BREEDING

Abdurakhmonov, I. (ed). 2016. Plant Genomics. Janeza Trdine, 9 51000, Rijeka, Croatia.

Acevedo, E., Fereres, E. 1993. Resistance to abiotic stresses. In M. D. Hayward, N. O. Bosemark and I. Romagosa (eds.). Plant Breeding: Principles and Prospects. Chapman and Hall, London.

Acquaah, G. 2007. Principles of Plant Genetics and Breeding. Blackwell Publishing, USA, UK and Australia. pp 569.

Allard, R. W. 1960. Principles of Plant Breeding. John Wiley and Sons, New York, London and Toppan Company, Tokyo. pp 485.

Alfanso, R 2007. Rice Biotechnology. Presentation during the PhilRice R & D. March 13-15, 2007. Biotech Mentor's Kit. 2003. Produced by ISAAA, PCARRD and SEARCA-BIC.

Barnawal, V. K., Mikkilineni, V., Zehr, U. B., Tyagi, A. K., Kapoor, S. 2012. Opinion Paper-Heterosis: Emerging ideas about hybrid vigour. Journal of Experimental Botany. doi:10.1093/jxb/ers291: pp 1-6.

Beale, M. H., Sussman, M. R. 2018. Metabolomics of *Arabidopsis thaliana*. Annual Plant Reviews online 157-180.

Belhaj, K., Chaparro-Garcia, A., Kamoun, S., Patron, N. J., Nekrasov, V. 2015. Editing plant genomes with CRISPR/Cas9. Current Opinion in Biotechnology 32: 76-84.

Blumwald, E., Grover, A., Good, A. G. 2004. Breeding for Abiotic Stress Resistance: Challenges and Opportunities. In New Directions for a Diverse Planet-Proceedings of the 4th International Crop Science Congress, 26 Sep-1 Oct 2004, Brisbane, Australia. Published on CDROM. Website www.cropscience.org.au

Borlaug, N. E. 1958. First Int. Wheat Genet. Symp. Univ. Manitoba, Winnipeg, Canada. p 12.

Briggs, F., Knowles, P. F. 1967. Introduction to Plant Breeding. Reinhold Publishing Corporation, New York, Amsterdam, London. pp 426.

Brim, C. A. 1966. A modified pedigree method of selection in soybeans. Crop Sci. 6: 220.

Collard, B. C. Y., Jahufer, M. Z. Z., Brouwer, J. B., Pang, E. C. K. 2005. An introduction to markers, quantitative trait loci (QTL) mapping and marker-assisted selection for crop improvement: The basic concept. Euphytica 142: 169-196.

Collard, B. C. Y., Mackill, D. J. 2008. Marker-assisted selection: An approach for precision plant breeding in the twenty-first century. Philos. Trans. R. Soc. Lond. B. Biol. Sci. 363: 557-572.

Cong, L., Ran, F. A., Cox, D., Lin, S., Barretto, R., Habib, N., Hsu, P.D., Wu, X., Jiang, W., Marraffini, L.A., Zhang, F., 2013. Multiplex genome engineering using CRISPR/Cas systems. Science 339 (6121): 819-823.

Coruzzi, G. M., Burga, A. R., Katari, M. S., Gutiérrez, R. A. 2018. Systems biology: principles and applications in plant research. Annual Plant Reviews online 3-40.

Crop Biotech Update. 2014. www.isaaa.org/kc/cropbiotechupdate.

Dhillon, B. S., Singh, A. K., Lather, B. P. S., Srinivasan, G. 2004. Advances in hybrid breeding methodology. In H. K. Jain and M. C. Kharkwal (eds.). Plant Breeding-Mendelian to Molecular Approaches. Narosa Publishing House, New Delhi. pp 419-450.

Drăghici, S. 2016. Statistics and data analysis for microarrays using R and bioconductor. Chapman and Hall/CRC.

Falconer, D. S. 1960. Introduction to Quantitative Genetics. Oliver and Boyd, Edinburgh and London.

Frankel, O. H., Bennett, E. (eds.). 1970. Genetic Resources in Plants-Their Exploration and Conservation. IBP Handbook No. 11. Blackwell Sci. Publ. Oxford.

Frankel, O. H., Hawkes, J. G. (eds.). 1975. Crop Genetic Resources for Today and Tomorrow. CIBP No. 2. Cambridge University Press.

Furbank, R. T., Tester, M. 2011. Phenomics-technologies to relieve the phenotyping bottleneck. Trends in Plant Science 16(12): 635-644.

Garcia, V., Salanoubat, M., Choisne, N., Tissier, A. 2000. An ATM homologue from *Arabidopsis thaliana*: complete genomic organisation and expression analysis. Nucleic Acids Research 28(8): 1692-1699.

Gardner, C. O. 1961. An evaluation of mass selection and seed irradiation with thermal neutrons on yield of corn. Crop Sci. 1: 241-245.

Gepts, P., Hancock, J. 2006. The future of plant breeding. Crop Sci. 46:1630-1634.

Gupta, P. K. 2007. Pyamiding of genes/QTLs for crop improvement using marker-assisted selection (MAS). In: V. L. Chopra, R. P. Sharma, S. R. Bhat, and B. M. Prasanna (eds.). Search for New Genes. Academic Foundation, New Delhi. pp 145-171.

Gupta, P. K., Kumar, J., Mir, R. R., Kumar, A. 2010. Marker-assisted selection as a component of conventional plant breeding. Plant Breeding Reviews 33: 145-217.

Grover, A., Kapoor, A., Kumar, D., Shashidhar, H. E., Hittaalmani, S. 2004. Genetic improvement for abiotic stress responses. In H. K. Jain and M. C. Kharkwal (eds.). Plant Breeding-Mendelian to Molecular Approaches. Narosa Publishing House, New Delhi. pp 167-193.

Hanson, J. 1985. Procedure for Handling Seeds in Genebanks (Practical Manual for Genebanks No. 1). IBPGR, Rome.

Harlan, J. R. 1975. Crops and Man. American Society of Agronomy, Madison, USA.

Harlan, J. R., de Wet, J. M. J. 1971. Towards a rationale classification of cultivated plants. Taxon. 20: 509-517.

Henderson, I. R., Jacobsen, S. E. (2007). Epigenetic inheritance in plants. Nature 447(7143): 418.

Hoisington, D. M. *et al.* 1999. Plant genetic resources: What can they contribute towards increased crop productivity? Proc. Natl. Acad. Sci. USA. 96: 5937-5943.

Hong, J., Yang, L., Zhang, D., Shi, J. 2016. Plant metabolomics: an indispensable system biology tool for plant science. International Journal of Molecular Sciences 17(6): 767.

ISAAA. 2017. Global Status of Commercialized Biotech/GM Crops in 2017: Biotech Crop Adoption: Surges as Economic Benefits Accumulate in 22 Years. ISAAA Brief No. 53. ISAAA, Ithaca, NY.

Jain, H. K., Kharkwal, M. C. (eds.). 2004. Plant Breeding- Mendelian to Molecular Approaches. Narosa Publishing House, New Delhi.

Johnson, R. 1978. Practical breeding for durable resistance to rust diseases in self-pollinating cereals. Euphytica 27: 529-540.

Kaul, M. L. H. 1998. Male sterility: classification and concept. In S. S. Banga and S. K. Banga (eds.). Hybrid Cultivar Development. Narosa Publishing House, New Delhi. pp 17-45.

Kinman, M. L., Sprague, G. F. 1945. Relation between number of parental lines and theoretical performance of synthetic varieties in corn. J. Amer. Soc. Agron. 37: 341.

Komari, T., Ishida, Y., Hiei, Y. 2004. Plant transformation technology. In Paul Christou and Harry Klee (eds.). Handbook of Plant Biotechnology. pp 233-261. John Wiley and Sons ltd.

Kuckuck, H., Kobabe, G., Wenzel, G. 1991. Fundamentals of Plant Breeding. Springer-Verlag Berlin, Heidelberg. pp 236.

Kumar, A., Dixit, S., Ram, T., Yadaw, R. B., Mishra, K. K., Mandal, N. P. 2014. Breeding high yielding drought-tolerant rice: genetic variations and conventional and molecular approaches. Journal of Experimental Botany. Published Sep 2014, doi: 10, 1093/jxb/eru363.

Lamesch, P., Berardini, T. Z., Li, D., Swarbreck, D., Wilks, C., Sasidharan, R., Muller, R., Dreher, K., Alexander, D. L., Garcia-Hernandez, M., Karthikeyan, A. S., 2011. The Arabidopsis Information Resource (TAIR): improved gene annotation and new tools. Nucleic Acids Research 40(D1), pp.D1202-D1210.

LeClerg, E. L., Leonard, W. H., Clark, A. G. 1962. Field Plot Technique, Second Edition. Burgess Publishing Co. Minneapolis, Minnesota, USA.

Leebens-Mack, J. OneKp Capstone Wiki [Internet]. 2015. Available from: https://pods.iplantcollaborative.org/wiki/display/iptol/OneKP±Capstone±Wiki[Accessed: 2015-11-10]

Li, C. C. 1955. Population Genetics. The University of Chicago Press, Chicago and London.

Mariani, C., Beuckeleer, M. D., Truettner, J., Leemans, J., Goldberg, R. B. 1990. Induction of male sterility in plants by chimeric ribonuclease gene. Nature 343: 737-741.

Mardis, E. R. 2017. DNA sequencing technologies: 2006-2016. Nature Protocols 12(2): 213.

Michael, T. P., Jackson, S. 2013. The first 50 plant genomes. The Plant Genome 6: 1-7. DOI: 10.3855/plantgenome 2013.03.001.

Michael, T. P., VanBuren, R. 2015. Progress, challenges and future of crop genomes. Curr. Opin Plant Biol. 24: 71-81. DOI: 10.1016/j.pbi.2015.02.002.

Minai Tehrani, A., Jafarzadeh, N., Gilany, K. 2016. Metabolomics: A state of the art technology for better understanding of male infertility. Andrologia 48(6): 609-616.

Mohan, M., Nair, S., Bhagwat, A., Krishna, T. G., Yano, M., Bhatia, C. R., Sasaki, T. 1997. Genome mapping, molecular markers and marker assisted selection in crop plants. Molecular Breeding 3: 87-103.

Ortiz, R. 2002. Germplasm enhancement to sustain genetic gains in crop improvement. In J. M. M. Engels *et al.*,(eds.). Managing Plant Genetic Diversity. CABI Publishing, Wallingford, UK. pp 275-290.

Ram, Hari Har. 2012. Vegetable Breeding: Principles and Practices. Third Revised and Enlarged Edition. Kalyani Publishers, Ludhiana. pp 791.

Ram, Hari Har. 2019. Crop Breeding and Biotechnology (Second Revised and Enlarged Edition). Kalyani Publishers, Ludhiana. pp 747.

Ram, Hari Har, Yadava, Rakesh. 2007. Genetic Resources and Seed Enterprises-in 2 Parts. New India Publishing Agency, New Delhi. pp 959.

Ram, Hari Har, Upadhyay, Rupa, Dubey, R. K., Mandal, B. C. 2017. Vegetable Seed Production. Kalyani Publishers, Ludhiana. pp 734.

Ramalingam, A., Kudapa, H., Pazhamala, L. T., Weckwerth, W., Varshney, R. K. 2015. Proteomics and metabolomics: two emerging areas for legume improvement. Frontiers in Plant Science 6: 1116.

Reddy, B. V. S., Green, J. M., Bisen, S. S. 1978. Genetic male sterility in pigeon pea. Crop Sci. 18: 362-364.

Reynolds, M. P., Borlaug, N. E. 2006. Applying innovations and new technologies for international collaborative wheat improvement. J. Agri. Sci. 144: 95-110.

Ricroch, A. E., Henard-Damave, M. C. 2015. Next biotech plants: new traits, crops, developers and technologies for addressing global challenges. Crit. Rev. Biotechnol. Early online:1-16. DOI:10.3109/07388551.2015.1004521.

Ricroch, A. E., Clairand, P., Harwood, W. 2017. Use of CRISPR systems in plant genome editing: toward new opportunities in agriculture. Emerging Topics in Life Sciences 1(2): 169-182.

Roane, C. W. 1973. Trends in breeding for disease resistance in crops. Annual Review of Phytopathology 11: 463-486.

Saxena, K. B., Wallis, E. S., Byth, D. E. 1983. A new gene for male sterility in pigeon pea (*Cajanus cajan*). Heredity 51: 419-421.

Saxena, K. B., Faris, D. G., Reddy, L. J., Sharma, D., Reddy, B. V. S., Gupta, S. C., Green, J. M. 1986. Prospects for hybrid pigeon pea. In B. Napompeth and S. Subhdrabandhu (eds.). Proc. Fifth International Congress of the Society for Advancement of Breeding Research in Asia and Oceania. Kasetsart Univ., Bangkok, pp 379-388.

Scheben, A., Batley, J., Edwards, D. 2017. Genotyping by sequencing approaches to characterize crop genomes: choosing the right tool for the right application. Plant Biotechnology Journal 15(2): 149-161.

Sharma, H. C. 1995. How wide can a wide cross be? Euphytica 82: 43-64.

Siddiq, E. A., Govila, O. P., Banga, S. S. 2004. Heterosis in crop improvement. In H. K. Jain and M. C. Kharkwal (eds.). Plant Breeding-Mendelian to Molecular Approaches. Narosa Publishing House, New Delhi. pp 451-472.

Simmonds, N. W. (ed.). The Evolution of Crop Plants. Longman, New York.

Singh, R. K., Gregoria, G. B., Jain, R. K. 2007. QTL mapping for salinity tolerance in rice. Physiol. Mol. Biol. Plants 13 (2): 87-99.

Singh, R. P., Rajaram, S. Saini, R. G., Huerta-Espino, J. , William, M. 2004. Genetics of host-pathogen interaction and breeding for durable resistance. In H. K. Jain and M. C. Kharkwal (eds.). Plant Breeding-Mendelian to Molecular Approaches. Narosa Publishing House, New Delhi. pp 145-166.

Sneep, J., Hendriksen, A. J. T. (eds.). 1979. Plant Breeding Perspectives. Centre for Agricultural Publication and Documentation, Wageningen.

Stansfield, W. 1969. Theory and Problems of Genetics. McGraw-Hill Book Company, New York, San Francisco, Sydney, Toronto, Mexico, and Panama. pp 281.

Sundaram, R. M., Rajendrakumar, P. 2008. Gene tagging and mapping: Basic concepts. In Proc. International Workshop-cum-Training Programme on Marker Assisted Selection. August, 3-9. Barwale Knowledge and Study Centre, Jalna. pp 4-16.

Tabien, R. 2000. Biotech for Agriculture. Presentation during the PhilRice Farmers' Forum, July 17, 2000.

Thiel, T., Michalek, W., Varshney, R., Graner, A. 2003. Exploiting EST databases for the development and characterization of gene-derived SSR-markers in barley (*Hordeum vulgare* L.). Theoretical and Applied Genetics 106(3): 411-422.

Trethowan, R. M., Reynolds, M. P., Ivan Ortiz-Monasteriio, J. 2007. Genetic basis of green revolution in wheat production. Plant Breed. Rev. 28: 39-58.

Van der Maesen, L. J. G. 1990. Pigeonpea: Origin, history, evolution and taxonomy. In Y. L. Nene, S. D. Hall and V. K. Sheila (eds.). The Pigeonpea. CABI.

Van Der Plank, J. E. 1968. Disease Resistance in Plants. Academic Press, New York and London. pp 206.

Vasal, S. K., Singh, N. N., Dhillon, B. S., Patil, S. J. 2004. Population improvement strategies for crop improvement. In H. K. Jain and M. C. Kharkwal (eds.). Plant Breeding-Mendelian to Molecular Approaches. Narosa Publishing House, New Delhi. pp 391-406.

Vavilov, N. I. 1951. The Origin, Variation, Immunity and Breeding of Cultivated Plants. Translated from the Russian by K. S. Chester. Chronica Botanica Vr. 1/6.

Virmani, S. S., Pandey, M. P., Singh, I. S., Xu, W. J. 2004. Classical and molecular concepts of heterosis. In H. K. Jain and M. C. Kharkwal (eds.). Plant Breeding-Mendelian to Molecular Approaches. Narosa Publishing House, New Delhi. pp 407-418.

Williams, W. 1964. Genetical Principles and Plant Breeding. Blackwell Scientific Publications, Oxford. pp 504.

Witcombe, J. R., Hollington, P. A., Howarth, C. J., Reader, S., Steele, K. A. 2008. Breeding for abiotic stresses for sustainable agriculture. Philos. Trans. R. Soc. B. Biol. Sci. 363 (1492): 703-716.

Xu, Y. 2010. Molecular Breeding. CABI.

Xu, Y., Crouch, J. H. 2008. Marker-assisted selection in plant breeding: from publication to practice. Crop Sci. 48: 291-307.

Yang, W., Duan, L., Chen, G., Xiong, L., Liu, Q. 2013. Plant phenomics and high-throughput phenotyping: accelerating rice functional genomics using multidisciplinary technologies. Current Opinion in Plant Biology 16(2): 180-187.

Yelagandula, R., Osakabe, A., Axelsson, E., Berger, F., Kawashima, T. 2017. Genome-wide profiling of histone modifications and histone variants in *Arabidopsis thaliana* and *Marchantia polymorpha*. In Plant Genomics, pp. 93-106. Humana Press, New York, NY.

Zhou, M., Sng, N. J., LeFrois, C. E., Paul, A. L., Ferl, R. J. 2019. Epigenomics in an extraterrestrial environment: organ-specific alteration of DNA methylation and gene expression elicited by spaceflight in *Arabidopsis thaliana*. BMC Genomics 20(1): 205.

www.seednet.gov.in

II-GENETICS

Braun, W. 1965. Bacterial Genetics. Second Edition. Saunders, Philadelphia.

Brewbaker, J. L. 1964. Agricultural Genetics. Printice-Hall, Englewood Cliffs, N. J.

Burnham, C. R. 1962. Discussions in Cytogenetics. Burgess, Minneapolis.

Carlson, E. A. 1966. The Gene: A Critical History. W. B. Saunders, Philadelphia.

Hayes, W. 1964. Genetics of Bacteria and Their Viruses. John Wiley, New York.

Li, C. C. 1955. Population Genetics. Univ. of Chicago Press, Chicago.

Medvedev, Z. 1969. The Rise and Fall of T. D. Lysenko. Columbia University Press, New York.

Peters, J. A. (ed.). 1959. Classic Papers in Genetics. Prentice-Hall, Englewood Cliffs, N. J.

Serra, J. A. 1965. Modern Genetics. Academic Press, New York.

Singh, R. J. 2017. Plant Cytogenetics, Third Edition. Taylor and Francis Group, LLC. pp. 528.

Sinnott, E. W., Dunn, L. C., Dobjhansky, T. 1958. Principles of Genetics. International Student Edition. McGraw-Hill Book Company, Inc. New York, Toronto, London.

Snustad, D. P., Simmons, M. J., Jenkins, J. B. 1997. Principles of Genetics. John Wiley and Sons, Inc. New York.

Stansfield, W. D. 1969.Theory and Problems of Genetics. McGraw-Hill Book Company, New York.

Stebbins, G. L. 1950. Variation and Evolution in Plants. Columbia University Press, New York.

Stern, C. 1973. Principles of Human Genetics. W. H. Freeman, San Francisco.

Strickberger, M. W. 1968. Genetics. The Macmillan Company, New York.

Swanson, C. P. 1957. Cytology and Cytogenetics. Printice-Hall, Englewood Cliffs, N. J.

Index

FSC
www.fsc.org
MIX
Papier | Fördert gute Waldnutzung
FSC® C083411

Zeitfracht Medien GmbH
Ferdinand-Jühlke-Straße 7
99095 Erfurt, Deutschland
produktsicherheit@kolibri360.de